MICROBIAL DIVERSITY AND ECOSYSTEM FUNCTION

CAB INTERNATIONAL is an intergovernmental organization providing services worldwide to agriculture, forestry, human health and the management of natural resources.

The information services maintain a computerized database containing over 2.7 million abstracts on agricultural and related research with 150,000 records added each year. This information is disseminated in 47 abstract journals, and also on CD-ROM and online. Other services include supporting development and training projects, and publishing a wide range of academic titles.

The four scientific institutes are centres of excellence for research and identification of organisms of agricultural and economic importance: they provide annual identifications of over 30,000 insect and microorganism specimens to scientists worldwide, and conduct international biological control projects.

International Mycological Institute
An Institute of CAB INTERNATIONAL

The Institute, founded in 1920, and employing over 70 staff:
- Is the largest mycological centre in the world, and is housed in a complex specially designed to support its requirements;
- Carries out research on a wide range of systematic and applied problems involving fungi (including lichens and yeasts), bacteria, and also on the preservation of fungi, the biochemical physiological and molecular characterization of strains, and on crop protection, environmental and industrial mycology, food spoilage, public health, biodeterioration and biodegradation;
- Provides an authoritative identification service, especially for microfungi of economic and environmental importance (other than certain human and animal pathogens), and for plant pathogenic bacteria and spoilage yeasts;
- Undertakes extensive computerized and indexing bibliographic work, including the preparation of 7 serial publications;
- Supports advice and undertakes a wide variety of project and culture work on crop protection, environmental, food spoilage and industrial topics.
- Offers training in pure and applied aspects of mycology and bacteriology, both at the Institute and overseas.

The Institute's dried reference collection numbers in excess of 355,000 specimens representing about 31,500 different species, and a genetic resource collection holds more than 17,000 living isolates by a variety of the most modern methods. The library receives about 600 current journals, and has extensive book and reprint holdings reflecting the Institute's interests.

International Mycological Institute
Bakeham Lane
Egham
Surrey TW20 9TY
UK

Tel: (01784) 470111
Telex: 9312102252
Telecom Gold/Dialcom: 84:CAU009
Fax: (01784) 470909
E-mail: CABI-IMI@CGNET.COM

For full details of the services provided, please contact the Institute Director.

Microbial Diversity and Ecosystem Function

Proceedings of the IUBS/IUMS Workshop held at Egham, UK, 10–13 August 1993 in support of the IUBS/UNESCO/SCOPE 'DIVERSITAS' Programme

Edited by

D. Allsopp
International Mycological Institute
Egham, UK

R.R. Colwell
University of Maryland
Biotechnology Institute
USA

and

D.L. Hawksworth
International Mycological Institute
Egham, UK

CAB INTERNATIONAL
in association with
United Nations Environment Programme (UNEP)

CAB INTERNATIONAL
Wallingford
Oxon OX10 8DE
UK

Tel: +44 (0)1491 832111
Telex: 847964 (COMAGG G)
E-mail: cabi@cabi.org
Fax: +44 (0)1491 833508

© CAB INTERNATIONAL 1995. All rights reserved. No part of this publication may be reproduced in any form or by any means, electronically, mechanically, by photocopying, recording or otherwise, without the prior permission of the copyright owners.

A catalogue entry for this book is available from the British Library.

ISBN 0 85198 898 9

Typeset by Solidus (Bristol) Limited
Printed and bound in the UK at the University Press, Cambridge

Contents

Foreword ix

Acknowledgements xi

PART I: THE MICROBIAL CONCEPT

1 **The Microbial Species Concept and Biodiversity** 3
R.R. Colwell, R.A. Clayton, B.A. Ortiz-Conde,
D. Jacobs and E. Russek-Cohen

2 **The Microorganisms: A Concept in Need of
Clarification or One Now to be Rejected?** 17
G.A. Zavarzin

PART II: THE EXTENT OF MICROBIAL DIVERSITY

3 **Described and Estimated Species Numbers: An
Objective Assessment of Current Knowledge** 29
P.M. Hammond

4 **Approaches to the Comprehensive Evaluation of
Prokaryote Diversity of a Habitat** 73
J.M. Tiedje

5   Identifying and Culturing the 'Unculturables':
    A Challenge for Microbiologists   89
    N. Ward, F.A. Rainey, B. Goebel and E. Stackebrandt

PART III: THE IMPACT OF MICROORGANISMS ON GLOBAL ECOLOGY AND NUTRIENT CYCLING

6   A Neglected Carbon Sink? Biodegradation of Rocks   113
    W.E. Krumbein

7   Lichens in Southern Hemisphere Temperate
    Rainforest and Their Role in Maintenance of
    Biodiversity   125
    D.J. Galloway

8   Mineral Cycling by Microorganisms: Iron Bacteria   137
    D.B. Johnson

9   The Potential Importance of Biodiversity in
    Environmental Biotechnology Applications:
    Bioremediation of PAH-contaminated Soils
    and Sediments   161
    P.H. Pritchard, J.G. Mueller, S.E. Lantz and D.L. Santavy

PART IV: MICROORGANISMS AND ECOSYSTEM MAINTENANCE

10 Bacterial Diversity and Ecosystem Maintenance:
    An Overview   185
    A. Bianchi and M. Bianchi

11 Ecological Role of Microphytic Soil Crusts in Arid
    Ecosystems   199
    S.D. Warren

12 The Diversity of Microorganisms Associated with
    Marine Invertebrates and Their Roles in the
    Maintenance of Ecosystems   211
    D.L. Santavy

| 13 | Fungi, a Vital Component of Ecosystem Function in Woodland
A.D.M. Rayner | 231 |

PART V: MICROORGANISMS IN EXTREME ENVIRONMENTS

14	Molecular Biology of Alkaliphiles T. Hamamoto and K. Horikoshi	255
15	Thermophilic Fungi in Desert Soils: A Neglected Extreme Environment J. Mouchacca	265
16	Biodiversity of the Rock Inhabiting Microbiota with Special Reference to Black Fungi and Black Yeasts C. Urzì, U. Wollenzien, G. Criseo and W.E. Krumbein	289

PART VI: INVENTORYING AND MONITORING MICROORGANISMS

17	Statistics, Biodiversity and Microorganisms E. Russek-Cohen and D. Jacobs	305
18	Traditional Methods of Detecting and Selecting Functionally Important Microorganisms from the Soil and the Rhizosphere F.A.A.M. de Leij, J.M. Whipps and J.M. Lynch	321
19	Problems in Measurements of Species Diversity of Macrofungi E. Arnolds	337
20	Inventorying Microfungi on Tropical Plants P.M. Kirk	355
21	Viral Biodiversity M.H.V. Van Regenmortel	361
22	Exploration of Prokaryotic Diversity Employing Taxonomy J. Swings	371

| 23 | International Biodiversity Initiatives and the Global Biodiversity Assessment (GBA)
V.H. Heywood | 381 |

PART VII: THE RESOURCE BASE IN MICROBIOLOGY

24	Living Reference Collections H. Sugawara and Jun-Cai Ma	389
25	Dried Reference Collections as a Microbiological Resource D.W. Minter	403
26	Microorganisms, Indigenous Intellectual Property Rights and the Convention on Biological Diversity J. Kelley	415
27	Extent and Development of the Human Resource E.J. DaSilva	427
28	Biodiversity Information Transfer: Some Existing Initiatives and How to Link Them B. Kirsop and V. Canhos	439
29	Indigenous Rhizobia Populations in East and Southern Africa: A Network Approach N.K.N. Karanja, P.L. Woomer and S. Wangaruro	447
30	Progress in the Synthesis and Delivery of Information on the Diversity of Known Bacteria J.G. Holt, M.I. Krichevsky and T. Bryant	455

Index 467

Foreword

The last five years have witnessed a renewed interest in microorganisms as a component of biological diversity. This revival and acknowledgement of the significance of the microbial dimension to the conservation and sustainable use of ecosystems has resulted in microorganisms being incorporated into broad international initiatives, and to specific international programmes being established for them. This workshop is a component of one of them, DIVERSITAS.

The concept of DIVERSITAS, a programme addressing scientific issues related to the nature and significance of biodiversity, originated from a meeting of experts organized by the International Union of Biological Sciences (IUBS) in Washington, DC, in June 1989. This programme has been developed in collaboration with the Scientific Committee on Problems of the Environment (SCOPE) and the United Nations Educational, Scientific and Cultural Organization (UNESCO), and now contains six themes:

- Ecosystem function of biodiversity
- Origins, maintenance and loss of biodiversity
- Inventorying and monitoring biodiversity
- Biodiversity of wild relatives of domesticated organisms
- Microbial biodiversity
- Marine biodiversity

The last two topics were envisaged as cross-cutting the first four, but meriting fast-track attention because of their neglect. In order to kick-start the microbial strand in this major initiative, IUBS and the International Union of Microbiological Societies (IUMS) developed the subprogramme MICROBIAL DIVERSITY 21 at a workshop in Amsterdam in September

1991. That occasion reviewed the current state of knowledge on the diversity of algae, bacteria, filamentous fungi, lichen-forming fungi, yeasts, protozoa, and viruses, and developed a 14-point action statement.

The IUBS/IUMS/SCOPE workshop convened at Egham in August 1993, supported by the Commission of the European Communities (CEC), United Nations Environment Programme (UNEP), and CAB INTERNATIONAL, took the complementary approach of focusing on microbial diversity and its relevance to ecosystem function. There is a tendency for fungi and other microorganisms to be viewed by macroecologists as a 'black box', performing crucial processes, but with little attention being paid to the actual species involved, their numbers, or their susceptibility to biodiversity loss or gain. Some biologists have been cognizant of this importance, but they have been the exception rather than the rule. For example, as early as 1927 J.B.S. Haldane drew attention to the ability of a single pathogen to change an oak forest with wild pigs to a pine forest with ants. As far as I am aware, there has been only one major attempt to synthesize what is known of the effects of diseases in the maintenance of natural (as opposed to agricultural) ecosystems, edited by J.J. Burdon and S.R. Leather in 1990. Further, the widespread nature of mutualisms involving fungi and bacteria is too often passed over, from corals to mycorrhizas and ruminant guts, and the magnitude of the microbial contribution to nutrient cycling and atmospheric composition is rarely analysed from the organismal level.

The workshop aimed to provide both a chart as to where we are now in these various seas of ignorance, and some pointers towards directions that can be taken to strengthen our knowledge base so that the microbial dimension of biodiversity can be more objectively assessed than is currently possible.

During the Egham workshop, it was suggested that an inter-union IUBS/IUMS Committee on Microbial Diversity be established, comprising representatives from each of the pertinent scientific members and commissions of each union. This recommendation was endorsed by both unions in 1994, and the inaugural meeting of the Committee was held at UNESCO headquarters in Paris in September 1994. Among the actions decided at that meeting was to seek support for an international workshop to be held at Egham in April 1996 to address the question of how to develop the most effective protocols to inventory microbial groups in soil and water.

The scientific and practical tasks ahead in microbial diversity are daunting, but at least fundamental issues are now being aired and actions to address them developed rather than passed over. If the level of international interest and support received by the Egham workshop is an indicator, the prospects of our attaining a better understanding of microbial diversity and its significance during this decade are encouraging.

Professor David L. Hawksworth
President, International Union of Biological Sciences (IUBS)

Acknowledgements

This workshop would not have been possible without the support of the following organizations, which was greatly appreciated by all participants:

Commission of the European Communities (CEC) – DG XII
CAB INTERNATIONAL
International Union of Biological Sciences (IUBS)
International Union of Microbiological Societies (IUMS)
Scientific Committee on Problems of the Environment (SCOPE)
United Nations Environment Programme (UNEP)

Meeting facilities provided by the Royal Holloway University of London are also gratefully acknowledged.

In addition, the organizers wish to record their gratitude to the following staff of the International Mycological Institute (IMI) for facilitating the smooth running of the workshop: Stephanie Groundwater, Marilyn S. Rainbow and Barbara J. Ritchie.

THE MICROBIAL CONCEPT I

The Microbial Species Concept 1
and Biodiversity

R.R. COLWELL, R.A. CLAYTON, B.A.
ORTIZ-CONDE, D. JACOBS AND E. RUSSEK-COHEN

University of Maryland Biotechnology Institute, 4321 Hartwick Road, Suite 550, College Park, Maryland 20740, USA.

Significance of Microbial Biodiversity

As E.O. Wilson observed, 'Biological diversity must be treated more seriously as a global resource, to be indexed, used, and above all, preserved' (Wilson, 1988). As habitat destruction and species extinction rates accelerate, biotechnology is identifying new ways to use information and products from the biota. Biodiversity management is crucial, yet there is little baseline information about its amount and distribution.

Management of biodiversity, from microorganisms to humans, is a major task currently being addressed by many national and international biodiversity programmes. Our world comprises a 'stockroom' of biodiversity. This stock of world genetic resources must be utilized wisely and managed carefully. Management of microbial biodiversity is a serious task to which we, as microbiologists, must set ourselves, since we find the microorganisms either on the fringe of biodiversity programmes, or left out entirely. Microbial biodiversity provides the foundation for biotechnology, i.e., the basis for new product discovery, bioremediation, and genetic manipulation of organisms for new commercial products and processes. Thus, retention and conservation of microbial genomic variation for future applications is not only an obligation of the highest order, but also of great importance for the global economy.

The advances made in molecular biology during the past three decades allow previously unforeseen applications of biotechnology and, more recently, have highlighted the value of microbial genetic biodiversity, especially in bioprospecting. A case in point is PCR technology, which was greatly facilitated by the discovery of a heat-stable DNA polymerase in the hot springs archaean, *Thermus aquaticus*. A significant aspect of biotechnology is that biodiversity is its linchpin and genetic characterization of species is a basic prerequisite. The first step in exploiting biotechnology is assessment of biodiversity and exploring genetic variability. The second is recognizing the necessity to preserve genetic information and to make a powerful argument for conservation, since every future need for any given microbial genetic resource cannot be anticipated.

The actual 'prospecting' for biodiversity is a third component, often leading to utilization of genetic resources. A good example is the study of biodiversity for purposes of chemical prospecting currently in progress in Costa Rica, South America, as a result of a contractual arrangement between Merck Co., a pharmaceutical company, and the Costa Rican government (Laird, 1993). In this contractual agreement, the bulk of royalties go to conservation, with conservation management being undertaken by the Costa Rican government, using its share of the income from the financial arrangement with Merck. The result is that biodiversity is maintained and simultaneously chemical prospecting is accomplished.

Inventory of Microbial Species

The total number of extant species on earth is unknown, even to its order of magnitude. Estimates of species biodiversity range from between three and five million species to 1.4 billion species. Approximately 1.4 million species have been named (Wilson, 1988). Multicellular eukaryotes are better studied than prokaryotes, yet it is estimated that less than 15% of the eukaryote species have formally been named, although some taxa are well characterized. For example, 45,000 species of vertebrates are described, representing an estimated 90% of extant vertebrate species. Insects, on the other hand, comprise 1 million described species, representing between 1% and 12% of extant species (Hammond, 1992). Hawksworth (1991) reported 1.5 million species of fungi, as a working estimate. The 69,000 described fungal species represent less than 5% of extant species for that group.

For the prokaryotes, it is difficult even to establish a rationale for estimating the number of species, other than counting described species in bacterial nomenclature updates published in the *International Journal of Systematic Bacteriology* (3800 species as of 1993), which gives a listing of all taxonomically valid names under the International Code of Nomenclature of Bacteria (Sneath *et al.*, 1992). Some investigators question the notion that microorganismal biodiversity can be enumerated, given the difficulty of delineating species in these groups. Viruses, especially, are viewed by many biologists as 'pieces of genomes', i.e., non-living biological entities. For those adhering to this view, viral 'species' do not exist.

An organized taxonomy for the viruses (Pringle, 1993) was presented to 4000 virologists attending the 1993 International Congress of Virology in Glasgow, Scotland. The International Committee on Taxonomy of the Viruses (Francki et al., 1991) listed over 5500 species of viruses. This includes 1500 recognized viruses of vertebrates, 2000 recognized viruses of plants, 1000 of insects, and 2000 of bacteria. Hammond (1992) has estimated that there are 500,000 extant viral species. Thus, at best, about 1% of virus species are described.

In the case of the bacteria, there are ca. 3800 characterized species, including archaeal and eubacterial species. Between 300,000 and 1,000,000 species of bacteria are estimated to exist globally (Hammond, 1992), a conservative estimate. In a 1965 review, MacLeod stated that less than 1% of all bacterial species in the sea have been cultured. Work by DeLong (1992), Giovannoni et al. (1990), Fuhrman et al. (1992, 1993), and others, based on nucleic acid sequence data, has shown this to be a reasonable estimate. A study by Scandinavian scientists revealed that, in a single gram of soil, as many as 5000 species of bacteria may exist (Torsvik et al., 1990).

Viable but Non-culturable Bacteria

Culturable bacteria represent only a small fraction of the prokaryotic biota. These species include important pathogens of humans, animals, and plants, as well as sources of many useful products and processes (for example, the waste treatment processes of most major cities of the world). In addition to these known bacteria, there is evidence of an enormous reservoir of uncultured bacteria, for example, those discovered solely by their ribosomal RNA sequences (Chapter 5). There are also the viable but non-culturable stages of Gram-negative bacteria (Xu et al., 1982; Roszak and Colwell, 1987), as well as the spore, or resting stages of Gram-positive bacteria. A reservoir of microbial genomes exists in the natural environment, of which only a small fraction is active at any given time. Cells in the viable but non-culturable (VBNC) stage are not metabolically inert, but can synthesize protein (Chowdhury et al., 1993), as well as take up substrate (Roszak and Colwell, 1987). When present in very large numbers, in the ocean, for example (Giovannoni et al., 1990), even resting stage cells metabolize large amounts of nutrients.

In the case of viable but non-culturable bacteria, Gram-negative bacterial cultures, under conditions adverse to growth and cell division, undergo conversion into non-culturable forms. The number of culturable cells decreases with time of incubation (Grimes et al., 1986; Fig. 1.1). Without direct detection techniques, the assumption was that such cultures were dead. However, the use of nucleic acid probes and PCR technology have confirmed the presence of large numbers of metabolizing, but not

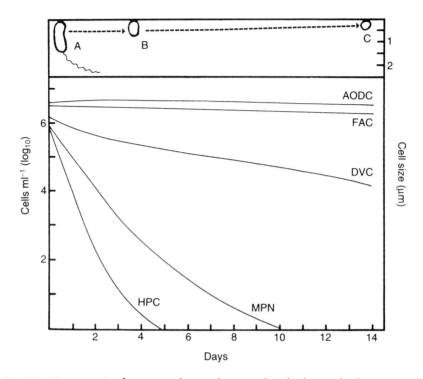

Fig. 1.1. Enumeration of enteric pathogens by several methods. Results shown provide evidence for viable but non-culturable organisms. AODC, Acridine Orange direct count; FAC, fluorescent antibody count; DVC, direct viable count: reproduced with permission from Grimes et al. (1986).

dividing, cells. Thus, an extensive array of non-culturable microorganisms needs to be characterized and catalogued. Characterization of habitats where VBNC bacteria may be present in large numbers (i.e., biofilms) will potentially provide a rich source of new taxa. Nucleic acid techniques make it possible to survey the vast numbers of previously unknown taxa. Thus, microorganisms that have never been cultured, either because appropriate techniques are not available or because the organisms are encountered primarily in a dormant stage, require reconsideration of cataloguing methods for microbial diversity. This factor must be addressed when undertaking estimates of bacterial biodiversity or preparing inventories, since mere enumeration by culture may yield misleading results.

Microbial Species Concept

An argument has been made that biodiversity estimates are meaningless for bacteria, because the species concept for bacteria is neither completely developed nor applicable, especially in the case of natural environmental isolates. However, history has shown that it is possible to develop a useful species concept for the bacteria. There are strong arguments supporting the usefulness of characterizing and cataloguing microbial species, with the driving forces behind speciation of microorganisms based on molecular genetic, environmental, and ecological data, i.e., a polyphasic approach (Colwell, 1970). It is instructive to review the history of microbial systematics, because it places in perspective the macrobiologist's view of the world of microorganisms. Although the species concept differs for animals, plants, and microorganisms, especially viruses, it is not irrelevant to the bacteria.

Haeckel, the German philosopher–biologist, coined the term 'phylogeny' to describe the genealogy of all organisms (Haeckel, 1866). His schema divided organisms into three Archephyla: Protista, Plantae, and Animalia. Organisms lacking nuclei, the bacteria and blue–green algae, were placed in Moneres, within Protista. Haeckel considered the Moneres to be the ancestors of all other living organisms (Haeckel, 1866). Morphology was the key feature in placing the Moneres in this scheme. Orla-Jensen (1909) was among the first to add biochemical activities to morphological features in the study of bacterial classification and evolution. Biochemical activity continues to dominate the phenotypic characterization of bacteria.

Since 1975, microbial taxonomy has turned to genetic approaches to expand understanding of prokaryotic relationships. Descriptions of cultured species now include DNA base composition, DNA/DNA hybridization, genomic fingerprinting, and nucleic acid sequence data. Before this, prokaryotic taxonomy was essentially phenotypically based and descriptive, or 'alpha taxonomy'. The transition period, 1960–1975, led to 'beta systematics', i.e. phylogeny. Some systematists seem to consider nucleic acid fingerprinting and sequencing the 'omega of systematics', a premature conclusion since there are surely surprises in new information to be gleaned in the future from, as yet unknown, new methodologies. One need only review the literature to observe the hubris of earlier microbiologists who believed that they, indeed, had discovered the key unlocking the mysteries of microbial systematics.

The taxonomic divisions for the microorganisms are now based on a polyphasic approach: phenotypic characterization, nucleic acid sequence analysis, DNA/DNA homology, etc. The criteria to delineate a prokaryotic species was established by consensus of an international committee on the systematics of bacteria. It was concluded that approximately 70% DNA/DNA hybridization was the level of relationship for defining a bacterial species (Wayne et al., 1987). Twenty years earlier, from numerical taxonomy estimates, it had been concluded that a phenotypic similarity of ca. 60–65% could be used to define the genus level relationship and a phenotypic similarity of ca. 75%, the species level relationship (Colwell and Liston, 1961). Thus, it is interesting to note that nucleic acid homology data fall roughly within the same range of similarity to define genus and species for the bacteria.

As technology has made it feasible, nucleic acid analyses of similarity have expanded to include DNA/rRNA hybridization, and later, 5S rRNA, 16S rRNA, and 23S rRNA sequencing. With the launching of the Human Genome Initiative, nucleic acid sequencing of complete genomes of prokaryotes has come into the realm of possibility (for example, *Haemophilus influenzae* by The Institute of Genomic Research (TIGR) in collaboration with Johns Hopkins University). TIGR is undertaking nucleic acid sequencing of a number of microorganisms, as an extension of their work on the Human Genome Initiative. The complete sequencing of the smallpox virus genome has already been accomplished (Massung et al., 1994).

This added genetic dimension will take microbial systematics to another level, from the monothetic, phenotypic approach of the earlier half of this century, i.e., microbial systematics based on phenotype, to a truly polyphasic approach, including molecular genetic data and a more complete integration of phenotypic, chemotaxonomic, and genetic information. Progress is such that we can now accomplish identification of species using gene probes targeted to specific nucleic acid sequences, without recovering the organisms into pure culture, based on a comprehensive, data-rich classification (Schmidt et al., 1991; Stackebrandt, 1991). Fluorescent monoclonal antibody probes used directly in the environment to react selectively with a target species are increasingly used to answer many questions about microbial community ecology. The Ribosomal Database Project (Larsen et al., 1993) provides aligned small subunit ribosomal RNA sequences for over 2500 operational taxonomic units, against which unknown RNA sequences can be compared, and provisionally identified. Thus, microbial systematics has moved from a phenotypic taxonomy to a polyphasic taxonomy that takes into account DNA base composition, DNA/rRNA and DNA/DNA hybridization, nucleic acid sequences and ecological data in population-based systematics.

Bacterial Species Definitions

There is no official definition of a species in microbiology, and the applicability of the species concept to bacteria is refuted by some (Cowan, 1971). A bacterial species is generally considered to be a collection of strains that show a high degree of overall similarity and differ considerably from related strain groups with respect to many independent characteristics (Staley and Krieg, 1984; Stanier *et al.*, 1986). A widely accepted bacterial species definition is 'a collection of strains showing a high degree of overall similarity, compared to other, related groups of strains'. This comprises a relatively vague definition, but it 'works'. It has evolved from the earlier contributions of van Niel (1943), Stanier (Stanier and van Niel, 1941), and, more recently, Sneath (1984), and Wayne *et al.* (1987). However, a bacterial species, maybe a taxospecies, is defined as a group of organisms (strains, isolates) with mutually high phenotypic similarity that form a phenetic cluster (Sneath and Sokal, 1973); a genospecies as a group of organisms capable of genetic exchange; a genomic species as a group showing high DNA/DNA hybridization values; and a nomenspecies as a group that bears a binomial name (Sneath, 1984; Stackebrandt and Goodfellow, 1991). For example, because most of the members of the family Enterobacteriaceae are capable of genetic exchange, they are considered one genospecies even though this family consists of many taxospecies (Jones and Sackin, 1980). Unfortunately, the distinction between genospecies and genomic species has become blurred because genospecies is now frequently used in the literature to refer to organisms belonging to the same DNA/DNA hybridization group (Stackebrandt, 1991). Despite the variability of the species definition, the biological reality represented by these varying concepts is describable and worthy of study and conservation.

The genus *Vibrio* presents a good example of how the bacteriological species concept has changed in practice. The name was first used by Pacini (1854) to describe rod-shaped cells associated with the disease cholera. Subsequently, microbiologists continued to characterize the genus phenotypically: Gram-negative rods, oxidase positive, fermenting sugars without the production of gas, etc. During the 1960s, the base composition of the genus was determined to be 41–49% G+C, with type species *V. cholerae*, measuring 47–49% G+C. In the 1980s and 1990s, 5S rRNA sequences were obtained for all the named strains available to date. A simple match coefficient based on the primary structure and cluster analysis using average linkage provides a dendrogram showing relationships among species of the genus *Vibrio* and some related organisms (Fig. 1.2). Among these species, a Nelson consensus tree based on 5S data shows a slightly different but comparable arrangement (Fig. 1.3). DNA homology data do not yield results fully consonant with a phylogenetic tree based on nucleic acid sequences, the latter being less 'noisy', in terms of arrangements of the *Vibrio* species within the genus (Fig. 1.4). A summary tree, with all the data aligned, provides a reasonable analysis of species relationships within the genus, allowing clustering of strains within species (data not shown). All the data sets clearly confirm that the *Vibrionaceae* is a family separate from the *Aeromonadaceae* (MacDonell and Colwell, 1985; Colwell *et al.*, 1986). Even with a full characterization of the group, including phenotypic data,

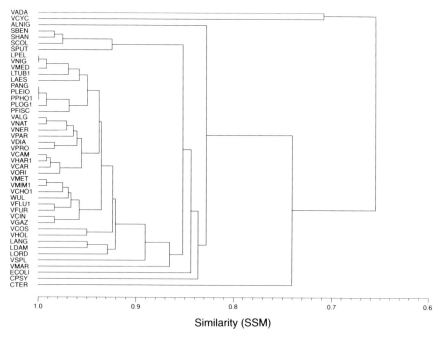

Fig. 1.2. 5S rRNA S_{SM}/UPGMA phenogram for 45 vibrios and relatives. Dendrogram constructed using simple match similarity (S_{SM}) values clustered by unweighted pair group with arithmetic averages (UPGMA). The S_{SM} values were calculated using 121 of the 134 aligned sequence positions from the 5S rRNA data set. The cophenetic correlation coefficient (r_{CS}) is 0.9764 for this tree. Strain abbreviations are as follows, with numbers being those of the American Type Culture Collection (ATCC, Parklawn Dr., Rockville, MD): ALNIG (*Alteromonas nigrifaciens* no. 19375), CPSY (*Colwellia psychroerythrus* no. 27364), CTER (*Comamonas terrigena* no. 14636), ECOLI (*Escherichia coli* K12), LAES (*Listonella aestuariana* no. 35048), LANG (*L. anguillarum* no. 19264), LDAM (*L. damsela* no. 33539), LORD (*L. ordalii* no. 33509), LPEL (*L. pelagia* no. 25916), LTUB1 (*L. tubiashii* no. 19109), PANG (*Photobacterium angustum* no. 25915), PFISC (*P. fischeri* no. 7744), PLEIO (*P. leiognathi* no. 25521), PLOG1 (*P. logei* no. 29985), PPHO1 (*P. phosphoreum* no. 11040), PSFL (*Pseudomonas fluorescens* no. 13525), SBEN (*Shewanella benthica* no. 43992), SCOL (*S. colwelliana* no. 39565), SHAN (*S. hanedai* no. 33224), SPUT (*S. putrefaciens* no. 8071), VADA (*Vibrio adaptatus* no. 19263), VALG (*V. alginolyticus* no. 17749), VCAM (*V. campbellii* no. 25920), VCAR (*V. carchariae* no. 35084), VCHO1 (*V. cholerae* no. 14035), VCIN (*V. cincinnatiensis* no. 35912), VCOS (*V. costicola* no. 33508), VCYC (*V. cyclosites* no. 14635), VDIA (*V. diazotrophicus* no. 33466), VFLU1 (*V. fluvialis* no. 33809), VFUR (*V. furnissii* no. 35016), VGAZ (*V. gazogenes* no. 29988), VHAR1 (*V. harveyi* no. 14126), VHOL (*V. hollisae* no. 33564), VMAR (*V. marinus* no. 15381), VMED (*V. mediterranei* no. 43341), VMET (*V. metschnikovii* no. 7708), VMIM1 (*V. mimicus* no. 33653), VNAT (*V. natriegens* no. 14048), VNER (*V. nereis* no. 25917), VNIG (*V. nigripulchritudo* no. 27043), VORI (*V. orientalis* no. 33934), VPAR (*V. parahaemolyticus* no. 17802), VPRO (*V. proteolyticus* no. 15338), VSPL (*V. splendidus* no. 33125), and VVUL (*V. vulnificus* no. 27562).

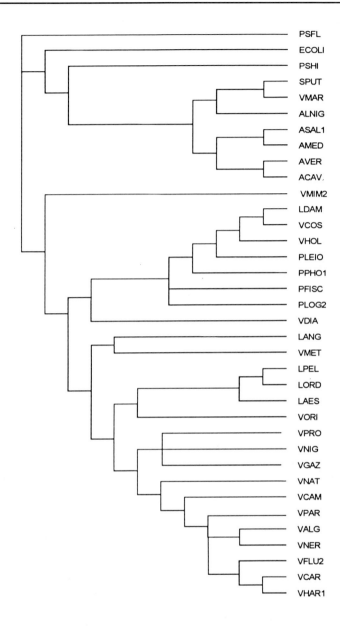

Fig. 1.3. 5S rRNA cladogram showing evolutionary relationships between 46 vibrios and their relatives. The tree was constructed from 59 characters of the 5S rRNA data set using Hennig86 with *Pseudomonas fluorescens* ATCC no. 13525 (PSFL) as an outgroup. It is a Nelson consensus tree representing over 100 equally most parsimonious trees of length 452. See legend of Fig. 1.2 for strain abbreviations.

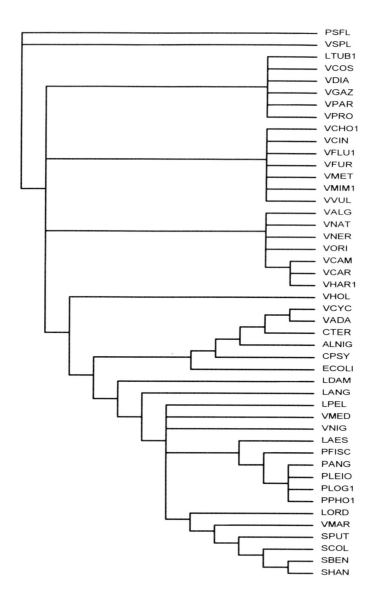

Fig. 1.4. Global consensus cladogram of 36 taxa. Cladogram showing evolutionary relationships between 36 vibrios and their relatives. The tree was constructed from 165 characters obtained by combining the 5S rRNA and 16S rRNA data sets, and using Hennig86 with *Pseudomonas fluorescens* ATCC no. 13525 (PSFL) as an outgroup. See legend of Fig. 1.2 for strain abbreviations.

ribosomal RNA sequences and DNA/DNA hybridization, it is not a simple process to analyse and combine the data sets. This is a research programme currently in progress in our laboratory.

In summary, microbial diversity is, indeed, a highly significant component of a sustainable biosphere, if not its very foundation. Microorganisms, even though invisible to the naked eye, are a source of extraordinary diversity, with unlimited potential. Both the sources and the potential should be categorized and inventoried. Significant microbial diversity exists that is, as yet, to be defined and remains unexplored. No international programme for assessment of biodiversity can be complete without the inclusion of microorganisms, the very foundation of all extant life on our planet!

Acknowledgements

This work was supported, in part, by a grant from the Office of Naval Research no. ONR N0014-86-K-0197, The Thrasher Research Fund and The Institute for Genomic Research, Rockville, MD, USA.

References

Chowdhury, M.A.R., Hill, R.T. and Colwell, R.R. (1993) The metabolic activities of viable but non-culturable microorganisms in the non-culturable state. Abst. Q-170, p. 377, 93rd General Meeting of the American Society for Microbiology, May 16-20, 1993. Atlanta, Georgia, USA.

Colwell, R.R. (1970) Polyphasic taxonomy of bacteria. In: Iizuka, H. and Hasegawa, T. (eds) *Culture Collections of Microorganisms*. Proceedings of the International Conference on Culture Collections. Oct. 1968, Tokyo. University of Tokyo Press, Tokyo, Japan, pp. 421-436.

Colwell, R.R. and Liston, J. (1961) Taxonomic relationships among the pseudomonads. *Journal of Bacteriology* 82, 1-14.

Colwell, R.R., MacDonell, M.T. and De Ley, J. (1986) Proposal to recognize the family Aeromonadaceae fam. nov. *International Journal of Systematic Bacteriology* 36, 473-477.

Cowan, S.T. (1971) Sense and nonsense in bacterial taxonomy. *Journal of General Microbiology* 67, 1-8.

DeLong, E.F. (1992) Archaea in coastal marine environments. *Proceedings of the National Academy of Sciences of the USA* 89, 5685-5689.

DeLong, E.F., Wickham, G.S. and Pace, N.R. (1989) Phylogenetic strains: ribosomal RNA-based probes for the identification of single cells. *Science* 243, 1360-1363.

Francki, R.I.B., Fauquet, C.M., Knudson, D.L., Brown, F. and International Committee on Taxonomy of Viruses (1991) Classification and nomenclature of viruses: fifth report of the International Committee on Taxonomy of Viruses for the Virology Division of the International Union of Microbiological Societies. *Archives of Virology*, Suppl. 2.

Fuhrman, J.A., McCallum, K. and Davis, A.A. (1992) Novel major archaebacterial group from marine plankton. *Nature* 356, 148–149.

Fuhrman, J.A., McCallum, K. and Davis, A.A. (1993) Phylogenetic diversity of subsurface marine microbial communities from the Atlantic and Pacific Oceans. *Applied and Environmental Microbiology* 59, 1294–1302.

Giovannoni, S.J., Brigtschgi, T.B., Moyer, C.L. and Field, G.K. (1990) Genetic diversity in Sargasso Sea bacterioplankton. *Nature* 345, 60–63.

Grimes, D.J., Attwell, R.W., Brayton, P.R., Palmer, L.M., Rollins, D.M., Roszak, D.B., Singleton, F.L., Tamplin, M.L. and Colwell, R.R. (1986) The fate of enteric pathogenic bacteria in estuarine and marine environments. *Microbiological Sciences* 3, 324–329.

Haeckel, E. (1866) *Generele Morphologie der Organismem*, vol. 2. Georg Reiner, Berlin.

Hammond, P. (1992) Systematics and diversity: species inventory. In: Groombridge, B. (ed.) *Global Biodiversity: Status of the Earth's Living Resources*. A report compiled by the World Conservation Monitoring Center. Chapman & Hall, London, pp. 17–39.

Hawksworth, D.L. (1991) The fungal dimension of biodiversity: magnitude, significance, and conservation. *Mycological Research* 95, 641–655.

Jones, D. and Sackin, M.J. (1980) Numerical methods in the classification and identification of bacteria with especial reference to the Enterobacteriaceae. In: Goodfellow, M. and Board, R.G. (eds) *Microbiological Classification and Identification*. Academic Press, London, pp. 73–106.

Laird, S.A. (1993) Contracts for biodiversity prospecting. In: Reid, W.V., Laird, S.A., Meyer, C.A., Gamez, R., Sittenfeld, A., Janzen, D.H., Gollin, M.A. and Juma, C. (eds) *Biodiversity Prospecting: Using Genetic Resources for Sustainable Development*. World Resources Institute, Washington DC, USA, pp. 99–130.

Larsen, N., Olsen, G.J., Maidak, B.L., McCaughey, M.J., Overbeek, R., Macke, T.J., Marsh, T.L. and Woese, C.R. (1993) The ribosomal database project. *Nucleic Acids Research* 21 (Suppl.), 3021–3023.

MacDonell, M.T. and Colwell, R.R. (1985) Phylogeny of the Vibrionaceae, and recommendation for two new genera, *Listonella* and *Shewanella*. *Systematic Applied Microbiology* 6, 171–182.

MacLeod, R.A. (1965) The question of the existence of specific marine bacteria. *Bacterial Review* 29, 9–23.

Massung, R.F., Liu, L.I., Qi, J., Knight, J.C., Yuran, T.E., Kerlavage, A.R., Parsons, J.M., Venter, J.C. and Esposito, J.J. (1994) Analysis of the complete genome of smallpox variola major virus strain Bangladesh-1975. *Virology* 201, 215–240.

Orla-Jensen, S. (1909) Die haptlinien des naturlichen bakteriensystems. *Zentralblatt für Bakteriologie, Parasitenkunde, Infektions-krankheiten und Hygiene* 2, 22, 305–346.

Pacini, F. (1854) Osservazione microscopiche e deduzioni patologiche sul Cholera Asiatico. *Gazzetta Medica Italiana Toscana Firenze* 6, 405–412.

Pringle, C.R. (1993) Virus taxonomy update. *Archives of Virology* 133, 491–495.

Roszak, D.B. and Colwell, R.R. (1987) Survival strategies of bacteria in the natural environment. *Microbiology Reviews* 51, 365–379.

Schmidt, T.M., DeLong, E.F. and Pace, N.R. (1991) Analysis of a marine picoplankton community by 1GS rRNA gene cloning and sequencing. *Journal of Bacteriology* 173, 4371–4378.

Sneath, P.H.A. (1984) Bacterial nomenclature. In: Krieg, N.R. and Holt, J.G. (eds) *Bergey's Manual of Systematic Bacteriology*, vol. 1. Williams & Wilkins, Baltimore, pp. 19–23.

Sneath, P.H.A. and Sokal, R.R. (1973) *Numerical Taxonomy: The Principles and Practice of Numerical Classification*. W.H. Freeman, San Francisco.

Sneath, P.H.A., Lapage, S.P., Lessel, E.F., Skerman, V.B.D., Seeliger, H.P.R. and Clark, W.A. (eds) (1992) *International Code of Nomenclature of Bacteria*. American Society for Microbiology, Washington, DC.

Stackebrandt, E. (1991) Unifying phylogeny and phenotypic diversity. In: Balows, A., Truper, H.G., Dworkin, M., Harder, W. and Schleifer, K.-H. (eds) *The Prokaryotes*, 2nd edn, vol. 1. Springer-Verlag, New York, pp. 19–47.

Stackebrandt, E. and Goodfellow, M. (eds) (1991) Introduction. *Nucleic Acid Techniques in Bacterial Systematics*. John Wiley and Sons, Chichester, UK, pp. xix–xxix.

Staley, J.T. and Krieg, N.R. (1984) Classification of procaryotic organisms: an overview. In: Krieg, N.R. and Holt, J.G. (ed.) *Bergey's Manual of Systematic Bacteriology*, vol. 1. Williams & Wilkins, Baltimore, pp. 1–4.

Stanier, R.Y. and van Niel, C.B. (1941) The main outline of bacterial classification. *Journal of Bacteriology* 42, 437–466.

Stanier, R.Y., Adelberg, E.A. and Ingraham, J.L. (1986) *The Microbial World*, 5th edn. Prentice-Hall, Englewood Cliffs, NJ, 689 pp.

Torsvik, V., Salte, K., Sorheim, R. and Goksoyr, J. (1990) Comparison of phenotypic diversity and DNA heterogeneity in a population of soil bacteria. *Applied and Environmental Microbiology* 56, 776–781.

van Niel, C.B. (1943) Biochemistry of microorganisms. *Annual Review of Biochemistry* 12, 551–586.

Wayne, L.G., Brenner, D.J., Colwell, R.R., Grimont, P.A.D., Kandler, O., Krichevsky, M.I., Moore, L.H., Moore, W.E.C., Murray, R.G.E., Stackebrandt, E., Starr, M.P. and Truper, H.G. (1987) Report of the *ad hoc* committee on reconciliation of approaches to bacterial systematics. *International Journal of Systematic Bacteriology* 37, 463–464.

Wilson, E.O. (1988) The current state of biological diversity. In: Wilson, E.O. and Peter, F.M. (eds) *Biodiversity*. National Academy Press, Washington, DC, pp. 3–18.

Xu, H.-S., Roberts, N.C., Singleton, F.I., Attwell, R.W., Grimes, D.J. and Colwell, R.R. (1982) Survival and viability of non-culturable *Escherichia coli* and *Vibrio cholerae* in the estuarine and marine environment. *Microbiology and Ecology* 8, 313–323.

The Microorganisms: A Concept in Need of Clarification or One Now to Be Rejected?

G.A. ZAVARZIN

Institute of Microbiology, Russian Academy of Sciences, Prospekt 60 let Octobra 7/2, Moscow 117312, Russia

Conventions

Conventions signed in 1992 in Rio de Janeiro make for a new international legal environment and there is a need to accommodate to it both by industrialists and scientists. The scientific community is slow and ignorant of changes in spite of the fact that scientific data presented via the mass media influenced public opinion which in turn compelled governments to act.

The Convention on Biological Diversity introduces a new tripartite formula 'plants, animals, and microorganisms' which supersedes the archaic 'plants and animals' concept for living organisms. This formula makes no attempt to clarify our understanding of what is meant by the term 'microorganisms'. The tripartite formula is supposed to include all kinds of living organisms; however, it was supplemented by inclusion of genetic resources, which are defined in a qualitative way, including not only organisms but 'parts thereof' carrying genetic information.

The crucial decision of the Convention is acknowledgement of national sovereignty over genetic resources on the territory under national jurisdiction. It brings a number of considerations.

Attention was drawn to the 'countries of origin', a concept which seems to be ridiculous for microbes whose country of evolutionary origin could be Pangaea. Is the country of origin for microbial species the laboratory where it was isolated from the dust?

The fastest group to react to the Convention on Biological Diversity were ornithologists who include more than half of all UK environmentalists.

The concept of migratory species was one of the most discussed scientific issues. Are microbes 'migratory' since some fungi and bacteria are distributed all over the planet by aerosols?

Attention was concentrated on living organisms inhabiting terrestrial environments under national jurisdiction. The question of the international oceans was dropped as well as organisms there. Might microbes be regarded as international as fishes?

Botanists followed ornithologists in activity, indicating the importance of tropical plants for pharmaceutical biotechnology; they noted the market value of agricultural biodiversity but did not mention its role in the struggle against famine. However, there was a need to stress the fact that antibiotics are produced by microbes.

The immediate problem for microbiologists regarding national jurisdiction is the fate of culture collections which have international roles in maintaining type strains but in fact are supported by a few active governments and organizations. Major service culture collections can only function as international with free access for scientific purposes. But type cultures do bear genetic information which might be used for biotechnological purposes. Contradictions would appear to be solved for patent strains involved in intellectual property rights under the Budapest Treaty, but how should contradictions be solved now when a culture is a national, practically governmental, property? That affects the Bacteriological Code which has living type cultures. The agreement between international scientific rules adopted by non-governmental unions is perhaps a question for a future Conference of the Parties and its scientific committee.

A solution might be found in the substitution of viable type cultures of bacteria by the sequences of their nucleic acids, and there is little doubt that supporters of this approach will be found in the laboratories. It should be noted that the concept of microbial biodiversity was outside the classic phylogenetic way of thinking as until recently there was no approach to the phylogeny of prokaryotes. After Woese (1987), bacterial phylogeny has the most solid basis but the consortium of life and earth sciences has moved in another direction: that is the sustainability of the biosphere, not the origin of the species, as the main problem for humanity.

What might be the result of substitution of the functional concept by a phylogenetic one based on nucleic acid sequences? Vast practical perspectives are obvious. Phylogeny will develop, but at the expense of functional uniqueness which depends on more peripheral adaptive genetic information than ribosomal functions. Identification of bacteria in their natural habitats will make a breakthrough in the coming decade. Molecular approaches advance the identification and quantification of bacterial species, but eliminate morphophysiological genera of the past which were operational units with probable functions in the ecosystem. Another approach might be to estimate a quantity of a certain enzyme, e.g.

methanemonooxygenase, say in 1 ml of pond water.

The contribution of microbes to biology was paradoxical. In descriptions of plant communities, species composition and their frequency make up the prime factual material, ecophysiology being a rather recent invention. Since characteristics of bacteria were based on their functions, the shortcomings in identification of bacterial species of non-medical significance turned out to be an advantage in understanding their role in nature as biogeochemical agents. The concept of microbial ecology was formulated about a century ago by Winogradsky (1952). He ascribed functional bacterial diversity as constructing an 'organism' (the word 'system' was not in use at that time) which is responsible for turnover of the matter on the Earth. For some reason he did not include this fundamental paper in his *Microbiologie du sol*. However, the idea, intrinsic to the Russian science of the late 19th century (which influenced the world no less than literature or music), was inherited by V.I. Vernadsky for his biogeochemical cycles which focused more on the accumulative role of organisms such as corals or molluscs rather than their catalytic role. The concept of a universal biological system became a paradigm of modern ecology which encouraged politicians in Rio de Janeiro to accept a Climate Convention (which is in fact a 'convention on greenhouse gases', many produced naturally mainly by microbes). The Summit concerns the interaction of human society and the life-support system; it is anthropocentric. However, it deals not only with partial problems but with the system as a whole.

Public Adoption

Here we can return to criteria which drive the international community to integrated action. Governmental agreements should be induced by public consensus corrected by a suitable technical approach and supported by scientific knowledge. However, public opinion is only a reflection of impressions gained from the media which for obvious reasons is concentrated on animals since according to the old definition 'animals are those beings which move' and thus only they are actors for movies. Landscape remains in the background of the scene. The invisible creatures, by definition, cannot be actors in visual entertainment. They appear on the screen as the invisible causes of human suffering and public opinion is surely turned against preservation *in situ* (in line with WHO opinion about human pathogens).

The natural scientist has an inverse scale of priorities: terrestrial animals are at the top of the trophic pyramid, but basic processes occur on the level of plants and invisible organisms are responsible for biogeochemical cycles which determine the sustainability of the biosphere. It is a peculiar fact that the Climate Convention which is dealing with the stability of the biosphere includes for consideration forests as sinks for CO_2 but has no direct interest in the decomposition of organic matter as a source of CO_2. This source from the terrestrial environment under national jurisdiction exceeds the industrial one on the global scale by about 10 times. Even for an industrialized country such as Russia – one of the main CO_2 emitters – soil respiration exceeds industrial emission by not less than three times. The contribution of animals, mainly inhabitants of the soil, is about 5% on the global scale. The main input is made by invisible organisms – microbes.

There is a myth supported by the popular media about tropical forests as the 'lungs of the planet'. However, oxygen in the atmosphere is stoichiometrically equivalent to the organic carbon remaining undecomposed by microbes. The immediate conclusion is that in the terrestrial environment the regions which are the sources of global O_2 (and sinks of CO_2) could easily be mapped on the maps of the organic carbon content in the soil. These maps indicate boreal regions, with high precipitation and groundwater levels, as global sources of oxygen due to the inability of microbes to decompose lignocellulosic material under waterlogged, anaerobic conditions. Another indication on 'lungs' is the seasonal oscillation of CO_2 in the atmosphere: these oscillations are absent in the oceanic Southern Hemisphere and most obvious in the continental Northern Hemisphere.

Tropical regions are obviously the habitat of biological diversity. Boreal regions are poor in diversity (except for mycorrhizal and lichenized fungi); the large plains of Eurasia are covered by monotonous taiga and forested bogs, but they are responsible for the biogenic gas cycle in the atmosphere and for sustainability of the Earth system. The contradiction between sites of maximal macrobiological diversity and sites of large-scale cycling leads to questions concerning common statements on the role of biodiversity in sustainability. Are the coral reefs the most important sites for contributions to life in the ocean? From the microbiological point of view the problem is not in high species diversity but in completeness (or incompleteness) of the trophic system within the community. The heretical question is: is excessive biological diversity the cause of instability?

It is accepted that atmospheric oxygen (and what is less noted, combined nitrogen) is the prerequisite for eukaryotic multicellular life. It appeared on Earth in late Proterozoic (vendian) time more than 0.5 gigaannum (Ga) ago. This is a time when macroorganisms are first observed in the fossil record and the Phanerozoic history of plants and animals begins. This makes a very definite border in time between micro- and macro-life: all that was before the first animals belongs to the field of invisible life to which an old term 'Cryptozoan' was applied. Cryptozoan is equivalent to 'invisible to the naked eye', that is, microbes. Cryptos (colloquial for communists in the *Oxford English Dictionary*) formed an interdependent sustainable community. This 'crypto-life' produced 100 m-high fossil structures as stromatolitic belts on the ecotones of the ancient continents and microscopy studies have discovered trichomic microorganisms in 3.5 Ga Archaean rocks in the beginning of the fossil record. This 3 billion year time span of exclusively microbial life makes the term 'microorganisms' unavoidable in Natural Sciences. That was a true era of the 'Microbial World'.

The diversity of life was obviously not as great in microbial Precambrian time as it is now, but which biosphere is more sustainable – 3 Ga or 0.5 Ga old? Microbes proved that they can sustain the Earth system without millions of invertebrates but that is not so for a system without microbes. The conclusion is that the thesis 'biodiversity is a prerequisite for sustainability' advocated by zoologists can only be accepted with most serious limitations which make the thesis equivocal.

Preliminaries to Definition

Both in time and in the biospheric system microbes contribute to a very definite and easily delineated domain. The conclusion is that the concept of 'microbe' is unavoidable in Natural Sciences just as it became unavoidable in the legal field. There is a need in usage of somewhat imprecise terms which cover broad and yet undefined fields. 'Group', meaning 'a set of objects', is a good example. These words are needed to indicate subjects which might be discussed further in more detail. Attempts to give precise definitions diminish the ability to communicate initial diffidence. A trivial language precedes a scientific one.

The relationship between the object and the word is not at all straightforward. The medieval idea was that objects have an inherent transcendentality which is manifested in interactions with other objects but their intrinsic value remains unexhausted by any definition. The medieval thinkers might have had a better insight into the problem since they were not preoccupied by the partiality of the objects as is so for reductionalists. The concept of 'genus' is derived from the logic of Thomas of Aquinas and designates the essence of the object whereas the 'specific' is accidental, non-essential (*aliquid*). The taxonomic language of Linnaeus arose from this logic.

The genus is a central concept in diversity. There is a regularity or system of regularities (logos) and all beings (objects, events) realize themselves if they do not contradict this system; they are allowed and fit into the space of logical possibilities. Fitness to logos is an essence of each object. (*Est veritas in omnibus, quae sunt entia.*) It is expressed in most general features inherent to the object (the second meaning comes from genealogy indicating the origin). The cognitive human mind in contradistinction to logos is not a universal system and can comprehend only a part of essence.

This part might be expressed in words but it remains partial and incomplete. The word-concept is an operational element in cognition. It is intuitive and comprehends more than is in its definition or given by descriptive adjectives. Our goal here is much simpler: it is to delineate borderlines between microbes (i.e. 'microorganisms') and other beings ('macroorganisms') in a non-contradictory way. To classify objects means to divide a set of objects into classes (subsets) which seem to us to be equivalent. The reason for classification is to find such a mode of division which reflects our preliminary ideas (prejudice, intuition). From this statement it is evident that there are many classifications and each object might be classified in different ways according to which characteristics seem to be most important in the study. Nevertheless, classification is correct if the object cannot belong to more than a single class (subsets do not overlap). Cognition of the object is substituted by a simpler problem: understanding the difference between objects. But this difference has to be defined.

Toward the Definition

The colloquial term 'microbes' or the more usual 'microorganisms' indicates dimensions as the main difference between two groups of living beings. Textbooks in microbiology do not emphasize dimensions as an important differentiating character, indicating that there are large microbes and small animals. Some *Rotifera* are smaller than some *Infusoria*. In the ecosystem they fulfil the similar function of micropredators in spite of one being multicellular and the other unicellular. We can regard dimensions as a non-essential feature. However, this is wrong. Dimensions determine molecular diffusion as a main transport process outside the cell which is extremely important for osmotrophic organisms. For instance, root hairs are in the same size range as mycorrhiza. Physical transport processes in the habitat neglected by biochemists are determinative not only for ecology but for morphogenesis which gives advantage in certain habitats. For instance, crenophilic microbes escape from diffusional limitations by living in streams. The minute microbial world has another set of limitations as compared to the visible one. From the above we obtain at least three features for definition of microorganisms: dimensions, cellularity, osmotrophy. Not a single one is exclusive.

Another cryptic definition is in the word 'organism'. Root hairs are parts of organisms. Gametes or eggs represent only a stage in the life cycle of an organism. So in the definition of 'microorganism' one might include the words 'minute during all its life cycle'. There immediately appears a notion about the fruiting bodies of fungi and even some prokaryotes such as myxobacteria. However, these can be considered as 'masses of minute organisms'. The word organism is more important to delineation than the lower size limit.

Dealing with genetic resources, understood in an exclusively qualitative way, the Convention on Biological Diversity regards not only organisms but 'parts thereof' bearing genetic information such as cell cultures or nucleic acids. That brings viruses into the scene. The definition of life as heredity is obviously incorrect since life is neither an entity nor even a sum of entities, it is a quality or feature. The word 'heredity' is misleading since it implicates life by itself. By substituting 'heredity' with 'reproducibility' you come to a definition of snow flakes or lattice crystallization. Heredity means the development of a similar being from something which is already alive. Viruses, plasmids and other molecular carriers of heredity belong to the domain of Life Sciences, they fit exactly into 'molecular biology' but they cannot be included in the concept of 'organism'. An organism is something which is organized from components. If viruses are included, then the thesis of cellular structure should be excluded from the definition of 'organism' which has 'cell biology' as its strongest study field. 'Some organisms have cellular structure' sounds suspicious in spite of the fact that some parts of organisms, including humans, may contain non-cellular components. Note that viruses (and virology) is commonly used with the disjunctive 'and'. The pathological origin (including phytopathology) of this concept is as obvious as in the case of 'microbiology and immunology' which belongs to the reaction of a macroorganism and not necessarily to microorganisms.

We are definitely on the safe side if we state that 'microbes are cellular organisms'. Immediately there are the statements that 'all unicellular organisms are microorganisms' or 'organisms which during their entire life cycle remain unicellular belong to microorganisms'. 'Microorganism' is a morphological concept and morphology is regarded nowadays as unimportant in spite of the absence of a rational explanation of its origin. The main reason is that it does not conform to phyletic origin as is exemplified by numerous parallelisms in shape and structure.

Microorganisms are built from primitive prokaryotic cells or composite eukaryotic ones. The border between pro- and eukaryotes is more essential than the border between microorganisms and plants and animals since it includes a set of intrinsic differences on a subcellular level. There is no need to list all these differences as they may be found in textbooks since the time when Stanier *et al.* (1957) introduced the concept of prokaryotes. That is one of the reasons why the concept of 'microorganism' is regarded as chimeric (not more than a eukaryotic cell itself). Another reason is the versatility of microorganisms which is discussed below.

Multicellular microorganisms, both pro- and eukaryotic, might have distinctive morphology (e.g. trichomic or mycelial) and they are regarded as single organisms in spite of low intercellular integrity. Difficulties begin in the borderline cases between multicellular 'colonies' and tissues. Beings whose body is built up from tissues are not microorganisms. There is a great difference between some plant tissues in which cells are potentially able to develop into organisms and animal tissue which is too differentiated. The definition of tissue as 'any coherent substance from which the bodies of plants and animals is built' is a circular one. The word 'tissue' when applied to the macroscopic structures formed by microorganisms is used metaphorically and terms such as 'pseudoparenchyma' may be added. These macroscopic structures are built by microorganisms from non-differentiated cells (which is not exactly true but could be adopted). Differentiation of the cells in the body of an organism might be one of the most promising distinguishing features between macro- and microorganisms.

The versatility of microorganisms is all-embracing. All types of nutrition are found among microorganisms. That makes it impossible to find here differentiating characters but it forces us to use the collective term for this set of living beings. The most obvious result is that only microorganisms can develop an autonomous community, that is, a community where cycles of matter are closed. The validity of this concept is confirmed by our knowledge of the Precambrian biosphere which was a true microbial world. The bacteria were the prime contributors and all other organisms, including eukaryotic microbes, were superimposed on the system formed by bacteria during Archaean to Middle Proterozoic. The true evolution in time disregards cladistic concepts of the genealogical tree; it resembles a trophic pyramid in spite of all the species diversity on its top. With microbes we are moving in a non-Darwinian domain.

The true difficulty begins when the concept of microorganism is to be applied in a phylogenetic context. There are some phylogenetic groups which fall entirely within the borders of microorganisms. *Archaea* and *Bacteria* (what a pity to use this term in a scientific way when it was so convenient as a colloquialism for *Prokaryotae*!) belong entirely to the microorganisms. Some phyla of unicellular *Eukaryotae* belong entirely to the microorganisms. *Fungi* are traditionally included in microorganisms in spite of representing the longest-living beings on the Earth. As lichenized fungi are macroscopic *'lichens'* are on questionable ground between old-style cryptogamists and modern mycologists. Victory is on the side of the mycologists but the role of symbiogenesis in evolution makes all situations with eukaryotes and some lower protists unsure. Many of them represent stable consortia and tend to exhibit sequential symbiosis.

The real difficulty is with some algae. Primitive forms are unicellular, advanced ones are multicellular, and the most differentiated ones form a 'thallome' which is equivalent to tissue. The usual understanding was that members of the *'Thallophyta'* do not belong to microorganisms. In this phylum there is a transition from microorganisms to macroorganisms and the absence of a sharp border in reality indicates the evolutionary origin and the need for both terms. I am convinced that the next breakthrough in biodiversity is in the field of various lower eukaryotes which form a bush land before genealogical trees of plants and animals.

The concept of 'microorganisms' is not essentially phylogenetic. It is important in practice, it is important in ecology, but it only partly coincides with phylogenetic classification – the only one which is regarded to be 'scientific' (i.e. subject to the formulation and testing of hypotheses). I do not accept this view. Classification might be scientific in certain fields of science but any classification can be universal. As a strong exaggeration it might be assumed that phylogenetic classification is not scientific (non-essential) to ecology or community formation, where concepts to be expressed in classification concern interactions between organisms and their environment and which produce a sustainable system with primary importance of trophic relations.

Definition

The definition which is needed should be as close to consensus as possible. The simplest way of constructing it is concentric: to come from the obvious nucleus to diffuse additional characters and to diffident outer circles which make the definition descriptive.

Microorganisms are living beings invisible to the naked eye (except when developing in large masses). Most microorganisms are osmotrophic, few are holozoic. Diffusion is the main limitation to dimensions and thus the most essential feature.

The concept of microorganisms is mainly morphological, it has limited phylogenetic significance. Microorganisms are cellular living beings (organisms) built either from primitive prokaryotic or from composite eukaryotic cells. All prokaryotic organisms belong to microorganisms; some eukaryotes belong to microorganisms. All unicellular eukaryotic organisms belong to microorganisms; some of the multicellular ones belong to microorganisms also. Multicellular ones might have a distinct morphology and in spite of diminished integrity are considered as a single organism (e.g. trichomic, mycelial) sometimes with differentiated cells. However, multicellular ones never have differentiated tissues as have even minute plants and animals. Those multicellular organisms which build a 'thallome' are in the arbitrary region and their ascription either to microorganisms or to plants depends on tradition. This might be a matter of uncertainty.

Physiologically, microorganisms are the most versatile of organisms and cover all known types of nutrition which has allowed them to build an autonomous community which constitutes a system of biogeochemical cycles determining sustainability of the biosphere up to the present time. Microorganisms are persistent over time. The evolutionary microbial world was exclusive from the beginning of the fossil record until the late Precambrian.

The colloquial name for microorganisms is 'microbes' and it is the least limiting word to use.

References

Stanier, R., Doudoroff, M. and Edelberg, E.A. (1957) *The Microbial World*, 1st edn. Prentice-Hall, Englewood Cliffs, p. 8.

Winogradsky, S. (1952) Sur la classification des bactéries. *Annales de l'Institut Pasteur* 82, 125–128.

Woese, C.R. (1987) Bacterial evolution. *Microbiological Reviews* 51, 221–227.

THE EXTENT OF MICROBIAL DIVERSITY II

# Described and Estimated Species Numbers: An Objective Assessment of Current Knowledge

3

P.M. HAMMOND

Department of Entomology, The Natural History Museum, Cromwell Road, London SW7 5BD, UK

Introduction

How accurate an understanding do we have of the overall extent of global species richness? This question – the subject of the present chapter – and the allied question of how the diversity of life on Earth is distributed are of direct relevance to current concerns over the threats that this diversity faces. However, the nature of these threats, their possible effects (e.g. extinctions, reduction in genetic diversity), and how to combat them (conservation efforts) will not be addressed here. Equally, the functional significance of particular segments of biodiversity – community and ecosystem roles, questions of functional redundancy and so on (di Castri and Younès, 1990; Grassle *et al.*, 1991; Solbrig, 1991; Hawksworth, 1992; Hawksworth and Ritchie, 1993) – lies beyond the scope of the present discussion.

The political dimensions of these topics have been dealt with by McNeely (1992), practical conservation issues by many others (e.g. Soulé, 1986; Reid and Miller, 1989), and ethical questions involved in the conservation of biological diversity have also received extensive comment (see Wilson, 1992 for a reasoned approach). Other aspects of the scientific and political agendas that have emerged in response to the 'biodiversity crisis' are treated by Wilson (1985a, b), Wilson and Peters (1988), May (1988, 1991, 1992a), McNeely *et al.* (1990), Ehrlich and Wilson (1991), Solbrig (1991) and Kristensen (1993) among others. The general area of enquiry has been particularly well served by the publication of a comprehensive source-book of information and analysis (Groombridge, 1992), and a well-rounded and eminently readable account of the diversity of life on Earth (Wilson, 1992).

© 1995 CAB INTERNATIONAL. *Microbial Diversity and Ecosystem Function* (eds D. Allsopp, R.R. Colwell and D.L. Hawksworth)

Nevertheless, with theory and explanation (Hutchinson, 1959; Van Valen, 1973; Janzen, 1977; Mound and Waloff, 1978; Cracraft, 1985; Williamson, 1988; Stevens, 1989; Marzluff and Dial, 1991) continuing to run well ahead of data and documentation, our understanding of the actual dimensions of biological diversity and appreciation of how it is distributed remain poor. As a consequence, except for the best-known groups of large animals and plants, few firm conclusions may be reached with respect to current extinction rates (Wilson, 1992). Both the extent to which the planet's biological diversity is curently being eroded and the scale of future losses likely to result from threats that are already posed are extremely uncertain, although these questions, as might be expected, are the subject of much speculation (Jenkins et al. in Groombridge, 1992).

My aims here are to review current knowledge of the numbers of different kinds of organisms that have been formally recognized to date and the number of as yet undescribed kinds that might exist, leaving to one side the at least equally interesting question of what these organisms do. If the focus is on those groups of organisms known or suspected to contribute most to the world's biological diversity at the species-level, this is not intended to imply that the less species-rich are of less significance, either in terms of their study or of conservation. Some of the most vulnerable areas (e.g. islands, freshwater systems) and groups (e.g. large animals) are not especially rich in species. In such instances, however, even small-scale interference may lead to the loss of important species and unique genetic resources.

Although species-richness is only one aspect of the diversity of life (Solbrig, 1991), for sound theoretical as well as operational reasons species play a pivotal role (Stanton and Lattin, 1989) in the measurement of biodiversity. The merit of employing 'biological' species concepts, wherever this is possible, as a basis for describing the elements of biodiversity has been well argued by Wilson (1992). Of course, many organisms, more especially microorganisms such as bacteria, unicellular algae and viruses, are not sexual and cannot be defined on the basis of interbreeding criteria. Nevertheless, the reality is that the asexual world is for the most part just as well subdivided into (at least potentially) easily defined biological taxa as is the sexual world (Templeton, 1989). If bacterial species, for example, are often less easy to define than metazoan species, it is not so much the lack of sex but a poor understanding of how phenotypic variation manifests itself in populations of these organisms that is the problem. A greater emphasis on demographic exchangeability as a criterion for species recognition, as in the 'cohesion species concept' proposed by Templeton (1989), can help in establishing rough equivalence of the units defined as species in asexual and sexual organisms. Contemporaneous rather than historical criteria are largely the most appropriate to employ in species definitions used in measurements of current biological diversity. However, it may be sensible as well as convenient to apply the notion of the independent evolutionary unit, a common element in 'evolutionary' species concepts, to arrays of parapatric semispecies, 'chromosomal races' and so on, for which there is little evidence of any independent long-term evolutionary future, and treat them as single species.

The preceding comments on species concepts have made no mention of the extent of genetic differences between species. Genetic distance is known to vary very substantially, with very low average differences between, for example, most bird and mammal species, and much greater average differences in a number of other groups, including many microorganisms. Although the extent of DNA homology may provide a useful rule of thumb for determining which individuals and populations belong to the same species in poorly known groups such as bacteria (Trüper, 1992), such measures cannot be expected to have a uniform biological significance, even within one particular group of organisms. From the point of view of comparing patterns of species-richness between animals, plants, fungi and microbes (but not, of course, from many other points of view) the absolute extent of genetic variation within and between species is at best something of a red herring. In groups such as the bacteria where genome heterogeneity within species clearly varies greatly (see cases discussed by Bull *et al.*, 1992), too great an emphasis on DNA homology indices could easily encourage the splitting of taxa that would best be regarded as belonging to a single species, if sensible comparisons of microbial and, for example, metazoan species-richness are to be made.

After reviewing the question of how many species have already been described and discussing recent estimates and conjectures as to how many actually exist, both topics covered more extensively by Hammond (1992), I shall take the opportunity to explore a few of the many issues arising from recent discussions of the dimensions of global species-richness, concluding with brief comment on the practical tasks of estimating species-richness in poorly known and speciose groups (Hammond, 1994).

Although the meeting for which this chapter was prepared is devoted to microorganisms, other major groups of organisms provide the basis for most of my observations. However, as the estimation of microbial species-richness is still in its infancy, it may be hoped that future endeavours will be informed by considering critically the methods that have been employed in dealing with other groups. In particular, object lessons furnished by work on other groups may help to avoid some of the more extreme examples of handwaving and hyperbole that questions of species-richness seem especially liable to engender.

Described Species Numbers

The number of species of organisms known to science remains rather inexactly known. Indeed, striking differences can be seen in the figures, varying from less than 1.4 million to more than 1.8 million, for the number of described species currently regarded as valid provided in recent high profile publications on biodiversity. Although this imprecision is clearly of a quite different order from that with respect to our understanding of how many additional species (those as yet unaccounted for by description) exist, it is perhaps more surprising. After all, although already named species undoubtedly represent a small and extremely biased sample of the biota, they do also represent a record of human endeavour, something that our anthropocentric world tends to consider as especially worthy of documentation!

Table 3.1. Estimated numbers of described extant species in major groups (those known or expected to contain in excess of 100,000 species), according to two recent works.

	Wilson (1992)	Hammond (1992)	Probable accuracy
Viruses	1,000	5,000	High
Bacteria	4,800	4,000	High
Fungi	69,000	70,000	High
Protozoans	30,800	40,000	Moderate
Algae	26,900	40,000	Moderate
Higher plants	248,400	268,000	High
Nematodes	12,000	15,000	Low
Molluscs	50,000	70,000	Low
Crustaceans	35,000[a]	40,000	Low
Arachnids	73,400	75,000	Moderate
Lepidoptera	112,000	150,000	Moderate
Coleoptera	290,000	400,000	Moderate
Hymenoptera	103,000	130,000	Moderate
Diptera	98,500	120,000	High
Other insects	147,500	150,000	Moderate
Total (including other groups)	1,413,000	1,700,000	Moderate

[a] Wilson (1992) gives 50,000 as the figure for 'Other arthropod Classes' (i.e. including 'Myriapoda' etc. as well as Crustacea), so the figure of 35,000 for crustaceans is an approximation.

The most recent estimate of the number of described species of organisms that incorporates relatively up-to-date information on a number of the groups that contribute most to the total is approximately 1.7 million (Hammond, 1992). This figure was arrived at by consulting relevant specialist opinion, critically reviewing published counts and estimates, and then making adjustments where necessary to bring figures up to date or to counteract obvious biases. The individual estimates for the more speciose groups and an indication of how accurate they might be expected to be are given in Table 3.1, together with a series of mostly earlier estimates for the same groups provided by Wilson (1992) for comparison.

The level of accuracy of the figures provided here for numbers of described species is not uniformly related to the size of the group or the proportion probably so far described. Accurate figures are already available for bacteria, as well as for birds and mammals, and are available or soon will be for most embryophytic plants. The major imprecisions concern invertebrate animal groups such as molluscs, nematodes and arthropods. Although the estimates given by Hammond (1992) for all of these groups involved a good deal of consultation, substantially improved counts for some, e.g. the major insect groups, could be made without too much difficulty. Generally speaking, insects are well catalogued and the Insecta parts of the *Zoological Record* routinely list newly recorded synonymies as well as newly described taxa. On the other hand, accurate figures for groups such as the Mollusca, for which new synonymies are not listed in abstracting journals and for which existing catalogues are patchy, could be obtained only by extensive recourse to the primary literature.

'Lifecount' and 'Lifelist' Proposals

Apart from representing a very small and biased sample of the biota, and thus seriously limiting any use to which it might be put, e.g. for predicting the actual number of species that occur in nature (see below), the existing partial inventory of life is deficient in its own terms. For all but a few major groups of organisms the process of stock-taking – of compiling and maintaining accurate catalogues of described species and of keeping a regular tally of the number of species recognized as valid – has received no more than episodic attention. Although some may be prepared to devote considerable efforts to the production of scholarly catalogues for the groups that fall within their own special areas of interest, taxonomists are generally reluctant to invest their time in the compilation of more comprehensive lists or in making described species counts. Such reluctance undoubtedly stems in part from awareness of the deficiencies that any comprehensive list of described organisms is sure to exhibit. The existing inventory of described species incorporates the results of much poor taxonomy and includes a great amount of unrecognized synonymy. Together with the arbitrariness that arises from inconsistent application of variable and sometimes inadequate species concepts, these shortcomings dictate that any comprehensive database of described species will be extremely uneven in reliability.

Although the value of accurate stock-taking as seen from the viewpoint of the individual taxonomists who work with poorly inventoried groups of organisms may be debatable, the compilation of a complete, centralized and computerized list of the described biota has been recognized (e.g. Solbrig, 1991) as an obvious first requirement if the task of biodiversity inventorying is to be taken seriously. The broader case for constructing such a list has been put eloquently by May (1986, 1988, 1990, 1991). In practice, it is recognized that this stock-taking process might best proceed by two separate steps: first the compilation of accurate group by group counts of species currently regarded as valid, to be followed by the larger effort of constructing a database listing these species and, as a minimum, indicating the place that each fits into the taxonomic hierarchy (Solbrig, 1991). Although originally envisaged, when coining the working titles 'Lifecount' and 'Lifelist' in 1990 (P.M. Hammond, unpublished), as a stepping-stone to and pilot project for the production of a full 'Lifelist', an accurate tally of described species at family level ('Lifecount') may have considerable value in its own right. This practical task could be achieved with relatively little difficulty and, for most groups, to a high level of accuracy at relatively low cost.

Undescribed Species Numbers

In assessing current knowledge of and views concerning the number of species of organisms that remain to be described we are faced with the not inconsiderable task of sifting out the hyperbole and the handwaving from data that are in fact pertinent. In the sense of establishing what the currently conceivable range of magnitudes of as yet undescribed species-richness is, the task is not so formidable, but making any sense of this, making an objective assessment of the arguments and conjectures is a much more intimidating task. The best one may hope for is a relatively informed review of current thinking, inevitably biased by one's own standpoints on a variety of issues, from speciation processes and species concepts, to the relative significance of past events as opposed to contemporary conditions, and the nature of biases in the past and current descriptive effort.

Placing to one side the merely hyperbolic and, as much as possible, the more ill-founded conjectures, rough and ready estimates of the possible overall species-richness of the major groups of organisms were assembled by Hammond (1992) (Fig. 3.1), giving a 'working figure' total of around 12.5 million species. The figures provided for viruses, bacteria and algae are frankly speculative, whereas those for fungi, protozoans and nematodes also

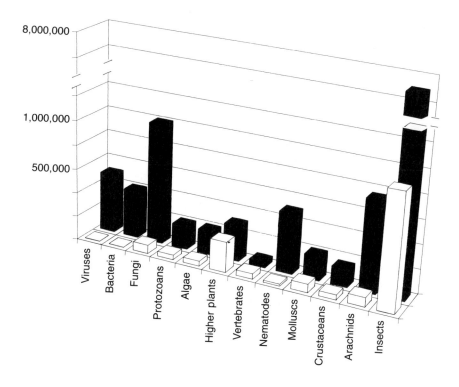

Fig. 3.1. Numbers of described species (open columns) and possibly existing species (dark columns) for those major groups of organisms expected to contain in excess of 100,000 and (for comparison) for vertebrates (after Hammond, 1992).

remain very insecurely based. On the other hand, the extent of the contribution made by the major arthropod groups, including the insects, is supported by evidence from a variety of sources (Hammond, 1990). In the light of available data on numbers of already described species (around one million), local species-richness in the moist tropics, and local : regional species-richness ratios, credible arguments for any substantial lowering of the working figure of 8 million for insects are difficult to envisage. Upper bounds to the possible species-richness of any poorly inventoried group, of course, are much more difficult to predicate with confidence. Nevertheless, for any upwards revision of the insect figure by a factor of more than two or so to be justified, quite radical changes in current understanding (albeit sketchy) of crucial parameters such as turnover rates in the tropics, would be required.

The Record to Date

Among the sources of evidence that might be tapped as a basis for any estimation of the number of species currently inhabiting the earth, the first and most obvious is what has been done so far in terms of description: our existing partial inventory of life. Is it possible to detect patterns in what has been done already that give us a good idea of the scale of the remaining descriptive task? Unfortunately, no. It has been well demonstrated that past and present trends in description do not furnish us with a suitable basis for prediction, at least with respect to the groups of organisms most likely to contain the greater part of biological diversity (Frank and Curtis, 1979; Simon, 1983; Hammond, 1992). Indeed, trend lines may be uninformative or misleading even in groups of organisms, such as birds, for which the descriptive task is nearing completion (Hammond, 1992).

Current description rates and the way in which rates have varied from year to year over the past decade or so (both considered in some detail by Hammond (1992)) illustrate these points well. First, 'growth' rates (the number of new species described in relation to the number already described) mostly differ remarkably little between major groups, with such variation as there is correlating very poorly with educated guesses as to the numbers of species that remain to be described. Modest rates of growth of around 0.5% are shared by some extremely well-inventoried groups (e.g. mammals) and a range of much less well-known groups (e.g. molluscs, beetles), whereas rather higher rates are exhibited by relatively well-known groups such as reptiles and amphibia (1.17%) and fish (1.22%). Although

Table 3.2. The constancy of description rates of new species in animal groups for which more than 1000 species are described per annum, and for all animal groups. Data from *Zoological Record* and Hammond (1992).

	per annum (1979–1988)			
	Highest	Lowest	Mean	SD
Arachnida	1,610	1,145	1,350	162.0
Hemiptera	1,277	903	1,103	117.8
Diptera	1,303	899	1,048	111.8
Hymenoptera	1,496	851	1,196	228.3
Coleoptera	2,843	1,960	2,308	287.0
All insects	8,021	6,758	7,221	342.2
All animals	12,365	10,912	11,599	371.7

SD, standard deviation.

the highest growth rates of all are confined to groups which may reasonably be assumed to be among the least well-known (e.g. fungi 2.43%, nematodes 2.43%) and perhaps indicate some investment here in 'catching up', other poorly known groups (e.g. Diptera, Hymenoptera) have substantially lower rates of growth. Second, absolute description rates, especially in the more sizeable groups, vary remarkably little over time scales of one or two decades (Table 3.2), suggesting that the capacity of the taxonomic 'system' to generate and handle descriptive work is relatively fixed, and unresponsive to short-term influences. The number and type of journals and other vehicles for the publication of descriptive taxonomy are probably among the more important immediate limiting factors.

Although past and current description rates are unlikely to provide any direct indication of the proportion of species remaining undescribed in any of the most speciose groups, careful scrutiny of how rates vary from group to group and with time may be of some value, by revealing or highlighting obvious biases in the way that taxonomic efforts have been apportioned. In turn, an appreciation of these biases may alert us to some of the potential pitfalls to be encountered in the process of estimating overall species-richness, as well as suggesting just where to look for major sources of as yet undisclosed diversity.

First Principles and Processes

What place do general rules with respect to body size, species–area relationships, food-web structure, trophic links, parasite : host ratios, and so on have in the estimation of major species-richness patterns? Although playing a vital role in providing conceptual frameworks and explanation for emerging patterns, such general rules and associated empirical relationships do not, generally speaking, provide a sound basis for actually predicting how many species are likely to be found, for example, in an area of a given size or in a given trophic group. Data available in the literature are inevitably strongly biased towards those which are easy to gather: i.e. they mostly concern large organisms and accessible (generally small) assemblages. Fresh relevant data are needed, but as these involve large assemblages and speciose but poorly known groups, they are hard won and thus always in short supply.

Food webs

Food web structure, a topic that has received a good deal of attention from ecologists (see Cohen *et al.*, 1990), provides an example. Although studies have begun to establish some general structural rules, the question of the numbers of species and overall numbers of links involved in webs of various types remains poorly documented. Although it is feasible to establish the number of links between the species in the simplest of webs, e.g. in a 'container' habitat such as a pitcher plant, rather accurately, as only a few species are involved, few attempts have been made to establish the full species composition of less confined webs, especially those of terrestrial rather than marine or freshwater habitats. Many terrestrial food webs may be expected to involve large numbers of saprophages and polyphagous predators that may feed on hundreds of prey species. Without taking into account pertinent data (currently unavailable) on webs of this type, food-web theory cannot be expected to provide any basis for predicting overall species numbers in any but a few situations. Equally, although interesting generalizations have emerged from the study of food webs involving plants, herbivores and their parasitoids (Hawkins and Lawton, 1987; Hawkins, 1993), most of the data available concern parasitoid loads in individual host species rather than host specificity (Memmott and Godfray, 1993). The overall ratio of parasitoid species to phytophagous insect species in Britain, with the best documented insect fauna in this respect, is in the region of 1 : 1, whereas the mean number of parasitoid species per host species for 2188 British hosts for which Hawkins (1993) assembled parasitoid records is around 5 or 6. Clearly, immensely detailed host specificity information would be needed in addition for parasitoid load data of this type to be used to predict overall numbers of parasitoid species from known numbers of potential herbivore hosts at this or any other scale. Systematic sampling of the full spectrum of herbivores and their parasitoids from an assemblage, without reference to specificities and loads, offers a simpler and more direct route to establishing the empirical relationship between local herbivore and parasitoid species numbers. A final caveat with respect to parasitoid–host links is in order. Although such links are undoubtedly numerous in some food webs, it is far from clear that they dominate terrestrial food webs as a whole, although this conclusion has been drawn (e.g. Lasalle and Gauld, 1993), with an element of circularity, from a consideration of patterns exhibited by a selection of food webs in which plant–herbivore–parasitoid systems are particularly well represented (Schoenly *et al.*, 1991).

Body size

The well-established relationship between body size of organisms, covering some 21 orders of magnitude (Table 3.3), and numbers of species also goes some way to explaining the approximate scale of global species-richness, if not providing, as yet, any firm basis for prediction. For the larger size classes, at least, a rather constant decrease in numbers of extant species with increasing size has been well demonstrated (Van Valen, 1973). However, the rules at the lower end of the spectrum, below body lengths of about 0.2 mm, remain poorly understood, although there have been suggestions (e.g. Andre *et al.* quoted by Stork, 1993) that it is reasonable to extrapolate

Table 3.3. Sizes of various organisms (after Schmidt-Nielsen, 1984). Each step represents a × 1000 increase in mass, so that the size range covers 21 orders of magnitude.

Organism	Size
Mycoplasma	10^{-13} g
'Average' bacteria	10^{-10} g
Tetrahymena (ciliate)	10^{-7} g
Large amoeba	10^{-4} g
Bee	10^{-1} g
Hamster	10^{2} g
Human	10^{5} g
Whale	10^{8} g

downwards. Interestingly, in their quest for life history correlates of taxonomic diversity, Dial and Marzluff (1988) and Marzluff and Dial (1991) found that early age at first reproduction rather than small body size *per se* (although the two are frequently associated) was more often a characteristic of speciose groups. Although we cannot be sure to what extent the apparent decline in species numbers at the very smallest end of the size spectrum is a real phenomenon (May, 1988) and at precisely what body size overall species-richness peaks, some fall-off at the lower end would appear inevitable. Characteristically, for any given taxonomic grouping, the greatest numbers of species are to be found close to the small end of their size spectrum, but not in the very smallest size class for the group. Such a spread around what is likely to be an optimum size for the functioning of a particular category of organism is not inherently surprising.

Processes

Can knowledge of processes (as well as of patterns) make a contribution to our understanding of the scale of global species-richness? The interplay between pattern and process may be intricate, but, where a real grasp of the way in which patterns of species-richness are generated is possible, this may be of considerable help. Often given no more than token attention, the historical dimension is particularly important to take into account (Cornell and Lawton, 1992). Whether a question of particular historical events, or, for example, the way in which cladogenesis operates, an appreciation of history and of process can help to guide us in judging which patterns are likely to be special cases and which of more general interest. Consideration of the likely rôles of major physical determinants of diversity, such as area, temperature, moisture and habitat architecture, along with ecosystem and habitat attributes such as stability, microclimates and so on may also be of benefit. Finally, not to be overlooked in any critical review of data concerning species-richness patterns are the intrinsic characteristics of the organisms in question (their optimum size, which resources they exploit and how, their ability to withstand desiccation, their population structure, tendency to speciate, etc.).

In sum, while providing much useful guidance, our understanding of the attributes of organisms and ecosystems, even when quite profound, does not in itself equip us to make reliable predictions with respect to the number of species awaiting description. This is generally because crucial data (on hyper-diverse but poorly known groups) are currently lacking and seemingly only very remotely accessible, and/or because the answers we seek are needed before the patterns from which we wish to extrapolate can be accurately determined.

Other Approaches

Extrapolation from well-known groups

A simple and direct method of predicting the likely overall species-richness of poorly known but undoubtedly speciose groups is to extrapolate from such reliable data as are already available, using ratios that are known to obtain for much better-known groups such as birds or vascular plants (Hammond, 1992, for examples). This approach depends, of course, on having some reliable and appropriate data for the group in question. For insects this is the case, with the British fauna, for example, known to contain roughly 22,000 species. If we take the number of British species of flowering plants (excluding 'microspecies' and most garden species) to be 2089 (Hawksworth, 1991), and a relatively conservative figure of 300,000 for flowering plant species overall, the number of British species of insects may be projected to a world total of some 3.16 million.

However, studies of well-known faunas and floras such as those of the British Isles demonstrate well how local species-richness of groups such as birds or vascular plants does not track that of others such as lichens or insects at all closely, and the same appears to be true in other parts of the world. Yen (1987), for example, found no correlation between vertebrate and beetle species numbers at 32 sites investigated in southeastern Australia (see Oliver and Beattie, 1993, for further examples).

In general, therefore, although providing a useful starting point, patterns exhibited by well-known groups, which, almost by definition, are atypical in some of the most crucial relevant respects, will at best give very approximate answers. Vascular plant species found in tropical regions as a whole represent at least two-thirds of the world total; mammal species-richness is

Table 3.4. Numbers of species of well-known groups of organisms recorded from various African countries (after Stuart and Adams, 1990). Such data are likely to provide a poor basis for predicting the relative species-richness of insects overall or of other hyper-diverse groups (see text).

	Birds	Mammals	Swallowtails	Plants
Kenya	1,067	314	30	7,500
Ethiopia	836	265	15	5,770
South Africa	774	283	15	20,300
Chad	496	131	7	1,600
Madagascar	250	105	13	11,000

about twice as high in tropical as opposed to temperate countries when countries of similar size are compared (Reid and Miller, 1989). There are no comparable data for species-rich groups of terrestrial invertebrates, but data for individual tropical and temperate sites (Hammond, 1990, 1992) suggest that, at this scale, tropical to temperate ratios are much higher (e.g. 5 (or more) : 1) for some of the more speciose groups. If, as might be assumed from the limited data available, turnover (the increase in species numbers as the area considered is enlarged) in groups such as insects is generally as great as in terrestrial vertebrates, these higher tropical to temperate ratios for single sites will also be reflected at larger (regional) scales.

Among the numerous factors likely to be responsible for marked differences in the species-richness patterns of vascular plants and at least some vertebrate groups on the one hand, and many terrestrial invertebrate groups on the other, special mention may be made of moisture levels at and above the soil surface. For example, the well-documented floristic richness of areas that are at least seasonally dry is not known to be paralleled by an equivalent richness of terrestrial invertebrate species, free-living fungi or microorganisms. Such differences between groups in species-richness patterns may be noted at all scales, including those of country-level and region. Figures for a range of African countries (and Madagascar) provided by Stuart and Adams (1990) provide a simple illustration (Table 3.4). The groups considered exhibit striking differences, both in relative species-richness per country and in the range of variation in species-richness. Differences between countries are most marked in the case of plants. The richest in plant species – South Africa – has 13.7 times more species than Chad, whereas the richest country for mammals (Kenya) has only three times as many species as the least rich (Madagascar). A consensus ranking of the five countries may be achieved using the bird, mammal and plant data by applying equal weight to each. Summing relative species-richness (species-richness of each country over mean species-richness for the five countries) gives a score for South Africa of 4.62, followed by Kenya with 3.80, Ethiopia with 3.05, Madagascar with 2.04 and Chad with 1.49.

Overall species-richness (including terrestrial arthropods, fungi and other speciose groups) for these countries, of course, remains a matter for speculation, but is unlikely to be predicted very well by these consensus figures for birds, mammals and plants. Data for some of the better-known groups of insects, for example, suggest that a lower relative species-richness of South Africa in comparison to Kenya, Ethiopia and Madagascar, and peak species-richness in Kenya rather than South Africa may be typical for the species-rich groups of terrestrial arthopods.

Biases

Are there any other short cuts to a reasonable answer as to the overall dimensions of global biodiversity? It is certainly feasible, for example, to simply sum the opinions of relevant taxonomic specialists, or use the proportion of previously described to as yet undescribed species in selected samples as a basis for extrapolation. However, in any attempt to employ either of these two approaches the influence of any past and/or continuing biases in the pattern of species description is likely to be great.

Various biases in the way that taxonomic effort is and has been apportioned between the major groups of organisms and between geographical regions were identified and discussed by Hammond (1992). Many derive from practical human concerns with crops or health, whereas others have more to do with taxonomic taste and fashion as well as what may be termed taxonomic apparency and taxonomic tractability. As might be expected, small organisms generally exhibit low taxonomic apparency, although this is not necessarily so. For example, some microscopic organisms such as diatoms are easy to collect and examine, when they reveal an intriguing variety of form, and have thus attracted a good measure of taxonomic interest throughout the period that adequate microscopes have been available. However, in general the largest, most accessible and most attractive of organisms have traditionally received and, to some extent continue to receive a disproportionate share of taxonomic attention. Biases of this type and how they have changed with time have been documented in some detail for British Coleoptera (Gaston, 1991b) and for Hymenoptera (Gaston, 1993). A further example is considered here.

How representative of the group as a whole are the species of beetles described in two of the earliest works (Linnaeus, 1758, 1761) adopting the binomial system of nomenclature? First, a very clear bias in favour of European species (the most accessible) is evident. However, on closer examination, a series of other biases are revealed. Linnaeus disproportionately described large species, attractively coloured species and those

Table 3.5. Taxonomic apparency of British Coleoptera at the time of Linnaeus (1758, 1761) by family. Families with 50 or more British species or for which Linnaeus described five or more British species are included. A total of 59 species of other families were described by Linnaeus. Figures for the number of British species, from a list maintained by P.M. Hammond, include introduced and 'doubtful' species.

Family	No. British species	Percentage recognized by Linnaeus (1758, 1761)	Predominant Habitat association[a]	Size class[b]
Staphylinidae	982	1.5	L / De (+)	1–6
Curculionidae	432	9.0	Pl (L / W / +)	2–6
Carabidae	360	12.8	L (+)	2–6
Chrysomelidae	256	20.3	Pl	2–6
Dytiscidae	114	8.8	A	2–7
Cryptophagidae	104	1.9	L / W / +	2–4
Nitidulidae	98	11.2	Pl / W / +	2–4
Leiodidae	95	1.1	De / L / W	2–4
Hydrophilidae	95	11.6	A / De (+)	2–7
Scarabaeidae	80	20.0	De / Pl (+)	3–6
Elateridae	72	25.0	Pl (W / +)	3–6
Ptiliidae	71	0.0	De / L / +	1–2
Cerambycidae	64	57.8	Pl / W	4–7
Scolytidae	60	8.3	W (+)	2–4
Lathridiidae	52	0.0	L / W (+)	2–3
Histeridae	50	8.0	De / W / +	2–5
Coccinellidae	46	43.5	Pl	2–5
Tenebrionidae	44	22.7	L (W / F / +)	2–6
Cantharidae	41	14.6	Pl	3–5
Dermestidae	32	18.7	De / Pl / W	2–5
Melyridae	25	20.0	Pl (W)	3–4
Silphidae	21	38.1	De / L	5–6
Attelabidae	21	28.6	Pl	3–4
Cleridae	14	35.7	W (De / Pl)	3–5

[a] Predominant habitat associations are given first, with other significant associations in parentheses. A, aquatic; De, dung, carrion, decaying fungi, etc.; F, large fruiting bodies of basidiomycete or ascomycete fungi; L, litter (including soil and the soil surface); Pl, on living plants; W, associated with wood of standing or fallen trees.
[b] Size classes are body lengths: 1, 0.5–1 mm; 2, 1–2 mm; 3, 2–4 mm; 4, 4–8 mm; 5, 8–16 mm; 6, 16–32 mm; 7, 32–64 mm. Median sizes are generally at or just below the middle of the range given for each family.

associated with plants. This is particularly clearly demonstrated by considering those North European species known to occur in the British Isles (Table 3.5). Although the modal body length for British beetle species is between 3 mm and 4 mm, only 16 of the 401 British species described by Linnaeus in these works are under 4 mm in length. In the nine major families with predominantly plant-associated species Linnaeus described 194 species (18.4%), whereas in the eleven major families in which species are mostly associated with the soil, litter or decaying matter he described only 117 (6.3%).

Taxonomists' views

However unsystematically they may have been formulated, the views of taxonomic specialists clearly have a valuable part to play in any attempt to gauge what proportion of the world's species remains to be described. Indeed, relevant taxonomists' opinions were taken fully into account (although by no means slavishly followed) in arriving at the working figures for global species-richness of the major groups of organisms (Fig. 3.1) provided by Hammond (1992). Fairly extensive polls of specialist opinion have formed the basis for species-richness estimates provided in several papers, e.g. those of Barnes (1989) for a number of major animal groups, Gaston (1991a) for insects, Winston (1992) for marine macrofauna and Andersen (1992) for eukaryotic algae. A lively exchange on the merits of this approach to estimating the likely number of as yet undescribed species (e.g. Erwin, 1991; Gaston, 1991c) followed Gaston's (1991a) paper. The way in which taxonomists are likely to arrive at their views on the proportion of species so far described and the consequences in terms of reliability have been discussed by Hammond (1992).

The way in which specialists, starting with much the same data, arrive at their conclusions is clearly not always the same as, in some instances, off-the-cuff guesstimates that they provide show little consensus. Figures for the Chromophyte Algae provided by a range of taxonomists, and quoted by Andersen (1992), for example, range from 100,000 or so to more than 10 million! More typically, however, there appears to be a common element in the adoption of a cautious approach, with an apparent unwillingness to go beyond what the taxonomist feels can be vouched for personally. Striking examples of the ultra-cautious approach are provided by some very early estimates, such as that of John Ray, 300 years ago (cited by Westwood, 1833), who considered that a reasonable global total for insect species might lie between 10,000 and 20,000. Counter-examples of course exist: Karsch's (1893) suggestion that the world's insect species number around 2 million is very much closer to the understanding that now exists, 100 years later. More modern estimates by taxonomic specialists (see Hammond, 1992, for examples) also suggest that only very close acquaintance with the group and/or region in question is sufficient for excessive caution to be dispensed with. Quite frequently, estimates for particular countries or regions (of which the estimator may have first–hand knowledge) are too high to be in line with views on world species-richness expressed by (sometimes the same) specialists, suggesting that these global estimates are too conservative. Of course, a lack of caution in dealing with narrowly based data sets may be responsible for generating estimates that are unjustifiably high. Extrapolations that pay no regard to how representative the data are and how observed patterns translate across scales are bound to be unsafe. Sample data on deep ocean sediment meiofauna, on pelagic microorganisms or forest canopy arthropods, for example, have been prone to treatment (although not necessarily by taxonomists) in this way, as have data on the number of parasites associated with particular hosts or the numbers of undescribed as opposed to described species encountered during taxonomic investigation of a little-studied group (see below).

One possible way for the taxonomist to systematize the basis of estimates is to take a manageable and generally smallish group and compare the number of species known before a thorough revision begins with the number recognized after it is completed. The ratio of already described to new species obtained will, of course, depend on how thoroughly new collections have been made, as well as on how thorough the previous descriptive effort had been. How typical the group in question is will depend on the biases in the descriptive record already discussed, and more directly on the group's pattern of representation in geographical regions and habitats. A recent example of employing a detailed taxonomic investigation to generate species-richness estimates, reported on in some detail by Uhlig (1991), illustrates the difficulty of interpreting results obtained in this way.

In Uhlig's study of the staphylinid beetle genus *Erichsonius* the number of species recognized before (115) and after revision (544) gives a ratio of described to undescribed species in the material available for study of around 1 to 4. Uhlig puts a detailed and plausible case that many more species of the genus await discovery, although it is probably fair to regard the 2000 or so given for these (furnishing a described to undescribed ratio of 1 to 25) as the maximum possible rather than an estimate of the most likely figure. From here on, the value of the results for extrapolation depends entirely on just how representative the genus examined is in terms of previous taxonomic attention. Extended to the animal kingdom as a whole a 1 to 25 ratio, using Uhlig's figures for described species, gives 31.5 million overall. Uhlig's own pragmatic approach is to suppose that the ratio of described to undescribed species found in *Erichsonius* is indeed more or less representative for the insects and mites but not other animal groups, for which he allows an approximately 50% increase over described species numbers. Thus Uhlig's conclusion, again using his own figures for described species, is that between about 4 million and 25 million animal species exist globally. Although a greater allowance than that given by Uhlig for undescribed species of such groups as nematodes, spiders, crustaceans might well be justified, the acceptability of the higher of his two figures largely depends on the reasonableness of a 1:25 described to undescribed ratio for the insects and mites as a whole. The evidence for such a high ratio (Gaston, 1991a, 1993; Hammond, 1992), including that from *Erichsonius*, remains uncompelling.

Described : undescribed ratios

Rather than as the by-product of revisional taxonomic work samples of a given, preferably sizeable, group may be taken or assembled for the specific purpose of assessing the proportion of undescribed species they contain (e.g. Casson and Hodkinson, 1991). The usefulness of this approach depends very largely on how representative the sample under consideration is with respect to the ratio of described to undescribed species that it contains. In fact, if we have a very good idea what a 'typical' sample in these terms is, the answer that we seek (the proportion of species that are undescribed globally) is already largely within our grasp. As the completeness (in any sense) of samples used for this type of exercise is of at most secondary importance, rather than use a more or less exhaustive set of samples from one site, it may be more profitable to take samples from a range of geographically well-separated sites. Another factor to be considered is the ease and accuracy with which it is possible to determine which species have already been described, and also the extent to which inaccuracies influence the answers obtained. As already noted (Hammond, 1992), samples of large but poorly known taxa (e.g. many major terrestrial arthropod groups) are likely to contain a very high proportion of undescribed species. In such instances apparently small errors will have a profound effect on the answers obtained. For example, if it is estimated that 95% of the species in a sample are undescribed when in reality only 90% are

Table 3.6. Representation of species in a comprehensive study of staphylinid beetles attracted to carrion-baited traps in Sarawak (Hanski and Hammond, 1987).

	Described		Undescribed		All
	No.	%	No.	%	No.
First 50% of samples	7	15.5	38	84.5	45
Remaining samples	11	29.2	28	70.8	39
All samples	18	20.6	66	79.4	84

Table 3.7. Variation in range size as a possible source of bias when using described to undescribed species ratios in samples from a single 'location' or subregion to predict regional (or global) species complements. In this hypothetical example each of 10 subregions that together comprise a region contains the same number of species (100). One of the subregions has been fully inventoried, with the finding that 30 (i.e. 30%) of the species there have been described previously. The number of described species known to occur in the region as a whole is 84, and thus a simple extrapolation would indicate that the region supports some $84 + (84 \times 7/3) = 280$ species. However, if already described species generally have broader ranges than undescribed species a very different answer may be obtained, as in the example below.

	Inventoried subregion	Increment for each of nine other subregions	Regional total
Described species			
Confined to one subregion	5	5	50
Present in five subregions	10	1	19
Present in all subregions	15	6	15
Total described spp.	30	6	84
Undescribed species			
Confined to one subregion	55	55	550
Present in five subregions	10	1	19
Present in all subregions	5	0	5
Total undescribed spp.	70	56	574
Overall total	100	62	658

new, projections based on numbers of already described species at regional or global levels will be twice as high as they should!

A number of other factors may come into play to make described : undescribed species ratios an unreliable basis for extrapolation. As already noted, the proportion of species in a sample that are already described depends in part on the vagaries of previous descriptive efforts, with biases that may be detectable or undetectable. However, in part also the proportion will depend on other variables, e.g. how samples are taken and their geographical spread. If we (reasonably enough) suppose that species already described are in some sense 'commoner' or more prone to be collected (Hodkinson and Hodkinson, 1993), then they may be over-represented in many types of sample, especially if samples are collected using 'traditional' techniques and are composed of relatively low percentages of the species present in the area(s) sampled. On the other hand, should the previous descriptive effort have been biased in favour of species that are more apparent (larger, for instance) than average, these may be under-represented in some samples, especially those of the standing-crop type. A simple illustration of this point is provided in Table 3.6, where additional species obtained after increasing sample size proved to be twice as likely to be already described as the species collected in the initial sample.

Further uncertainties with respect to scale attend any extrapolation from samples derived from a limited geographical area. For example, as illustrated in Table 3.7, a difference in average range size between described and undescribed species will make any extrapolation across scales unreliable. Realistically, ratios of described to undescribed species in individual samples may best serve as pointers to unexplored or previously under-estimated sources of species-richness, rather than as a basis for extrapolation.

The General Picture

A number of ecosystems (e.g. the deep sea, tropical forests and their canopies) and taxonomic or other groups (fungi, microorganisms, nematodes, mites, insects, parasites) might reasonably be expected to harbour particularly large portions of the world's as yet unassessed species-richness. However, after considering available data on these 'uncharted realms' in turn, the tentative (but scarcely novel) conclusion reached by Hammond (1992) was that we live in a world dominated in species terms by terrestrial arthropods, and that these species are concentrated much more in the moist tropics than elsewhere. However, the actual scale of terrestrial arthropod species-richness in the moist tropics, and the respective contributions of the major groups involved – Coleoptera, Diptera, Hymenoptera and Acari – are still very uncertain. In particular, a poor understanding of how the number of species increases as one moves from smaller to larger spatial scales remains a major impediment to any reliable estimation of regional or global species-richness in these groups. Of course, over and beyond such concerns, even greater uncertainties are faced with respect to nematodes, fungi, deep sea sediment faunas and, especially, with respect to microorganisms.

The Sea

In terms of volume, around 99% of the planet's supply of living space is provided by the oceans (Barnes and Hughes, 1982). The richest parts of the oceans, accounting for some 30% of productivity according to some estimates, are at their margins – the shallow waters adjacent to coasts and in bays and estuaries – although these areas amount to only 10% of the surface of the oceans, and around 0.5% of volume (Cherfas, 1990).

However, despite the great volume of potential living space only some

15% of described organisms are marine, and systematists working on most marine groups are reluctant to predict particularly large numbers of as yet undescribed marine species in most groups (Barnes, 1989; Winston, 1992). As a result of a poll of taxonomists working on many of the major marine groups (not, however, including most microorganisms, nematodes, crustaceans, or the fauna of the deep sea) Winston (1992) concluded that about half of marine taxa probably remain to be described, but noted that with all microorganisms and the deep sea taken into account, undescribed marine taxa might number one million or more. That marine species-richness may indeed be of this order obtains some support from studies of ocean-bottom samples, and is reinforced by recent findings concerning marine productivity. The discovery of a 'microbial loop' in the pelagic food chain makes it clear that old estimates of primary productivity in the oceans are too low (Fenchel, 1988), and it is now equally evident that traditional techniques of measurement have underestimated the amount of organic 'rain' that reaches the ocean floor. Systematic sampling of the deep ocean benthos began late and modern techniques remain relatively unsophisticated and expensive to use (Barnes and Hughes, 1982; Leblond, 1990; Winston, 1992). Nevertheless, some relatively substantial sampling programmes have now been completed, and one of these, in which a series of box-core samples (21 m^2 in total) were taken at depths of from 1500 to 2500 m in the Northwest Atlantic, has provided the basis for some controversial claims (Grassle, 1991; Grassle and Maciolek, 1992). The Northwest Atlantic samples were found to contain 90,677 individuals of macrofauna referable to 798 species, and somewhat similar results have been reported from elsewhere, for example by Poore and Wilson (1993) who found 800 invertebrate species in 10 m^2 of benthos from Bass Strait, Australia. On the basis of extrapolation from such samples, it has been argued that marine species are likely to number 10 million or more (Grassle, 1991), with most of these to be found in deep ocean sediments. In response, the inadequacy of the database from which these projections were made has been pointed out by Briggs (1991), who considered the figure of 10 million to have been reached by an act of 'statistical legerdemain', and further commented on by Hammond (1992) and by May (1992b), the latter concluding on general grounds that a more reasonable total for marine species might be around 0.5 million. Crucial to these arguments are the assumptions made about the pattern of species turnover on the ocean floor. This might reasonably be expected not to follow an even course, and to reflect both the degree of sediment heterogeneity and the spatial scale(s) at which this heterogeneity is mostly expressed. Lacking the architectural complexity that vascular plants give to terrestrial habitats, with less relief, no direct equivalent of moisture gradients, and with microclimatic uniformity the rule, the patchiness of the ocean floor is likely to differ markedly from that of (say) a forest. Although so far little documented, the dominant animal groups in deep ocean sediments, in terms of species-richness as well as abundance and biomass, may well be the interstitial meiofauna, principally nematodes and harpacticoid crustaceans. There are suggestions from a variety of ocean-bottom studies that much of the sediment heterogeneity experienced by these small interstitial organisms is at very small scales (from millimetres to metres). In addition, the results of box-core sampling indicate considerable species turnover at a larger scale, coincident with substantial changes of sediment type and/or depth. However, any great turnover at intermediate scales (hundreds of metres to kilometres), within what might be regarded as a single depth or sediment-type assemblage, has not been clearly demonstrated. At the largest scales (hundreds or thousands of kilometres), with fewer gross geographical barriers and less-marked gradients in temperature and other physical variables than on land, species turnover in the deep ocean benthos may be expected to be relatively low, with individual ranges of species very broad, mirroring those already known to be characteristic of many pelagic species. If this general picture reflects the way in which deep ocean sediment species are distributed even moderately faithfully, the high species-richness that evidently obtains at the smallest scales will not be expected to translate to especially great numbers of species at regional or global scales.

Terrestrial Arthropods

The major groups of terrestrial arthropods, and insects in particular, unarguably contribute substantially to overall global species-richness. Just how great this contribution is, although attracting considerable discussion and comment with a variety of views being expressed (Erwin, 1982; May, 1986; Gaston, 1991a; Hammond, 1992; Stork, 1993), remains controversial. However, the problem of estimating the approximate dimensions of global species-richness is, in fact, distinctly more tractable for insects than for many of the other hyper-diverse groups, including microorganisms. Inevitably, as with deep ocean benthic organisms, some extrapolations have been made from narrow and inappropriate data sets. These generally involve extremely unsafe assumptions, but a growing understanding of relevant patterns and relationships is beginning to furnish us with reference points against which the credibility of such estimates may be gauged. Among the areas of improved understanding are the nature and slope of latitudinal and altitudinal gradients in species-richness, the relative contributions made by the canopy and lower layers of a forest, and the proportions of species belonging to the various trophic groups and associated with various habitats and microhabitats (Fig. 3.2). Work on the size and composition of tropical assemblages (e.g. Barlow and Woiwod, 1989; Noyes, 1989; Hammond, 1990; Casson and Hodkinson, 1991) has contributed to the task of determining the relative species-richness of some of the major terrestrial

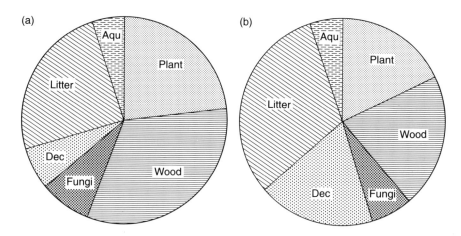

Fig. 3.2. Proportional representation of 'habitat-groups' among the beetle species found in a 500-ha area of relatively uniform moist tropical forest in Sulawesi (a) and in a 200-ha area of deciduous woodland at Burnham Beeches, UK (b). Data from Hammond (1990 and unpublished) and Hammond and Owen (1995). The habitat associations are: Aqu, aquatic; Dec, decaying matter such as dung, carrion, fallen fruit, etc.; Fungi, large fungus fruiting bodies; Litter, including soil and soil surface; Plant, non-woody parts of vascular plants; Wood, standing and fallen timber.

arthropod groups, but this remains an important area for which much better data are needed. As noted above, an improved understanding of the way in which species-richness of these organisms scales, especially in tropical regions, is an equally urgent requirement. Two topics of current interest and of direct relevance to the way estimates of terrestrial arthropod species-richness are viewed (insect–plant relationships and forest canopy assemblages) are discussed below.

Insects and plants

The bias in the historical pattern of description in favour of plant-associated and more particularly phytophagous species has already been noted. Undoubtedly fuelled by human interest in crops, and also influenced by the greater apparency of plant-associated as opposed to (say) soil-dwelling species, the bias may not have been sufficiently widely appreciated and apparently remains reflected in the accepted wisdom that considerably more than half the world's terrestrial arthropod species (as yet undescribed as well as described) are associated directly or indirectly with living plants (Strong *et al.*, 1984; Price, 1988). In fact, careful examination of well-known terrestrial arthropod faunas such as that of the British Isles and new data on tropical assemblages (e.g. Fig. 3.2) suggest that a much lower proportion of species (probably less than 50%) may be plant-associated, with perhaps half of these feeding directly on living plant tissues (P.M. Hammond, unpublished). Nevertheless, the relationship between terrestrial arthropods and living plants is an extremely important one and, with the pattern of occurrence of vascular plant species relatively well understood, suggests various possible approaches to estimating the magnitude of terrestrial arthropod species-richness. As already noted, at present no short cut to any sort of reliable figure is possible using specificities or loads, as the necessary data are unavailable and only remotely accessible. However, empirical relationships between plant and terrestrial arthropod species numbers, if these can be determined, offer more promise, particularly if plant-associated arthropods represent a constant proportion of the species present at any single site. But how robust, for areas of a similar size, is the relationship between insect and plant species-richness, and how does the relationship scale?

A careful examination of how insect and plant species numbers change with respect to one another at a regional scale has been conducted by Gaston (1992), who concluded that patterns vary substantially from one insect group to another and from region to region. He found no examples of insect taxa for which species-richness increases with area faster than that of plants. However, if latitudinal species-richness gradients are indeed steeper for insects than for plants, as suggested in an earlier section of this chapter, overall insect species-richness at the largest scales might be expected to increase more quickly than that of plants. Country to country and region to region differences in insect to plant species ratios undoubtedly involve a number of factors, but are likely to be heavily dependent on surface moisture. Gaston's (1992) data (his Table 1) incidentally go some way towards making this point. Regions that include extensive arid areas have very low insect to plant species ratios (e.g. 2.5:1 for the Canaries, 2+:1 for South Africa and 4:1 for Australia), whereas in regions with large areas of moist forest ratios of between 10:1 and 20:1 or more are usual. In fact, with a few exceptions (e.g. European countries), the figures provided by Gaston may be expected to underestimate insect:plant species ratios, most markedly in the case of China and certain island groups in the moist tropics. The figures given for Canada (Danks, 1979), where subsequent work has already shown those for Coleoptera, for instance, to be underestimates, and the figure of 3500 insect species given by Cogan (1984) for Seychelles and Aldabra, where recent studies indicate that this should be substantially higher, illustrate this point. Overall, when like in terms of moisture is compared with like, a latitudinal trend towards higher ratios in the tropics seems very probable. The ratio given by Gaston for Arctic North America is probably reasonably accurate at 6:1, as are those for North European countries (13:1 and 20:1, but including only native plants). No estimates are available for continental areas of the moist tropics, but distinctly higher ratios (of 25 or more:1) might be expected on the basis of what is known of relevant local insect species-richness patterns. All in all, insect to plant species ratios appear to offer a promising avenue of investigation, but the quality of the insect data used is clearly crucial. Even more important is the acquisition of reliable data from the areas of the moist tropics where most insect species are likely to occur. Conclusions that may be reached from data sets that are predominantly temperate and/or for islands, as Gaston (1992) notes, inevitably fail to reflect the larger global picture.

Forest canopies and other forest strata

The point that richness of tropical forest canopy arthropod assemblages of the scale suggested in some discussions (e.g. Erwin, 1983, 1988) cannot be sustained by current evidence was made at some length by Hammond (1992). Further evidence may be adduced in support of this view, for example from the studies by Mawdsley (unpublished) in Brunei, where similar differences between canopy and ground level samples (in abundance and species-richness) of beetles to those reported for Sulawesi (Hammond, 1990, 1992) appear to obtain. Although unsupported by extensive sampling at ground level the studies (Basset, 1992; Basset and Arthington, 1992) of arthropods associated with the Australian rainforest tree *Argyrodendron actinophyllum*, as well as data from temperate forest sites (e.g. Hammond and Owen, 1995), also support the view that the proportion of tropical forest species regularly inhabiting the canopy is not overwhelming. In addition, although relevant data are fragmentary, there is little to demonstrate that assemblages of other major groups of organisms, e.g. bacteria (Mitchell and Weller, 1992) and basidiomycete fungi (Ryvarden and Nunez, 1992) are at all rich at canopy level.

Fuller data available from the Sulawesi study (P.M. Hammond, N.E. Stork and Brendell, unpublished) also suggest that in the tract of lowland rainforest investigated most insects taken in the canopy by insecticide fogging are either tourists from lower levels or generalists that are to be found regularly at both canopy and lower levels. The significance to be attached to the Sulawesi data depends heavily on the completeness with which the canopy fauna proper (i.e. excluding tourists) has been sampled, rather than the form of species accumulation curves for canopy samples as a whole, which contain many vagrants as well as stratum generalists (see below) and canopy specialists. Adequate examination of this question in turn rests on the reliability with which vagrants or tourists can be identified. Although not as thoroughly investigated as temperate reference sites (e.g. Hammond and Owen, 1995), extensive and detailed information enabling tourists to be identified as such has been gleaned from the several thousand samples taken during the Sulawesi project (Hammond, 1990) and the more

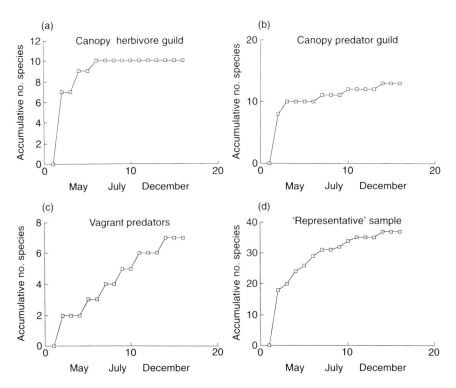

Fig. 3.3. Accumulation of beetle species in canopy fogging samples taken at different times of year at one site in a Sulawesi lowland rain forest (Hammond, 1990, and P.M. Hammond, N.E. Stork and Brendell, unpublished). (a) The 'canopy herbivore guild' (10 species) is composed of broad-nosed phyllophage weevils that are mostly canopy specialists; (b) the 'canopy predator guild' (13 species) is composed of arboricolous Staphylinidae of the genus *Palaminus*; (c) the 'vagrant predators' (seven species) are normally ground-dwelling species of Histeridae, Staphylininae and bembidiine Carabidae; (d) the 'representative' sample includes these 30 species together with seven saprophagous and normally ground-dwelling species of aphodiine Scarabaeidae and oxyteline Staphylinidae.

than one million individual Coleoptera they contain. Data for at least some guilds of canopy specialists (Fig. 3.3a, b) demonstrate clearly that for these groups the programme of fogging in the lowland forest was more than adequate to collect most if not all species present at any one site. If the same holds true on average for other canopy specialists and for generalists, the figures in Table 3.8 are likely to prove a more or less accurate reflection of the picture in the forest in question, bearing in mind, of course, that a number of true canopy species are unlikely to be taken by fogging (Adis *et al.*, 1984; Hammond, 1990, 1992).

Table 3.8. Proportional representation of beetle species in canopy fogging and other samples from a 500 ha area of lowland tropical forest in Sulawesi. Column 1: results for a selection of 51 well-inventoried groups (families or subfamilies); column 2: results for a larger selection of groups, some of them not so well inventoried; column 3: results (in part estimated) for all beetle species collected in the study area. On the basis of their occurrence in samples the 863 species in the well-inventoried group are considered to be ca. 5.5% canopy specialists, ca. 68.5% below canopy specialists, ca. 7.5% generalists (i.e. of regular occurrence in the canopy and at lower levels), and the remaining 18.5% of uncertain category.

Occurrence	1	2	3	Mean
n	863	1636	4815	
Canopy fogging	20.7	23.5	29.4	24.5
Other samples	92.1	89.2	90.7	90.6
Fogging samples only	7.9	10.8	9.9	9.4
Other samples only	79.3	76.5	70.6	75.5
Fogging and other samples	12.8	12.7	19.9	15.1

Further hints that the lower levels of a forest and the soil/litter layers in particular provide a plethora of resources for microorganisms, fungi and small animals are furnished by recent findings that levels of root production and turnover in forest trees are substantially higher than previously thought. Not only does most primary production (leaves, wood, fruit, etc.) bypass canopy herbivores and, generally speaking, end up in, on or near the forest floor, a substantial proportion (roots) also arises there. Figures for biomass, numbers of individuals and numbers of species – the latter at various spatial scales – of animals in forest soils all tend to corroborate the view that more animals of more kinds are to be found where most of the resources occur. Where seemingly contrary results are obtained, it is common to find that major resources present at ground level have beeen ignored or inadequately sampled, or scrutiny of the methods employed reveals that like is not being compared with like. The greater densities of insects and other macro-invertebrates reported by Paoletti *et al.* (1991) for suspended litter and soil around epiphytes in the canopy of a Venezuelan cloud forest, as opposed to the litter and soil of forest floor, for example, may be seen on examination to be a result of sampling vagaries. Forest floor samples were taken to the arbitrary depth of 10 cm, including layers with very little organic matter, and inevitably furnishing lower figures for the density of macroinvertebrates (per cm^2 of soil). Extraction of a much shallower layer of soil would have produced a very different result.

It may also be noted that vertical stratification in forests is not always marked, and a good part of many forests, especially closed forests, may be experienced as a continuum by generalist surface-active animals, including some insects. If their resources are generally distributed all layers of a forest may be utilized by some species that may be regarded as stratum generalists, whereas others may depend on different levels according to life stage, time of day or seasonal changes in feeding or resource availability. Certain types of resource are rare in the canopy, and it is no surprise that canopy assemblages include, for example, few species associated with carrion, dung or aquatic habitats. Of course, canopy samples, especially light trap catches, may contain many or even be dominated by ground-level, often aquatic or riparian, species. However, as trees are the most prominent feature of forest architecture it is only to be expected that many flying insects from the forest floor move through and land on them. Compared with lower strata, forest canopies with their greater exposure to weather (see various papers included by Hallé and Pascal, 1992), may represent unsuitably extreme environments for members of some insect groups. The numbers of species of such groups as basidiomycete fungi and lichens (D.J. Galloway, personal communication) to be found in the canopy may also be limited for this reason.

Certain caveats, however, are in order. Despite a heavy rain of plant and other organic material from above, some moist tropical forests, especially those with hard clay floors have a sparse litter layer and little development of humus, thus limiting resources available for some forest floor species. Tropical dry forests also provide fewer suitable resources for species of the forest floor, and seasonal flooding in inundation forests has obvious effects. However, in view of their limited area, it is likely that neither inundation forests nor dry forests make a very great contribution to tropical species-richness. Inundation forests represent in area less than 1% of all tropical forests and, within any large river system (e.g. Amazon or Orinoco), appear to support a relatively uniform fauna and flora.

Thoughts on other metazoan groups

The possible contribution of other undoubtedly species-rich but poorly known groups such as nematodes (Poinar, 1975, 1983) was discussed by Hammond (1992) who concluded that, in the absence of good data demonstrating substantial turnover at larger spatial scales, a suitable working figure for nematode species should lie in the hundreds of thousands rather than millions. Despite comments to the contrary (e.g. Briggs, 1991), the nematodes probably do not achieve their greatest species representation in strictly terrestrial habitats. In terms of free-living forms at least, nematodes in marine sediments, especially those of the deep sea (see above) may be expected to well outnumber those found on dry land. At all but the smallest spatial scales a relatively low turnover of both terrestrial and other free-living species is indicated by the few available data. For example, more than 200 nematode species have been found in a few cubic centimetres of estuarine mud in Britain, whereas only 700 or so species have been reported from the whole of the North Sea area. The number of parasitic species of nematodes is sure to be high, but evidence of massive species-richness of the order that, for example, many specific ties to individual species of arthropod would entail, is lacking.

Parasites

In view of their large body size and morphological complexity, it is not altogether surprising that many vertebrates and other large animals are known to support large numbers of obligately associated parasites and a range of other commensal organisms (Price, 1980). Plants and sessile colonial invertebrate animals too may be expected, and in some instances are known, to have high parasite loads, but even in these groups it remains to be demonstrated that the majority of species have many specific parasites dependent upon them. In general, small, active, unitary organisms are likely to have fewer obligately associated parasites (or parasitoids). In addition to morphological and behavioural defences, many small animals may not represent suitable resources for exploitation by potential parasites. Their suitability is likely to be limited not only by their size and food value but by their movements, population structure and the foraging ability of potential parasites. Other aspects of parasite biology, including the mode of infection, whether transfer between hosts is horizontal or vertical, whether parasites can survive away from their hosts, and so on, will also be of significance.

Despite the inadequacy of our knowledge of parasites in those groups of organisms, notably terrestrial arthropods, that we may be sure exhibit very great species-richness, the suspicion that these organisms may themselves be greatly outnumbered in species terms by their parasites and endosymbionts is fairly frequently voiced. For example, various parasitologists (quoted by Frank, 1982) have suggested that all insect species are subject to infections by microsporidians (Weiser, 1961), that most harbour one or several species of gregarines, fungi (Madelin, 1966) or nematodes. Certainly, it may reasonably be expected that bacteria, protists (Vickerman, 1992), nematodes and other parasites are of much more general occurrence in arthropods than the smattering of documented associations currently available indicate. However, firm evidence in these respects is lacking, and the degree of host-specificity exhibited by microbial parasites in particular remains unsure. Among the more likely sources of great, as yet undisclosed, richness of parasites in terrestrial arthropods are the tylenchid nematodes (Poinar, 1975) and ectoparasitic fungi belonging to the *Laboulbeniales*, as a relatively high degree of host-specificity is characteristic of both groups. Of course, insects and other terrestrial arthropods are defended against parasites in many ways, for example by secretions, grooming behaviour and mobility, although the latter may increase rather than decrease the chances

Table 3.9. Organisms reported as parasites or possible parasites of staphylinid beetles, a group with some 40,000 described species (after Frank, 1982). The host ranges given are those reported for individual parasite species; 'genera' indicates that recorded hosts belong to more than one genus of the same family, 'families' that recorded hosts belong to more than one family of the same insect Order, etc.

Parasite group	No. species	Host range	Category
Fungi			
Deuteromycotina	13	Orders (some)	Parasites
Ascomycotina			
Laboulbeniales	377	1 genus or group of genera (most)	Parasites
Zygomycotina			
Entomophthorales	1	Families	Parasite
Protozoa			
Gregarinida	4	Genera (most)	Parasites?
Microsporida	1	?	Parasite?
Nemata			
Dorylaimida	2	Orders/families	Parasitoids
Rhabditida	26	Families (many)	Phoretic?
Tylenchida	21	1 genus or group of genera	Parasites
Nematomorpha			
Gordioidea	1	Orders (most)	Parasite/parasitoid?
Arthropoda			
Acarina	4	Orders	Phoretic?
Hymenoptera	17	Subfamilies (some)	Parasitoids (mostly of larvae)

of infection by parasites that are themselves immobile (see below). The low levels of parasitism reported in some insect groups, notably in tropical forests, tentatively suggest that a diffuse population structure may often provide an adequate defence.

Essentially, although there are few direct hints of vast numbers of as yet unaccounted-for parasite species (Brooks and McLennan, 1993), the question of just how heavy parasite loads are in the species-rich but least well-investigated groups of insects and other terrestrial arthropods remains to be answered. This is well illustrated by examining what is known of parasites in the Staphylinidae (Table 3.9), a beetle family with some 40,000 described species and possibly as many as ten times or more that number awaiting description (Gaston, 1991a; P.M. Hammond, unpublished). A notable feature of the list of recorded parasites of Staphylinidae is the low number of hymenopterous parasitoids and the complete absence of parasitic Diptera. It may be noted that the suitability of Staphylinidae as parasitoid hosts is likely to be limited by their considerable mobility in both larval and adult stages, and the infrequency with which they are found on plants and in other exposed situations.

Laboulbeniales: *a case study*

As may be seen from Table 3.9, in number of species, laboulbeniaceous fungi exceed all other recorded parasites of Staphylinidae together. Because of the relative ease with which these ectoparasites may be detected, they offer a particularly good opportunity for determining patterns of host utilization in a major group of terrestrial arthropod parasites. Studies in progress (A. Weir *et al.*, unpublished), based on extensive matched quantitative samples of beetles and other terrestrial arthropods from moist tropical forest and temperate sites, are expected to furnish reasonable estimates for the number of laboulbeniaceous parasites that are typically associated with fully inventoried assemblages of staphylinid and other 'potential' host species.

Around 2000 or so species of this ascomycete fungus group have been described to date. All are obligate ectoparasites of Arthropoda, mainly insects, the great preponderance of records being for adult Coleoptera and Diptera. Although loads in natural populations of the hosts in which they occur may be between 5% and 50% (Frank, 1982), records are very clumped taxonomically, and records for some of the largest groups of terrestrial arthropods are virtually lacking. Transmission is by adhesive spores, normally by direct contact between host individuals (often during copulation). Opinions differ as to the prevalence of indirect transmission, although this may occur in some instances. Spores are short-lived, persisting away from the hosts for at most two weeks. Hulden (1983) identified seven factors (rephrased here) favourable, or in some cases essential (factors 1–3), to the existence of laboulbeniaceous parasites:

1. Hosts overwintering at least partially in the imaginal stage.
2. Overlapping generations of adults.
3. Mating between members of different generations.
4. Large and/or dense host populations.
5. Low isolation between host populations.
6. Stable host populations.
7. Warmth.

To these factors may be added: **8.** Moisture. The pattern of occurrence on particular hosts is also likely to be influenced by host cuticle characteristics, and host defences, including cleaning behaviour.

Generally, a high degree of host specificity is indicated, with most species largely restricted to particular genera or groups of closely related genera, but these patterns may be obscured by the tendency to occur on 'accidental' hosts. Persistence on such accidental hosts is generally limited, either by unsuitable host physiology, or lack of the conditions (e.g. overlapping generations) essential for transmission. However, many aspects of the biology of the parasites, including methods of transmission, the rôle of host defences and effects on host fitness, remain rather poorly understood (Scheloske, 1969; Benjamin, 1971; Hulden, 1983).

The occurrence of *Laboulbeniaceae* on terrestrial arthropods is very clumped, taxonomically speaking, with about 50% of recorded hosts belonging to just two beetle families: Carabidae and Staphylinidae. Although the fungi have been reported from several thousand beetle species these records are restricted to less than 30% of currently recognized beetle families and to just 12 of 24 recognized superfamilies. Some of the families lacking records, e.g. Eucnemidae, Cantharidae, Anobiidae, Melyridae, Mordellidae, Meloidae, Aderidae, Cerambycidae, Bruchidae, Scolytidae, contain large numbers of species (more than 20,000 species of Cerambycidae have been described to date). Despite the oft-emphasized specificity of the parasites, their geographical ranges tend to be large, as may be seen from various regional floras (e.g. Lee, 1986). This is particularly clearly demonstrated by data on the known European flora, in which 146 out of the 356 species and 22 subspecies reported (Santamaria *et al.*, 1991) have also been reported from regions outside of the Western Palaearctic, and 86 of these from truly tropical countries. Such broad distributions and this level of shared species between temperate and tropical regions are remarkable and quite unlike any pattern found in major groups of insects themselves. It is also striking that almost half (159 species) of the *Laboulbeniaceae* known from Europe have been recorded from a single country – Poland. Again the equivalent figure for a major insect group such as Coleoptera is much lower, probably less than 20%. There are few records of *Laboulbeniales* from more arid regions, including those with climates of a Mediterranean type.

Microorganisms

Not at all surprisingly, the pattern and scale of species-richness in the major groups of microorganisms is especially poorly understood. What then are we to make of such (largely anecdotal) evidence as is available in this regard? In the circumstances, the extent to which we expect exceptionally large numbers of as yet undescribed species to exist might reasonably depend less on individual reports than on our broad appreciation of microorganismal biology and the context in which it operates.

An important general characteristic of microorganisms is that they are essentially aquatic. Free-living non-marine forms therefore inhabit fresh water (in global terms very limited in extent), moist substrates, or the films of water, when these are of sufficient thickness, that coat soil and other particles. Dry land is essentially an alien environment and resistant resting stages are frequently required. Exploitation of and concomitant species-richness in terrestrial habitats, to the extent seen in speciose groups of multicellular organisms such as vascular plants, fungi, insects and arachnids, may not be, therefore, an obvious expectation. Indeed, many major groups are apparently poorly represented on land; for example, only 5% of described ciliate species are terrestrial and no terrestrial Foraminifera are known (Vickerman, 1992). It may also be of significance that microorganismal biomass in forest soils has been reported to be substantially lower in the tropics than in temperate regions.

However, in the marine realm, new methods of sampling have revealed the presence of whole groups of previously unknown pelagic algae and other microorganisms in the smallest size classes: the nanoplankton (2–20 µm), including both photosynthetic and heterotrophic forms from a variety of protistan groups, and the picoplankton (0.2–2 µm), including some chlorophytes as well as bacteria and cyanobacteria (Winston, 1992). However, despite suggestions that species-richness for marine algae (Andersen, 1992) and cyanobacteria (Whitton, 1992), for example, may be great, there are few good sample data to go on and extrapolation remains very unsafe. Ranges of many marine species may be expected to be extremely broad; in one study (D. Stackebrandt, personal communication) sediments taken from different oceans exhibited considerable similarities in the 'unculturable' bacteria that they contained.

Recent studies of what had been previously little explored habitats in microbial terms, including 'extreme' environments such as deep terrestrial subsurfaces (Fliermans and Ballewill, 1989), deserts, hot springs and hypothermal vents, have revealed the presence of a range of previously unknown microorganisms, but there is little to suggest that these will make an especially large overall contribution to microorganismal species-richness. Investigations of various soils and sediments (e.g. Ward *et al.*, 1990; Liesack and Stackebrandt, 1992) have also suggested that less than 20% of microorganisms are isolated by standard culture methods, but the proportion of such 'invisible' species or 'unculturables' that are previously unknown taxa remains unclear.

Suggestions that the species-richness of microorganisms lies in the range of tens of millions generally depend heavily on the view that endosymbionts of high host-specificity occur in virtually all insects and members of other speciose metazoan groups, and that symbiotic associations of bacteria with eukaryotic microorganisms are also not only frequent but mostly specific. If, as suggested by Bull *et al.* (1992), the diversity of microorganisms in large measure reflects 'obligate or facultative associations with higher organisms and [is] determined by the spatio-temporal diversity of their hosts or associates', the question of the dimensions of this diversity becomes very largely one of loads and specificities. For the moment, however, bearing in mind the patchiness in occurrence of most parasite groups (see above), there is little to suggest that, for example, mycoplasmas such as *Spiroplasma* species that are obligate associates (in the main) of insects are as ubiquitous or perhaps as specific as some have suggested (Whitcomb and Hackett, 1989).

Finally, what of range sizes of microorganisms? If the rather large geographical ranges reported for *Laboulbeniales* (see above), various of the better-known fungus groups (Hawksworth, 1991a; Oberwinkler, 1992), lichens, some protists and bacteria prove to be at all typical, then it is reasonable to expect species-richness of these groups to scale very differently from that of terrestrial arthropods. In other words, high local species-richness may not translate to especially high diversity at a global scale.

Patterns in View

It is difficult to escape the conclusion that a multitude of separate patterns needs to be taken into account in any attempt to gauge the overall dimensions of global organic diversity. Although many of these patterns are significantly and broadly shaped by temperature, moisture and the distribution and size of the various parts of the Earth's contemporary land surface, they are also influenced at every scale, including the very local, by history. Along with the intrinsic genetic, physiological and other characteristics that are their legacy, the groups of organisms that go to make up the greater part of present-day diversity carry with them the ghosts of interactions past, including reproductive relationships as well those involving, for example, competition or predation. The lottery of which groups became established where and when also may go some way to explaining the current richness of various floras and faunas. It may be unreasonable to expect assemblages, floras and faunas to have simple equivalents in all quarters of the globe. Even if some of the 'rules' are the same in, for example, two climatically and vegetationally similar areas of (say) Australia and Western North America (Westaby, 1988) or New Zealand and Europe, different decks of cards may have been dealt, with the consequence, if we decline to accept that all groups of organisms are equal and interchangeable, that different species-richness patterns will have emerged.

Returning to the question of threats to the biological diversity of the planet posed by human activities, it is clear that we remain woefully short of the data necessary to assess the scale of extinctions that result from major changes in land use and interference with pristine habitats. What should be evident, however, is that the various regions, ecosystems and landscapes of this world are of greatly varying fragility and differ greatly in their propensity for species loss. Once again, history may be expected to have been a major shaper of these differences (Ricklefs, 1989).

The relatively well-known patterns demonstrated by flowering plants and vertebrates already enable us to identify correlates of local and regional species-richness that are almost certain to be of more general significance. In addition, the corpus of biological and natural historical knowledge and experience is sufficiently great to give us some basis for judging the credibility of many species-richness patterns that are predicted from narrow and/or inappropriate databases. Nevertheless, vertebrate and flowering plant data are likely to be of little help in determining some of the major patterns (e.g. marine to terrestrial ratios in local and regional species-richness, the slope of latitudinal gradients) of global species-richness. This depends on the provision of good data concerning more speciose groups. Only these can provide us with a satisfactory understanding of the relative rôles played by geography, temperature, moisture and history.

In the short term, the most significant advances in documenting the dimensions of global biotic diversity are likely to come from a building-block by building-block approach in which extrapolative procedures are used to estimate the species-richness of hyper-diverse groups. For the very least-known of these, including microorganisms, extended anecdotes are also likely to prove useful, as long as these provide reasonable hints as to single-site species totals, turnover levels, or typical range sizes. For the more tractable groups, such as terrestrial arthropods, the compilation of single-site inventories of moist tropical forest and other species-rich sites is an essential step. Single-site data may be used both to calibrate sampling methods and to establish species-richness relationships across spatial scales for some of the more easily inventoried and hopefully typical (in terms of range sizes) groups. The practical problems of obtaining and calibrating appropriate sample data have been discussed, among others, by Coddington *et al.* (1991), Hammond (1990, 1992), Hammond and Harding (1991), Hammond and Owen (1994), Oliver and Beattie (1993), Pearson and Cassola (1991), and more fully by Hammond (1994), and the statistical treatment of sample data by Baltanas (1992), Palmer (1990, 1991) and Colwell and Coddington (1994).

In Conclusion

Some conclusions with respect to species-richness patterns and their investigation that derive largely from my own experience of working on large insect assemblages are listed below. Not all of these views are substantiated by (or even necessarily referred to in) the preceding discussion. Although, to a degree, they may thus be seen as a simple parade of prejudices, they are all based on critical appraisal of at least anecdotal evidence, and their validity is mostly relatively accessible to more systematic investigation.

With respect to all taxa:

1. High local species-richness is a necessary but by no means sufficient basis for concluding that global species-richness of a taxon is exceptionally high.
2. 'Extreme' environments and other newly discovered or investigated sources of fascinating novelties ('new frontiers') are unlikely to harbour vast numbers of previously unknown species.
3. It cannot be assumed safely that species-richness relationships (e.g. between taxa) that hold at one geographical scale hold at others. In some instances, ratios that obtain at different scales may be very disparate.
4. Latitudinal gradients in species-richness of major groups vary, with those for terrestrial arthropods, for example, likely to be steeper than those for vascular plants.

With respect to metazoans:

1. Although many parasites and endosymbionts of metazoans remain undetected and undescribed, evidence to suggest that overwhelming numbers remain to be discovered is lacking.
2. There is unlikely to be an average of one or more species of obligately associated mycoplasma, fungus, nematode or insect parasitoid for each species of terrestrial arthropod.
3. A smaller proportion of terrestrial arthropod species is associated with living higher plants than is frequently supposed.
4. At least as many terrestrial species are involved in food chains containing dead plant tissues as in those based on living green plants.
5. A high proportion of free-living nematode species-richness is expressed at very small scales.

With respect to microorganisms:

1. Relatively broad geographic ranges and low turnover at most scales are likely to be common in many groups.
2. Although 'unculturables' continue to provide a serious impediment to studies of microbial assemblages, they do not necessarily represent a massive hidden source of microorganismal diversity.

With respect to methods of investigation:

1. Extrapolation on the basis of simple ratios based on good sample data is the most practical and reliable approach to the estimation of species-richness in poorly known and speciose groups.
2. In judging the usefulness and validity of ratios the most pertinent criteria

are (i) the quality and extent of calibration, and (ii) whether they make biological sense.
3. Methods that ignore the effects of scale are unlikely to prove reliable.
4. Single-site species-richness data for a wide range of speciose groups are a desideratum. They not only provide the basis for calibrating sampling regimes, but also help to 'ground-truth' ratios such as numbers of parasite species to numbers of host species.
5. The numbers of species described to date, and previous trends in and current rates of description provide little in the way of guidance to the actual scale of species-richness in poorly known but hyper-diverse groups.
6. Biases in the past and present descriptive efforts, if correctly identified, may provide some indication of where the greatest poorly explored sources of species-richness are to be found.
7. Theoretical and other approaches that make no use of sample data on speciose groups may provide a framework for explaining species-richness patterns, but not a reliable basis for predicting the actual scale and major patterns of species-richness.

References

Adis, J., Lubin, Y.D. and Montgomery, G.G. (1984) Arthropods from the canopy of inundated and terra firma forests near Manaus, Brazil, with critical considerations of the Pyrethrum-fogging technique. *Studies on Neotropical Fauna and the Environment* 19, 223–226.

Andersen, R.A. (1992) Diversity of eukaryotic algae. *Biodiversity and Conservation* 1, 267–292.

Baltanas, A. (1992) On the use of some methods for the estimation of species richness. *Oikos* 65, 484–492.

Barlow, H.S. and Woiwod, I.P. (1989) Moth diversity of a tropical forest in Peninsular Malaysia. *Journal of Tropical Ecology* 5, 37–50.

Barnes, R.D. (1989) Diversity of organisms: How much do we know? *American Zoologist* 29, 1075–1084.

Barnes, R.S.K. and Hughes, R.N. (1982) *An Introduction to Marine Ecology*, Blackwell Scientific Publications, London.

Basset, Y. (1992) Host-specificity of arboreal and free-living insect herbivores in rain forests. *Biological Journal of the Linnean Society* 47, 115–133.

Basset, Y. and Arthington, A.H. (1992) The arthropod community associated with an Australian rainforest tree: abundance of component taxa, species richness and guild structure. *Australian Journal of Ecology* 17, 89–98.

Benjamin, R.K. (1971) *Introduction and Supplement to Roland Thaxter's Contribution towards a Monograph of the Laboulbeniaceae*. J. Cramer, Lehre.

Briggs, J.C. (1991) Global species diversity. *Journal of Natural History* 25, 1403–1406.

Brooks, D.R. and McLennan, D.A. (1993) *Parascript: Parasites and the Language of Evolution*. Smithsonian Institution Press, Washington.

Bull, A.T., Goodfellow, M. and Slater, J.H. (1992) Biodiversity as a source of innovation in biotechnology. *Annual Review of Microbiology* 46, 219–252.

Casson, D. and Hodkinson, I.D. (1991) The Hemiptera (Insecta) communities of tropical rain forest in Sulawesi. *Zoological Journal of the Linnean Society* 102, 253–275.

Cherfas, J. (1990) The fringe of the ocean – under siege from the land. *Science* 248, 163–165.

Coddington, J.A., Griswold, C.E., Silva, D., Penaranda, E. and Larcher, S.F. (1991) Designing and testing sampling protocols to estimate biodiversity in tropical ecosystems. In: Dudley, E.C. (ed.) *The Unity of Evolutionary Biology: Proceedings of the Fourth International Congress of Systematic and Evolutionary Biology.* Dioscorides Press, Portland, OR, pp. 44–60.

Cogan, B.H. (1984) Origins and affinities of Seychelles insect fauna. In: Stoddard, D.R. (ed.), *Biogeography and Ecology of the Seychelles Islands*, Monographiae Biologicae 55. Junk, The Hague, pp. 245–265.

Colwell, R.K. and Coddington, J.A. (1994) Estimating terrestrial biodiversity through extrapolation. *Philosophical Transactions of the Royal Society of London, B* 345, 101–118.

Cohen, J.E., Briand, F. and Newman, C.M. (eds) (1990) *Community Food Webs: Data and Theory.* Springer-Verlag, New York.

Cornell, H.V. and Lawton, J.H. (1992) Species interactions, local and regional processes, and limits to the richness of ecological communities. *Journal of Animal Ecology* 61, 1–12.

Cracraft, J. (1985) Biological diversification and its causes. *Annals of the Missouri Botanical Gardens* 72, 794–822.

Danks, H.V. (1979) Summary of the diversity of the Canadian terrestrial arthropods. *Memoirs of the Entomological Society of Canada* 108, 240–244.

Dial, K.P. and Marzluff, J.M. (1988) Are the smallest organisms the most diverse? *Ecology* 69, 1620–1624.

di Castri, F. and Younès, T. (1990) Ecosystem function of biological diversity. *Biology International Special Issue* 22, 1–20. IUBS, Paris.

Ehrlich, P.R. and Wilson, E.O. (1991) Biodiversity studies: science and policy. *Science* 253, 758–762.

Erwin, T.L. (1982) Tropical forests: their richness in Coleoptera and other arthropod species. *Coleopterists Bulletin* 36, 74–75.

Erwin, T.L. (1983) Tropical forest canopies, the last biotic frontier. *Bulletin of the Entomological Society of America* 29, 14–19.

Erwin, T.L. (1988) The tropical forest canopy: the heart of biotic diversity. In: Wilson, E.O. and Peters, F.M. (eds) *Biodiversity.* National Academy Press, Washington, DC, pp. 123–129.

Erwin, T.L. (1991) How many species are there?: revisited. *Conservation Biology* 5, 330–333.

Fenchel, T. (1988) Marine plankton food chains. *Annual Review of Ecology and Systematics* 19, 19–38.

Fliermans, C.B. and Ballewill, D.L. (1989) Microbic life in deep terrestrial subsurfaces. *BioScience* 39, 370–377.

Frank, J.H. (1982) The parasites of the Staphylinidae. *Bulletin (technical) of the Agricultural Experiment Stations, Institute of Food and Agricultural Sciences, University of Florida, Gainesville* 824, i–vii, 1–118.

Frank, J.H. and Curtis, G.A. (1979) Trend lines and the number of species of Staphylinidae. *Coleopterists Bulletin* 33, 133–149.
Gaston, K.J. (1991a) The magnitude of global insect species richness. *Conservation Biology* 5, 283–296.
Gaston, K.J. (1991b) Body size and the probability of description; the beetle fauna of Britain. *Ecological Entomology* 16, 505–508.
Gaston, K.J. (1991c) Estimates of the near-imponderable: a reply to Erwin. *Conservation Biology* 5, 564–566.
Gaston, K.J. (1992) Regional numbers of insect and plant species. *Functional Ecology* 6, 243–247.
Gaston, K.J. (1993) Spatial patterns in the description and richness of the Hymenoptera. In: LaSalle, J. and Gauld, I.D. (eds) *Hymenoptera and Biodiversity*. CAB International, Wallingford, pp. 277–293.
Grassle, J.F. (1991) Deep-sea benthic biodiversity. *BioScience* 41, 464–469.
Grassle, J.F. and Maciolek, N.J. (1992) Deep-sea species richness: regional and local diversity estimates from quantitative bottom samples. *American Naturalist* 139, 313–341.
Grassle, J.F., Laserre, P., McIntyre, A.D. and Ray, C.G. (1991) Marine biodiversity and ecosystem function. *Biology International Special Issue* 23, i–iv, 1–19.
Groombridge, B. (ed.) (1992) *Global Biodiversity, Status of the Earth's Living Resources*. Chapman & Hall, London.
Hallé, F. and Pascal, O. (eds) (1992) *Biologie d'une canopée de forêt équatoriale* – II. M.R. Communications, Lyon.
Hammond, P.M. (1990) Insect abundance and diversity in the Dumoga-Bone National Park, N. Sulawesi, with special reference to the beetle fauna of lowland rain forest in the Toraut region. In: Knight, W.J. and Holloway, J.D. (eds) *Insects and the Rain Forests of South East Asia (Wallacea)*. Royal Entomological Society of London, London, pp. 197–254.
Hammond, P.M. (1992) Species inventory. In: Groombridge, B. (ed.) *Global Biodiversity, Status of the Earth's Living Resources*. Chapman & Hall, London, pp. 17–39.
Hammond, P.M. (1994) Practical approaches to the estimation of the extent of biodiversity in speciose groups. *Philosophical Transactions of the Royal Society of London B* 345, 119–136.
Hammond, P.M. and Harding, P.T. (1991) Saproxylic invertebrate assemblages in British woodlands: their conservation significance and its evaluation. In: Read, H.J. (ed.) *Pollard and Veteran Tree Management*. Richmond Publishing, Slough, pp. 29–37.
Hammond, P.M. and Owen, J.A. (1995) The beetles of Richmond Park SSSI – a case study. *English Nature Science* (in press).
Hanski, I. and Hammond, P.M. (1987) Assemblages of carrion and dung Staphylinidae in tropical rain forests in Sarawak, Borneo. *Annales Entomologici Fennici* 52, 1–19.
Hawkins, B.A. (1993) Refugia, host population dynamics and the genesis of parasitoid diversity. In: LaSalle, J. and Gauld, I.D. (eds) *Hymenoptera and Biodiversity*. CAB International, Wallingford, pp. 235–256.
Hawkins, B.A. and Lawton, J.H. (1987) Species richness for parasitoids of British polyphagous insects. *Nature* 326, 788–790.

Hawksworth, D.L. (1991a) The fungal dimension of biodiversity: magnitude, significance and conservation. *Mycological Research* 95, 641–655.
Hawksworth, D.L. (ed.) (1991b) *The Biodiversity of Microorganisms and Invertebrates: Its Role in Sustainable Agriculture.* CAB International, Wallingford.
Hawksworth, D.L. (1992) Biodiversity in microorganisms and its role in ecosystem function. In Solbrig, O.T., van Emden, H.M. and van Oordt, P.G.W.J. (eds) *Biodiversity and Global Change.* IUBS, Paris, pp. 83–93.
Hawksworth, D.L. and Ritchie, J.M. (eds) (1993) *Biodiversity and Biosystematic Priorities: Microorganisms and Invertebrates.* CAB International, Wallingford.
Hodkinson, I.D. and Hodkinson, E. (1993) Pondering the imponderable: a probability-based approach to estimating insect diversity from repeat faunal samples. *Ecological Entomology* 18, 91–92.
Hulden, L. (1983) Laboulbeniales (Ascomycetes) of Finland and adjacent parts of the U.S.S.R. *Karstenia* 23, 31–136.
Hutchinson, G.E. (1959) Homage to Santa Rosalia, or why are there so many kinds of animals? *American Naturalist* 937, 117–125.
Janzen, D. (1977) Why are there so many species of insects? *Proceedings of the XV International Congress of Entomology*, Washington, DC, pp. 84–94.
Karsch, F. (1893) Wie viel Insectenarten giebt es? *Entomologische Nachrichten, Berlin* 19, 1–5.
Kristensen, N.P. (1993) Biodiversitetens dimensioner: kvantitet og 'kvalitet'. *Naturens Verden* 5, 163–179.
Lasalle, J. and Gauld, I.D. (1993) Hymenoptera: their diversity and their impact on the diversity of other organisms. In LaSalle, J. and Gauld, I.D. (eds) *Hymenoptera and Biodiversity.* CAB International, Wallingford, pp. 1–26.
Leblond, P.H. (1990) The role of cryptozoology in achieving an exhaustive inventory of the marine fauna. *La Mer* 28, 1–4.
Lee, Y.-B. (1986) Taxonomy and geographical distribution of the Laboulbeniales in Asia. *Korean Journal of Plant Taxonomy* 16, 89–185.
Liesack, W. and Stackebrandt, E. (1992) Unculturable microbes detected by molecular sequences and probes. *Biodiversity and Conservation,* 1, 250–262.
Linnaeus, C. (1758) *Systema Naturae: regnum animale* (10th edition). Holmiae.
Linnaeus, C. (1761) *Fauna Suecica*, Editio altera. Stockholmiae.
Madelin, M.F. (1966) Fungal parasites of insects. *Annual Review of Entomology* 11, 423–448.
Marzluff, J.M. and Dial, K.P. (1991) Life history correlates of taxonomic diversity. *Ecology* 72, 428–439.
May, R.M. (1986) How many species are there? *Nature* 32, 514–515.
May, R.M. (1988) How many species are there on earth? *Science* 241, 1441–1449.
May, R.M. (1990) How many species? *Philosophical Transactions of the Royal Society B* 330, 293–304.
May, R.M. (1991) Biodiversity and UK scientific research. In: Hill, J. (ed.) *Conserving the World's Biological Diversity: How can Britain Contribute?* Department of the Environment, London, pp. 14–22.
May, R.M. (1992a) Past efforts and future prospects towards understanding how many species there are. In: Solbrig, O.T., van Emden, H.M. and van Oordt, P.G.W.J. (eds) *Biodiversity and Global Change.* IUBS Monograph 8, IUBS, Paris, pp. 71–81.

May, R.M. (1992b) Bottoms up for the oceans. *Nature* 357, 278–279.
McNeely, J.A. (1992) The sinking ark: pollution and the worldwide loss of biodiversity. *Biodiversity and Conservation* 1, 2–18.
McNeely, J.A., Miller, K.A., Reid, W.V., Mittermeier, R.A. and Werner, T.B. (1990) *Conserving the World's Biological Diversity.* IUCN, WRI, Conservation International, WWF-US and World Bank, Gland.
Memmott, J. and Godfray, H.C.J. (1993) Parasitoid webs. In: Lasalle, J.L. and Gauld, I.D. (eds) *Hymenoptera and Biodiversity.* CAB International, Wallingford, pp. 217–234.
Mitchell, J.G. and Weller, R. (1992) Diversity of bacterial taxa in a rain forest canopy. In: Hallé, F. and Pascal, O. (eds) *Biologie d'une canopée de forêt équatoriale* – II. M.R. Communications, Lyon, pp. 39–40.
Mound, L.A. and Waloff, N. (eds) (1978) Diversity of insect faunas. *Symposia of the Royal Entomological Society of London* 9, i–ix, –204.
Noyes, J.S. (1989) The diversity of Hymenoptera in the tropics with special reference to Parasitica in Sulawesi. *Ecological Entomology* 14, 197–207.
Oberwinkler, F. (1992) Biodiversity amongst filamentous fungi. *Biodiversity and Conservation* 1, 293–311.
Oliver, I. and Beattie, A.J. (1993) A possible method for the rapid assessment of biodiversity. *Conservation Biology* 7, 562–568.
Palmer, M.V. (1990) The estimation of species richness by extrapolation. *Ecology* 71, 1195–1198.
Palmer, M.V. (1991) Estimating species richness: the second-order jackknife reconsidered. *Ecology* 72, 1512–1513.
Paoletti, M.G., Taylor, R.A.J., Stinner, B.R. and Benzing, D.H. (1991) Diversity of soil fauna in the canopy and forest floor of a Venezuelan cloud forest. *Journal of Tropical Ecology* 7, 373–383.
Pearson, D.L. and Cassola, F. (1991) Worldwide species richness patterns of tiger beetles (Coleoptera: Cicindelidae): indicator taxon for biodiversity and conservation studies. *Conservation Biology* 6, 376–391.
Poinar, G.O. (1975) *Entomogenous Nematodes. A Manual and Host List of Insect–Nematode Associations.* Brill, Leiden.
Poinar, G.O. (1983) *The Natural History of Nematodes.* Prentice-Hall, Englewood Cliffs, NJ.
Poore, G.C.B. and Wilson, G.D.F. (1993) Marine species richness. *Nature* 361, 597–598.
Price, P.W. (1980) *Evolutionary Biology of Parasites.* Princeton University Press, Princeton.
Price, P.W. (1988) An overview of organismal interactions in ecosystems in evolutionary and ecological time. *Agriculture, Ecosystems and Environment* 24, 369–377.
Reid, W.V. and Miller, K.R. (1989) *Keeping Options Alive: the Scientific Basis for Conserving Biodiversity.* World Resources Institute, Washington, DC.
Ricklefs, R.E. (1989) Speciation and diversity: the integration of local and regional processes. In: Otte, D. and Endler, J.A. (eds) *Speciation and its Consequences*, Sinauer Associates, Sunderland, MA, pp. 599–622.
Ryvarden, L. and Nunez, M. (1992) Basidiomycetes in the canopy of an African rain forest. In: Hallé, F. and Pascal, O. (eds) *Biologie d'une canopée de forêt équatoriale*

– II. M.R. Communications, Lyon, pp. 116–118.
Santamaria, S., Balazuc, J. and Tavares, I.I. (1991) Distribution of the European Laboulbeniales (Fungi, Ascomycotina). An annotated list of species. *Trebalis de l'Institut Botanic de Barcelona* 14, 1–123.
Scheloske, H.-W. (1969) Beitrage zur Biologie, Okologie und Systematik der Laboulbeniales (Ascomycetes) unter besonderer Berucksichtigung des Parasit-Wirt-Verhaltnisses. *Parasitologische Scriftenreihe* 19, 1–176.
Schmidt-Nielsen, K. (1984) *Scaling. Why is Animal Size so Important?* Cambridge University Press, Cambridge.
Schoenly, K., Beaver, R.A. and Heumier, T.A. (1991) On the trophic relations of insects: a food web approach. *American Naturalist* 137, 597–638.
Simon, H.R. (1983) Research and publication trends in systematic zoology. PhD Thesis, The City University, London.
Solbrig, O. (ed.) (1991) *From genes to ecosystems: a research agenda for biodiversity.* Report of a IUBS-SCOPE-UNESCO workshop, Harvard Forest, Petersham, MA. USA, June 27–July 1, 1991. IUBS, Cambridge, MA.
Soulé, M.E. (ed.) (1986) *Conservation Biology: The Science of Scarcity and Diversity.* Sinauer Associates, Sunderland, MA.
Stanton, N.L. and Lattin, J.D. (1989) In defense of species. *BioScience* 36, 368–373.
Stevens, G.C. (1989) The latitudinal gradient in geographical range: how so many species coexist in the tropics. *American Naturalist* 133, 240–256.
Stork, N.E. (1993) How many species are there? *Biodiversity and Conservation* 2, 215–232.
Strong, D.R., Lawton, J.H. and Southwood, T.R.E. (1984) *Insects on Plants. Community Patterns and Mechanisms.* Blackwell Scientific Publications, Oxford.
Stuart, S.N. and Adams, R.J. (1990) Biodiversity in Sub-Saharan Africa and its islands. Conservation, management and sustainable use. *Occasional Papers of the IUCN Species Survival Commission* 6, i–v, 1–242. IUCN, Gland.
Templeton, A.R. (1989) The meaning of species and speciation: a genetic perspective. In: Otte, D. and Endler, J.A. (eds) *Speciation and its Consequences.* Sinauer, Sunderland, MA, pp. 3–27.
Trüper, H.G. (1992) Prokaryotes: an overview with respect to biodiversity and environmental importance. *Biodiversity and Conservation* 1, 227–236.
Uhlig, M. (1991) Erforschungsstand und Forschungstrends in der Kurzfluglergattung *Erichsonius* Fauvel, 1874, - Uberlegungen eines Entomologen zur Zahl der Tierarten der Weltfauna (Coleoptera, Staphylinidae, Philonthini). *Verhandlungen des Westdeutschen Entomologen Tages* (1990), pp. 121–146.
Van Valen, L. (1973) Body size and numbers of plants and animals. *Evolution* 27, 27–35.
Vickerman, K. (1992) The diversity and ecological significance of Protozoa. *Biodiversity and Conservation* 1, 334–341.
Ward, D.M., Weller, R. and Bateson, M.M. (1990) 16S rRNA sequences reveal numerous uncultured microorganisms in a natural community. *Nature* 345, 63–65.
Weiser, J. (1961) Die Mikrosporidien als Parasiten der Insekten. *Monographien zur Angewandte Entomologie* 17, 1–149.
Westaby, M. (1988) Comparing Australian ecosystems with those elsewhere. *BioScience* 38, 549–556.

Westwood, J.O. (1833) On the probable number of species in the Creation. *Magazine of Natural History* 6, 116–123.
Whitcomb, R.F. and Hackettt, K.J. (1989) Why are there so many species of mollicutes? In: Knutson, L. and Stoner, A.K. (eds) *Biotic Diversity and Germplasm Preservation, Global Imperatives.* Kluwer, Amsterdam, pp. 205–240.
Whitton, B.A. (1992) Diversity, ecology and taxonomy of the Cyanobacteria. In: Mann, N.H. and Carr, N.G. (eds) *Photosynthetic Prokaryotes.* Plenum Press, New York, pp. 1–51.
Williamson, M. (1988) Relationship of species number to area, distance and other variables. In: Myers, A.A. and Giller, P.S. (eds) *Analytical Biogeography.* Chapman & Hall, London, pp. 91–115.
Wilson, E.O. (1985a) The biological crisis: a challenge to science. *Issues in Science and Technology* 3, 20–29.
Wilson, E.O. (1985b) Time to revive systematics. *Science* 230, 1227.
Wilson, E.O. (1992) *The Diversity of Life.* Belknap Press, Cambridge, MA.
Wilson, E.O. and Peters, F.M. (eds) (1988) *Biodiversity.* National Academy Press, Washington, DC.
Winston, J.E. (1992) Systematics and marine conservation. In: Eldredge, N. (ed.) *Systematics, Ecology and the Biodiversity Crisis.* Columbia University Press, New York, pp. 144–168.
Yen, A.L. (1987) A preliminary assessment of the correlation between plant, vertebrate and Coleoptera communities in the Victorian mallee. In: Majer, J.D. (ed.) *The Role of Invertebrates in Conservation and Biological Survey.* Department of Conservation and Land Management, Western Australia, pp. 73–88.

Approaches to the Comprehensive Evaluation of Prokaryote Diversity of a Habitat

4

J.M. TIEDJE

Center for Microbial Ecology, Michigan State University, East Lansing, Michigan 48824, USA.

Introduction

The question of what microbial types make up a community in soil, lake water, the rumen, or other habitats has challenged the microbiologist for nearly a century. We have a coarse level of information on the fungi, algae, protozoa, and bacteria that make up such communities based on microscopic observation and on information from culturable members. But it is also well known that we can culture perhaps only 0.1–5% of the microorganisms we see in most natural samples, and those that we can see by microscopy often do not exhibit sufficient characteristics to distinguish them from many other similar morphotypes. Thus, we only know a small fraction of the extant microbial diversity in most natural habitats. This is fundamentally the same question we face when asked what is the microbial diversity at an ecological reserve or study site, except that this new question is significantly larger because many microbial habitats make up large-scale sites. In the recent past, researchers avoided attempting to describe the microbial composition of a habitat because it was considered an intractable problem. However, with the entry of molecular biology tools into microbial ecology, along with the advances in microscopy, chemotaxonomy, automation, and computer databases and their analyses, it is now feasible to attack this question if done in a coordinated manner.

Value and Scope of a Comprehensive Microbial Diversity Study

The reasons that it is valuable to know the microbial composition of a habitat or site are many. Some strains undoubtedly produce novel and potentially important biotechnology products; many members likely play key roles in recycling nutrients and energy and sustain efficient agricultural, forestry or fishery ecosystems; some members detoxify hazardous wastes that are in our environment; and some members, if discovered, can provide new insight into how life is possible in unusual or extreme niches. Knowledge of the composition of communities should also lead to better management of microbial communities, such as improving efficiency of waste treatment systems, developing more efficacious biological control agents, and enhancing the feed efficiency of ruminant animals. Knowledge of microbial community composition and structure also provides a baseline measure against which the impact of environmental changes can be measured, such as the impact of a chemical spill, desertification, land farming of wastes and thermal pollution.

The methods developed and information learned in a programme aimed at understanding the structure and composition of microbial communities would directly contribute to one or more of these practical goals. One example of a project in which a comprehensive study of microbial diversity would play a unique and valuable role is an All Taxa Biodiversity Inventory, an ATBI (Janzen and Hallwachs, 1993). The purpose of such a study would be to inventory as comprehensively as feasible the biological species at a particular site. An ATBI would be unique in that biologists knowledgeable in all organismal groups would work in a coordinated manner at the same site and thus information between and among species could be evaluated in a manner not previously possible. Furthermore, the attempt to be comprehensive is new for microbiology. It would certainly lead to the discovery of new species and genera and possibly even new families. The general outcome of such a study would be that the 'face of biology' would be revealed for the first time in much the same manner that the 'face of the moon' has been revealed by space exploration (Colwell, 1993).

A comprehensive microbial diversity study offers several novel challenges to the microbiologist. First, a comprehensive study requires an effort to examine the microbial community at greater depth, i.e., to identify the rare community members that we currently ignore. Second, since all higher taxa are probably host to many novel microbes, some strategy must be devised on which higher organisms should be chosen for study of their microbial symbionts. Third, since most sites of interest to conservationists have many different soil and water communities, some criteria should also be developed on which samples would be most valuable for study. Fourth, since the information collected, whether it be on isolates, specimens, DNA sequences, or environmental characteristics will be massive, significant new thought must be given to efficiency of methods, automation, electronic data entry, and usable accessible databases.

In advance of any comprehensive diversity study, the microbiologist will also need to perfect methods of DNA recovery and cleanup from soils, sediments, and the intestinal tract of numerous organisms, develop selective culturing methods for the non-standard organisms, agree on methods that most accurately and quickly provide genus and/or species level of resolution, and perfect sampling schemes and methods for the many habitats of interest within a site. A serious deficiency for many of the microbial groups is the inadequate knowledge base of environmental species. This is needed to serve as a reference against which to compare new isolates or sequences. A second deficiency is the number of experienced taxonomists for various microbial groups. We can only attempt to counter the latter by involving all the microbial taxonomic expertise in the world. The current databases of bacteria used in automated identification systems are heavily dominated by clinical isolates, and it is a common mistake to attempt to identify new environmental isolates using a database dominated by clinical strains. This problem exists because the taxonomy of environmental isolates continues to be neglected. Microbiology research has traditionally focused on only the organisms of obvious economic interest, e.g., especially pathogens, but also organisms important in food, agriculture, and industry.

The microbial world includes the viruses, eubacteria, archaebacteria, filamentous fungi, yeasts, algae, protozoa, lichens, nematodes, and parasites. Each of these groups are studied by specialists trained in the tradition of that field and thus bring very different perspectives, questions, and approaches to a comprehensive study. Many of the challenges faced by working on these different groups of small organisms are common, because most groups have many unknown members, and thus it is difficult to rapidly identify new specimens or cultures. Furthermore, the delineation of species is not clear or easily done with most of these groups, and a large number of individuals and types is expected to be found in any natural habitat and therefore must be efficiently processed. Also, many of the organisms in these groups are associated with higher organisms or specialized niches with little past understanding of these interactions that would be helpful to guide an efficient study. Because of these common challenges, experts from these different fields could likely benefit by pooling their expertise to improve and coordinate an approach to any comprehensive diversity study.

Extent of Microbial Diversity

The ignorance of extant diversity appears to increase as organism size decreases. Mammals and vascular plants are the most comprehensively

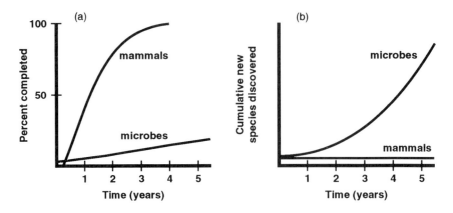

Fig. 4.1. The achievements of an ABTI may be different for different biological groups. In a well-funded ATBI, the species inventory for larger organisms may be nearly completed (a), whereas the microbial studies will result in discovery of many new species (b).

described, insects less so, and the microorganisms harbour the most undiscovered diversity. The evidence for this latter point comes from several lines. First, Hawksworth (1991) estimated on the basis of the ratio of fungi to plants that fungal species on Earth may be in the range of 1.5 million organisms but only about 5% have been described. Torsvik et al. (1990) measured the rates of reannealing of soil DNA and found that there were more than 4000 independent genomes of bacteria in a gram of Norwegian forest soil. Using the suggested definition of bacterial species (> 70% DNA homology, Wayne et al., 1987) and considering species-richness patterns, this would suggest that there are perhaps 20,000–40,000 bacterial species per gram of soil. The problem can be seen if this number is compared to the number of described species, which is only 3100, and most of these are not environmental organisms (Holt, 1984–1989). Further evidence for a high proportion of undiscovered prokaryote species comes from the increased rate of description of not only new species, but new genera, and the fact that most of the 16S rRNA sequences extracted from nature do not match sequences in the database of known organisms (Giovannoni et al., 1990; Schmidt et al., 1991; Liesack and Stackebrandt, 1992). Hence, all lines of evidence suggest that we have only begun to discover the extant microbial diversity on Earth.

There is not yet sufficient information from which to extrapolate the number of prokaryotic species on Earth. The evidence discussed above suggests that it is very high, at least on the spatial scale of grams of soil. What is not known is how different the soil community is at larger scales, e.g., the landscape, regional and global scales. May (1986) described a relationship which showed that diversity increases as body length decreases, at least down to the scale of the small organisms, i.e. < 1 cm length. For the small organisms, however, there may not be separate species in different climatic and geographic regions as there is for larger organisms, thus altering the size–diversity extrapolation to the smallest organisms. Although the extrapolation of microbial diversity to large geographic scales may not reveal an extensive increase in diversity, the already high diversity at small scales, the extensive range of very different microbial niches, and the microbial diversity that has been discovered in the extreme environments, e.g., high salt, high temperature, acid, alkaline, high pressure, suggest that the total microbial species could be in the range of 10^5 to 10^6 species. If this is the case, an attempt at a comprehensive microbial diversity study should be very productive in discovering new microbial diversity. The trends of discovery for a comprehensive diversity study, e.g., an ATBI, are suggested in Fig. 4.1. The expected diversity discovery for large organisms is expected to be low, as illustrated by the curve for mammals (Fig. 4.1b), since most are already known. The opposite is the case for microbes. Figure 4.1a shows that the expectations are also different, namely that an ATBI can be completed for mammals, but the more appropriate goal for the microorganisms is the discovery of new species (Fig. 4.1b). Thus, a major benefit of an ATBI is to reduce our ignorance about the microbial unknown in proportions that could not have been considered before. Figure 4.1b is drawn with an initial accelerating rate of microbial diversity discovery because of the effect caused by focusing on this question, the development of new methods, and the experience that will initially be gained. The later portion of the curve has been drawn linear, however, because it is suspected that the rate-limiting factor will likely be the number of experienced microbial taxonomists and their resources and not the extent of the microbial diversity that is available to be discovered. The ordinate of Fig. 4.1b is dimensionless because this second level of speculation would not be useful.

One Example of a Microbial Biodiversity Study, an ATBI

The challenge of comprehensive diversity study to the microbiologist is large and unique, and even more challenging than a project on the sequencing of a prokaryotic genome. To illustrate, we can draw on the example discussed at a workshop on planning an ATBI organized by Dan Janzen at the University of Pennsylvania (Janzen, 1993). This example assumes a site of $\geq 50,000$ ha on an undisturbed forest preserve in the tropics, and that the site contains a high level of taxonomic and ecological diversity. The study should describe all taxa at the level of species and identify the location where each species can again be found. The scientists should work in a coordinated manner, and both national and international scientists should be involved. In order that the goals of an ATBI are met, i.e. an inventory of taxa and their location, research on development of methods, development of new taxonomy, investigation of population dynamics, or investigating new physiological properties would not be goals of this ATBI. For microorganisms, however, research on method development, sampling strategies, rapid identification, and data handling are essential to maximize the scientific value of the microbial portion of an ATBI. Hence, there is a high priority for microbiologists to initiate this preparatory research immediately so that an ATBI or any comprehensive diversity study will have its maximum value.

Using this particular example, the objectives of the microbial phase of an ATBI (M/ATBI) might be:

1. to assess as comprehensively as possible the microbial taxa at the site;
2. to uncover patterns that relate taxa to their environment;
3. to conduct studies that draw on information from other taxa to produce a value-added product, a synthesis;
4. to substantially advance knowledge of microbial ecology and systematics;
5. to involve a broad spectrum of microbiologists in the study, both national and international, to increase expertise in microbial diversity.

Because it is impossible to complete a comprehensive inventory for microbes on such a site, some decisions that focus and limit the work are necessary. Microbiologists will probably have many opinions on what should be the priority efforts. A framework is offered here that attempts to ensure both the unique purpose and comprehensive nature of a comprehensive study. I suggest that the research plan can be crystallized by using a scheme in which there is a dynamic dialogue between the research question to be addressed and type of sample to be studied. The research question identification and prioritization attempt to ensure that the research chosen is of greatest value and raises the level of information to a level beyond that of an inventory. The focus on what samples are to be taken, e.g. which soil, leaf, mite, snake, etc., to study brings reality to the planning process as well as highlighting the unique character of such a study. It also identifies early which other biologists or environmental scientists need to be involved in the prioritization and planning. Once the research questions and types of samples are identified, then the specific approaches and methods to be used can be determined (Fig. 4.2).

An ATBI can be a vehicle to answer some of the most important and fundamental questions in microbial ecology, and these questions should

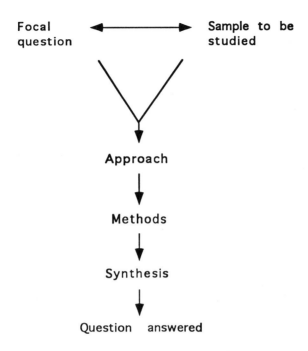

Fig. 4.2. Framework for prioritizing research and identifying an experimental plan for a comprehensive microbial biodiversity study.

shape the priorities discussed above. It is suggested that some of these questions are as follows.

1. Where are new species?
   (a) What is the relationship between environmental difference and genotypic difference? Another way to express this question is what is the microbial area species curve? If such a fundamental relationship were known, we would for the first time have a scientific basis to guide the discovery of microbial diversity. This information would also be of great value to the pharmaceutical and biotechnology industries, which are fed by biodiversity discovery.
   (b) What is the degree of specialization between microbes and their host animal, plant or insect? For pathogens we know there is considerable specialization which would suggest that there are also specialized (thus novel) symbionts which reside in these diverse niches.
2. What are the dominant but uncultured microbes in nature and how different are they from what we already know? We can recover DNA from soil in quantities one hundred times greater than that accounted for by the culturable organisms (Holben *et al.*, 1988). This DNA is presumed to reside

in live cells in order to have remained stable. What is the nature of these life forms? Are they injured forms of what we know, similar to what we know, very different from what we know, or a combination of these in a community or even in a genome?

3. What are the rare microbial forms and how different are they from what we know? In the past, we have ignored rare forms, unless these forms have economic value. This group should be a major source of new diversity discovery, but also the most difficult to discover.

4. Can genotypic (including phylogenetic), phenotypic and ecological information be synthesized to produce a greater understanding of microbial distribution? For example, can an *Arthrobacter* landscape be predicted? The optimum level of understanding would be to interrelate function, organisms and environment; a comprehensive study advances this goal.

5. Can the above information be used to advance the definition of a microbial species? The principle of polyphasic taxonomy has been endorsed by the prokaryotic microbiologists. Information from a comprehensive study aids this goal, thus sharpening the understanding of what is a microbial species.

Given the questions in this scheme of prioritization, then what are the samples? Some samples should reflect the comprehensive nature of an ATBI and therefore should include representative microbial diversity from the major microbial habitats, i.e., soils, plants, animals, arthropods, water, and sediments. However, each of these host environments has many types and component parts, and the challenge comes in prioritizing which of these is to be analysed. Some samples are driven by the specific question, e.g., how does the host reflect microbial diversity, which would suggest sampling along a gradient such as a transect of soil or a phylogenic line of insects.

Martinus Beijerinck, a famous Dutch microbiologist, established a century ago a fundamental paradigm for microbial ecology. He stated that, 'Everything [in bacteria] is everywhere, the environment selects.' I suggest that we are now in a position to refine this paradigm. For example, what genotypic level corresponds to 'everything'? The species, the 'ecovar', the DNA sequence? Also, at what geographic scale is 'everywhere'? The grain of sand, the soil aggregate, a square metre, a catena? Evaluation of this statement also raises the question of what we know about rates of microbial extinction, rates and range of microbial dispersal, and rates and range of gene transfer among microorganisms. An ATBI will give some insight into answering these questions and hence refine Beijerinck's paradigm.

Beijerinck's statement also recognizes the ecological principle that 'the environment selects' which members make up a successful community. Hence, there is expected to be a fundamental relationship between the degree of difference among environments and the degree of difference among the microbes occupying those different environments (Fig. 4.3).

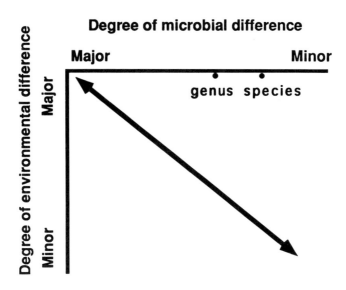

Fig. 4.3. Suggested relationship between the degree of microbial difference and the degree of difference between microbial habitats in which the organisms are found.

This relationship can be useful in considering the approach to an ATBI. As an example, some pathovars have a high degree of host specificity indicating a fine level of both environmental and genotypic difference (Fig. 4.3, lower right). It would be unmanageable to conduct the microbial phase of an ATBI at this level of resolution. At the other end of the spectrum, one might conclude that the Proteobacteria occupy all of the world's soils and waters, which is not a very useful level of resolution. What can be done, however, is to adjust the level of resolution to a position that is both meaningful and manageable. For an ATBI, the species level has been suggested, which is probably at the limit of what is manageable for microorganisms. From a practical perspective, however, many of the major economic properties of microorganisms, such as antibiotic production and pathogenesis, occur at a finer level of difference, e.g. at strain level. Figure 4.3 is also useful in that it emphasizes the equivalence between two key parameters, the organism and the environment. Both must be considered together in designing the specific experimentation. The figure also suggests some flexibility in design. For example, if the methodology is inadequate to adjust the level of genotypic resolution, the level of difference among environmental samples might be varied instead.

Strategy and Methods

The experimental scheme for evaluation of biodiversity should probably include at least four components: the appropriate experimental approach; a means to identify quickly if the organism or DNA sequence has been found before; a rapid coarse level of characterization, i.e. to identify to family or genus; and finer levels of resolution as necessary to confirm species or describe a new species (Fig. 4.4). An essential strategy of any such scheme is to efficiently reduce the number of organisms analysed by identifying the ones of unique value for intensive study. Figure 4.4 shows the approaches and possible methods that can be used in such a progression (left to right) to identify bacteria, fungi (including yeast), algae, and protozoa. The basic approaches are three: obtaining isolates, characterization of total community DNA, and using microscopy. With the new nucleic acid-based techniques it might be questioned, why bother with isolates? There are several reasons why isolates remain important. The level of biological characterization can be much greater with isolates, rare species are much more easily obtained if they can be selectively grown over the other community members, and cultures are necessary for evaluation and production of any products of practical value. Some investors in such a project, including the host country, are likely to want more than a DNA sequence, and isolation and culture characterization are an especially valuable way to involve a larger microbiology research community in the project. The isolation and culture approach should also contribute to the important goal of revitalizing microbial physiology and systematics on a broad scale. The DNA approach is to extract total community DNA from the natural sample, e.g. soil, gut contents, a micro-fungal community on a decaying log, and to identify the members of that community by DNA sequence or by use of specific probes or fingerprints without culturing the organisms. This approach has its greatest value with unculturable organisms. The three approaches are not equally applicable to all microbial groups or families within groups. For example, the microscopic approach is often a more valuable method for many fungi, algae, and protozoa, whereas culturing and DNA analysis are the priority methods for bacteria.

The approach may be coupled with a second strategy – fractioning the community into groups. This has the advantages of reducing the complexity of the community to be dealt with by subsequent methods; it allows group specific methods to be employed, e.g., chemotaxonomy, pigment analysis, specific nucleic acid or antibody probes; and the fractionation itself provides a valuable coarse level of information on structure and composition of the community. Methods for fractionating microbial communities by each of the approaches are available: e.g. culturing by selective growth of physiological or motility groups; separating DNA on the basis of its % G+C versus A+T (Holben et al., 1993) or by group-selective hybridization; and

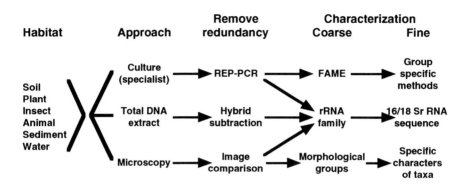

Fig. 4.4. General scheme for total diversity characterization.

separating cells by micromanipulation or by flow cytometry (Olson *et al.*, 1991). Groups sorted in this manner would then be subjected to redundancy reduction and the subsequent steps in Fig. 4.4.

The methods to be used to accomplish redundancy reduction and coarse and fine levels of characterization can be many, and new ones are continually being explored. The criteria for method selection for a comprehensive diversity study include the following: (i) provides the necessary biological information for its role in the analysis scheme; (ii) rapid; (iii) automated; (iv) provides direct electronic database entry; and (v) is well proven for comprehensive type of study. Not all methods of value may meet these criteria, e.g. the Gram-positive versus Gram-negative characterization for bacteria. Such methods are still valuable and should be included. The methods shown in Fig. 4.4 are the ones that meet most of these criteria, and are currently among the best available. The microbiologist typically focuses primarily on the biological value of the method – which is also important to this study – but much more attention must be given to the other criteria because of the large numbers of samples and organisms involved and the need to merge the data with other biological and environmental information in the electronic databases.

The scheme in Fig. 4.4 shows a flow of methods from left to right for a particular approach, but it is equally important that organisms and information obtained from one approach be transferred to the knowledge base developed for all approaches, as is illustrated by the non-horizontal arrows. For example, interesting specimens found by both culture and microscopy will also probably be analysed for information contained in their DNA. Furthermore, information from ribosomal sequence evaluation on uncultured organisms may be helpful in improving isolation and culturing strategies for these organisms.

A large number of methods have been used for microbial characterization and identification, but many of these do not meet the criteria for a comprehensive diversity study. Only those that seem to be most suitable for this type of study are mentioned here. Figure 4.4 lists most of these methods. Several of these and a few others deserve explanation and comment.

Community-level Methods

These are methods in which the total microbial community can be analysed for information about its structure and/or composition. The C_{ot} analysis determines the complexity, i.e. diversity, of DNA sequences in a DNA sample by measuring the rate of reannealing of single strands (Torsvik *et al.*, 1990). The more DNA sequence diversity in a sample, the longer it takes to reanneal. Cross-hybridization analysis is based on the same reannealing principle except that DNA from two communities is reannealed to determine how similar the two communities are to each other (Lee and Fuhrman, 1991). The % G+C analysis is based on separation of community DNA based on its % G+C content (Holben *et al.*, 1993). This is accomplished by using a dye that binds to A–T-rich regions, affecting density and allowing separation by density gradient separation. Since % G+C reflects taxonomy, a profile of the amount of each % G+C group is obtained. Image analysis is just beginning to be used on complex communities, but offers the potential to rapidly determine structure on communities that show some differences in morphological, density and/or fluorescence features. Cell sorting has been used very effectively in oceanography to determine structure and composition of those micro- and picoplankton communities (Olson *et al.*, 1991).

Redundancy Reduction Methods

The REP-PCR (repetitive extragenic palindromic sequences) method is based on using primers for PCR of conserved regions dispersed around the chromosome of most bacteria. After gel separation of the amplified fragments, one can detect siblings or very closely related organisms by noting matching band patterns (de Bruijn, 1992). This analysis requires only a colony and can be completed within 12 h. For redundancy reduction of community DNA, some type of hybrid subtraction method is needed to leave for analysis those sequences not previously seen. This method is yet to be adapted for community analysis.

Methods for a Coarse Level of Characterization

For microbial cultures, total cellular fatty acid analysis (FAME) has probably become the most used method for rapid taxonomic analysis (Sasser and Wichman, 1991; Stead *et al.*, 1992). The results are generally reproducible. The difficulty is that many of the novel strains do not grow on the media for which the database is built and the database, though extensive, lacks many environmental strains causing many isolates to be 'unidentifiable'. A second rapid commercial system suitable for environmental isolate identification is manufactured by BIOLOG Co. (Hayward, California). This system is based on which of 95 substrates in a 96-well plate the microbe oxidizes as detected by reduction of a tetrazolium dye. This method has also become widely used but has similar growth and database limitations that will limit its use in a comprehensive biodiversity study.

Identification of rRNA families has become a particularly attractive approach because of its phylogenic base, its comprehensiveness and its speed. The most attractive versions of this for a comprehensive study include amplified ribosomal DNA restriction analysis (ARDRA) (Vaneechoutte *et al.*, 1992). An advantage of this method is that the primers can be varied from 16S to 23S and from universal to group-specific, and the restriction enzymes can also be varied, thus allowing the method to be as comprehensive or specific as needed. Group-specific probes allow detection by fluorescent microscopy if labelled with a fluorescent probe (Amann *et al.*, 1991), or allow detection of DNA bound to a membrane (Devereux *et al.*, 1992).

Methods for Fine Level of Characterization

Today, the first method to be recommended would be the complete 16S rDNA sequence (Stackenbrandt and Goodfellow, 1991). The recent improvements in automatic sequencing and in the databases (Neefs *et al.*, 1991; Olsen *et al.*, 1991) make this feasible and the information obtained most useful. In a comprehensive study, a partial sequence would most likely be obtained first to determine if the organism (or a close relative) has been studied before. If not, the full sequence would be obtained. The REP-PCR and higher specificity version of ARDRA could also be used for a fine level of characterization.

Conclusion

A comprehensive microbial community analysis is a new concept to the microbiologist. Such studies have substantial basic and applied value. The methodology is ready to be adapted, perfected and automated for such a study. The expertise of the computer scientist, taxonomist, ecologist and engineer needs to be brought together to develop the infrastructure to carry this effort forward to the next stage.

Acknowledgements

The financial support of the US NSF to the Center for Microbial Ecology is acknowledged, BIR9120006 and DER-92 13435.

References

Amann, R., Springer, N., Ludwig, W., Görtz, H.-D. and Schleifer, K.-H. (1991) Identification *in situ* and phylogeny of uncultured bacterial endosymbionts. *Nature* 351, 161–164.
Colwell, R.K. (1993) Counting creatures great and small. *Science* 260, 620.
de Bruijn, F.J. (1992) Use of repetitive (repetitive extragenic palindromic and entrobacterial intergeneric consensus) sequences and the polymerase chain reaction to fingerprint the genomes of *Rhizobium meliloti* isolates and other soil bacteria. *Applied and Environmental Microbiology* 58, 2180–2187.
Devereux, R., Kane, M.D., Winfrey, J. and Stahl, D.A. (1992) Genus- and group-specific hybridization probes for determinative and environmental studies of sulfate-reducing bacteria. *Systematic and Applied Microbiology* 15, 601–609.
Giovannoni, S.J., Britschgi, T.B., Moyer, C.L. and Field, K.G. (1990) Genetic diversity in Sargasso Sea bacterioplankton. *Nature* 345, 60–63.
Hawksworth, D.L. (1991) The fungal dimension of biodiversity: magnitude, significance, and conservation. *Mycological Research* 95, 641–655.
Holben, W.E., Jansson, J.K., Chelm, B.K. and Tiedje, J.M. (1988) DNA probe method for the detection of specific microorganisms in the soil bacterial community. *Applied and Environmental Microbiology* 54, 703–711.
Holben, W.E., Calabrese, V.G.M., Harris, D., Ka, J.O. and Tiedje, J.M. (1993) Analysis of structure and selection in microbial communities by molecular methods. In: Guerrero, R. and Pedros-Alio, C. (eds) *Trends in Microbial Ecology: Proceedings of the Sixth International Symposium in Microbiological Ecology.* Spanish Society for Microbiology, pp. 363–370.
Holt, J.G. (ed.) (1984–1989) *Bergey's Manual of Systematic Bacteriology*, 4 vols. Williams & Wilkins, Baltimore.
Janzen, D.H. and Hallwachs, W. (1993) Highlights of the NSF-sponsored 'All Taxa Biodiversity Workshop', 16–18 April 1993, Philadelphia. Internet, Biological Systematics Discussion List (taxacom@harvard.bitnet).
Lee, S. and Fuhrman, J.A. (1991) Spatial and temporal variation of natural bacterioplankton assemblages studied by total genomic DNA cross-hybridization. *Limnology and Oceanography* 36, 1277–1287.
Liesack, W. and Stackenbrandt, E. (1992) Occurrence of novel groups of the domain bacteria as revealed by analysis of genetic material isolated from an

Australian terrestrial environment. *Journal of Bacteriology* 174, 5072–5078.
May, R.M. (1986) The search for patterns in the balance of nature's advances and retreats. *Ecology* 67, 1115–1126.
Neefs, J.-M., van de Peer, Y., de Rijik, P., Goris, A. and de Wachter, R. (1991) Compilation of small ribosomal subunit RNA sequences. *Nucleic Acids Research* 19, 1987–1998.
Olsen, G.J., Larsen, N. and Woese, C.R. (1991) The ribosomal RNA database project. *Nucleic Acids Research* 19, 2017-2021.
Olson, R.J., Zettler, E.R., Chisholm, S.W. and Dusenberry, J.A. (1991) Advances in oceanography through flow cytometry. In: Demars, S. (ed.) *Particle Analysis in Oceanography.* Springer-Verlag, Berlin, pp. 351–399.
Sasser, M. and Wichman, M.D. (1991) Identification of microorganisms through use of gas chromatography and high-performance liquid chromatography. In: Hausler, W.J. Jr, Herrmann, K.L., Isenberg, H.D. and Shadony, H.D. (eds) *Manual of Clinical Microbiology*, 5th edn. American Society of Microbiology, Washington, DC, pp. 111–118.
Schmidt, T.M., DeLong, E.F. and Pace, N.R. (1991) Analysis of a marine picoplankton community by 16S rRNA gene cloning and sequencing. *Journal of Bacteriology* 173, 4371–4378.
Stackebrandt, E. and Goodfellow, M. (1991) *Nucleic Acid Techniques in Bacterial Systematics.* John Wiley & Sons, New York.
Stead, D.E., Seilwood, J.E., Wilson, J. and Viney, I. (1992) Evaluation of a commercial microbial identification system based on fatty acid profiles for rapid accurate identification of plant pathogenic bacteria. *Journal of Applied Bacteriology* 72, 315–321.
Torsvik, V., Goksoyr, J. and Daae, F.L. (1990) High diversity in DNA of soil bacteria. *Applied and Environmental Microbiology* 56, 782–787.
Vaneechoutte, M., Rossau, R., de Vos, P., Gillis, M., Janssens, P., Paepe, N., de Rouck, A., Fiers, T., Claeys, G. and Kersters, K. (1992) Rapid identification of bacteria of the Comamonadaceae with amplified ribosomal DNA-restriction analysis (ARDRA). *FEMS Microbiology Letters* 93, 227–234.
Wayne, L., Brenner, D.J., Colwell, R.R., Grimont, P.A.D., Kandler, O., Krichevsky, M.I., Moore, L.H., Moore, W.E.C., Murray, R.G.E., Stackebrandt, E., Starr, M.R. and Trüper, H.G. (1987) Report of the Ad Hoc Committee on Reconciliation of Approaches to Bacterial Systematics. *International Journal of Systematic Bacteriology* 37, 463–464.

Identifying and Culturing the 'Unculturables': A Challenge for Microbiologists

N. WARD, F.A. RAINEY, B. GOEBEL AND
E. STACKEBRANDT

DSM – German Collection of Microorganisms and Cell Cultures, Mascheroder Weg 1B, 38124 Braunschweig, Germany

Introduction

The term 'unculturables' (or 'as yet uncultured', to be more precise) has been used in the past to describe the inability to cultivate and identify microorganisms that have been shown by light and electron microscopy to be present in natural samples. Recent advances in culturing techniques have permitted determination of the phylogenetic position of some obligately inter- and intracellular organisms (Amann *et al.*, 1991; Distel *et al.*, 1991; Embley *et al.*, 1992). However, the majority of microorganisms in the environment remain unidentified. In the past, the actual number of existing prokaryotic species was merely an estimate, barely supported by scientific experiments, and limited only by the courage of the microbiologist who provided this number. Today, we are able to show that the genomic variety in a natural sample is, indeed, significantly larger than even the greatest estimated number. In the context of worldwide concern about species loss, ongoing discussion about preservation of genetic variety and *in situ* preservation of the gene pool, and the role of microorganisms associated with more highly evolved organisms, the basis for a discussion of the relative importance of microorganisms, compared to animals and plants, will be strengthened when the estimated number of species of bacteria, Archaea, fungi (including yeasts), algae and protozoa is more scientifically sound.

This chapter addresses the following topics:

1. the methodologies applied to determine the presence of genetic variety in an environmental sample. Since most of these techniques have been

published (Liesack and Stackebrandt, 1992a, b; Fuhrman *et al.*, 1993; Muyzer *et al.*, 1993; Ward *et al.*, 1993) they will only be summarized and compared;
2. genomic variety of microorganisms in different environments, such as soil, water, eukaryotic host, e.g. a marine invertebrate, and a continuous bioleaching reactor;
3. the limited degree to which results of the molecular ecological studies can be correlated with actual biodiversity; and
4. the challenge of future biodiversity studies.

Strategies in Molecular Environmental Studies

Traditionally, microbiologists have attempted to recover as many different types of organisms from an environmental sample as time, funds and available methods allowed. Depending on the aims of the study, organisms were selected either on the basis of physiological properties (photosynthesis, N_2 fixation, sulphate or sulphur reduction, anaerobic respiration, H_2 production etc.), formation of products (antibiotics, metabolites), or morphological characters (in general, buds, prosthecae, and mycelia). Isolation of microorganisms in a given biotope usually resulted in recovery of a large number of strains. Allocation of these to described species was based on characters that were presumed to be sufficiently descriptive. However, because of the above-mentioned restrictions, most isolates were not compared to the type strains of described species using methods now available, such as chemotaxonomy, DNA–DNA hybridization, gene sequence comparison, or DNA probing. Obviously, neither the accuracy of the taxonomic conclusion, nor the genomic range of a species could be determined, and a large number of strains, potentially belonging to new taxa and/or with new properties of biotechnological and pharmaceutical importance, were probably discarded since their uniqueness could not be assessed. Today, progress in the development of new methods allows us, in principle, to determine not only the phylogenetic relatedness of any isolate to previously described taxa, but also the genomic variety of microbial communities.

Diversity of the Genome

Hybridization of DNA extracted from whole communities against that obtained from individual species, or other communities, allows the determination of the presence or absence of defined taxa, and the temporal and spatial distribution of defined and undefined community members, respectively (Sayler and Layton, 1990). The homogeneity of the microbiota of three different open ocean samples (Lee and Fuhrman, 1990), and the differences in microbial populations between coastal water and the open ocean have been suggested by DNA–DNA hybridization. DNA reassociation kinetics indicated the presence of approximately 4000 different prokaryotic genomes in 1 g of soil (Torsvik *et al.*, 1990). If such a result, indeed, reflects the extent of biodiversity, and each genome represents an individual species, the number of species in a single soil sample exceeds the number of all described prokaryotic species. This information clearly supports tapping the Earth's abundant, but, as yet unrecognized, supply of microbial physiological and genetic diversity.

Probing

Similar questions can be investigated by the use of taxon-specific, diagnostic oligonucleotide probes. The use of probes fulfils two functions: first, rapid screening of a library of cloned genes to reduce the sequencing effort. In this case, the probe sequence is designed either from the information of sequenced clones or from those of culturable organisms. However, for the reasons discussed below, it is unlikely that the numbers of different clones reflect the true abundance ratio of different microorganisms in the community under study and second, the use of oligonucleotide probes designed to bind to rRNA of strains in the environment. This approach has been used to verify the molecular identity of symbionts within their hosts and to determine the presence and relative abundance of organisms directly in the environment, either of a particular organism or group in total environmental rRNA, or by whole-cell hybridization of individual cells, the presence of which is indicated by analysis of the clone library.

The identification of clones and the determination of the relative abundance of microorganisms in environmental samples are often performed using dot blot or slot blot hybridization (Stahl *et al.*, 1988; Giovannoni *et al.*, 1990; DeLong, 1992; Devereux *et al.*, 1992; Liesack and Stackebrandt, 1992b). Precise quantification of a given microorganism is not yet possible to achieve, since different bacteria contain different numbers of rrn operons, and probably different numbers of ribosomes. Factors negatively affecting quantitative measurements are: masking of low-abundance homologous rRNAs by large amounts of heterologous rRNA; binding of probes to their target occurring with different efficiencies, since not all sites in rRNA are equally accessible to probes; and binding of the specific probe to environmental rRNA extracts that must be calibrated against the binding of reference probes to known rRNAs. It is likely that in the near future radioisotopes will be replaced by non-radioactive detection systems such as digoxigenin or biotin (Zarda *et al.*, 1991; Kessler, 1992; Hahn *et al.*, 1993). More precise quantification of individual microbes can potentially be achieved using fluorescent probes binding to intracellular rRNA within fixed whole cells directly in their natural environment (Yu and

Gorovsky, 1986; DeLong et al., 1989b; Stahl and Amann, 1991; Hicks et al., 1992; Manz et al., 1992). Cells can be counted directly and, if sampling permits, spatial distributions can be observed, even in mixed populations (Amann et al.,1992; Spring et al., 1992). Different probes can be labelled with different fluorescent dyes and it can be assumed that, in the future, it will be possible to visualize numerous targets in a single experiment (Ried et al., 1992).

Cloning and Sequencing

The methods applied to the determination of microbial genetic diversity of natural samples reflect the evolution of methods over the last 10 years. Although initially (Pace et al., 1985) the 5S rRNA of a low-diversity community was analysed by direct sequencing, more laborious methods were needed for samples with a higher degree of genomic variation. For example, profiling of low-molecular-weight RNA (5S, tRNA) was used to detect seasonal changes of microbial diversity in aqueous environments (Höfle, 1990), but the identity of individual taxa could not be determined, except for populations with low diversity. The use of denaturating gradient gel electrophoresis (DGGE) for the separation of PCR-amplified environmental 16S rDNA marked an important milestone (Muyzer et al., 1993), especially when this method is combined with high-resolution gels (eventually in a two-dimensional system), probing with taxon-specific oligonucleotides, and sequence analysis of the separated DNA fragments. The main breakthrough in the determination of genomic variety within natural samples came with the introduction of cloning strategies and the use of DNA that was isolated either from harvested cells (Giovannoni et al., 1990), or from the autochthonous population directly in the natural matrix (Liesack and Stackebrandt, 1992a, b). Cloning of rDNA or rcDNA is needed to single out individual sequences from the bulk of DNA restriction fragments or amplificates. Initially, rRNA and rDNA were considered to be equally useful, in their suitability as templates for sequence analysis, and for subsequent determination of biodiversity. However, the introduction of PCR-mediated amplification of rRNA genes and the recognition that reverse transcriptase may not faithfully transcribe the rRNA, shifted the preference of most workers to the amplification of rDNA. The rather laborious technique of shotgun cloning (Pace et al., 1986) was replaced by cloning only those DNA fragments shown to contain rDNA genes by hybridization (Unterman et al., 1989), whereas the current method involves cloning of PCR-amplified rRNA genes or partial fragments of these genes, preferably with the use of blunt-end cloning (see below).

The combined application of cloning and oligonucleotide probing facilitates the recognition of clones in clone libraries (Britschgi and Giovannoni, 1991; Liesack and Stackebrandt, 1992b). The strategy includes sequence analysis of randomly selected clones to determine the identity of the most abundant taxa represented in the library. Comparison of these sequences with those from databases allows the identification of diagnostic sequence regions that are suitable for the recognition of specific taxa, leading to the synthesis of probes. Dot blot hybridization between the probes and either clones or PCR-amplified cloned inserts allows rapid screening of libraries. This permits the determination of the total number of clones for which sequence information is available, and the detection of those clones that do not hybridize with any of the probes; the latter contain phylogenetic information of taxa that can be identified only by sequence analysis. Once a set of specific probes has been generated for the majority of taxa occurring in an environmental sample, the influence of methodological parameters on the composition of libraries, as well as fluctuations in the population due to seasonal changes, can be investigated much more rapidly than by sequence analysis (see below).

Sequence information obtained from nucleic acids isolated from an environmental sample allows detection of a significantly broader range of diversity than is achieved by isolation procedures alone. Four observations made from molecular ecology studies attracted the attention of microbial ecologists.

First, in an environment that contains a phylogenetically highly diverse population, hardly any of the clone sequences obtained have been identical to those of the recognized culture collection strains, a result similar to earlier studies employing culture methods alone, where the isolates could not be identified by reference to culture collection strains. Although reference sequences are not available for all described species, this finding was a confirmation of the assumption stated above that only a fraction of prokaryotic species are culturable. The available examples also showed that the as yet not cultured are not close relatives of the culturables. This hypothesis cannot be proved for genera with a high number of phylogenetically non-characterized species, such as *Streptomyces, Brevibacterium, Corynebacterium, Micromonospora, Nocardia, Pseudomonas* and other genera that are likely to occur in the environments analysed to date.

Second, in several cases, phylogenetic analysis revealed a group of related sequences, suggesting the presence of several members of the same, previously unknown, taxon. Although in certain cases the sequence differences were small and comparable to those observed in culturable members of a species or genus (Fuhrman *et al.*, 1993; Stackebrandt *et al.*, 1993), larger differences indicated membership in a higher taxon. Britschgi and Giovannoni (1991) pointed out that this phenomenon is not artefactual and caused by methodological problems, but is due to intrinsic properties of the sequences.

Third, only in rare cases were sequences of culturable strains isolated from the environment identical to those obtained from the genetic material that was analysed from the same environment. This finding emphasizes the previously documented bias of culture conditions that may select for a portion of the microbiota that is not predominant.

Last, where phylogenetic analysis of environmental nucleic acid sequences has indicated membership in a taxon of culturable organisms, in most cases attempts to cultivate those organisms have failed. This argument is often used to highlight the importance of molecular ecological studies. However, rDNA/RNA studies have demonstrated that even phylogenetic neighbours may exhibit distinct physiological differences which would exclude their growth on the same medium.

Molecular Biodiversity in Environmental Samples

Symbionts

The small number of microorganisms shown to be involved in symbiotic relationships with higher organisms has attracted the interest of molecular microbial ecologists, because the identities of the prokaryotic and the eukaryotic partners can be determined directly using PCR or cloning to recover selectively the rRNA genes of each partner. Probes to the recovered sequences can then be used for *in situ* hybridizations, in order to verify the intracellular location of the prokaryotes (Amann *et al.*, 1991; Distel *et al.*, 1991; Embley *et al.*, 1992; Finlay *et al.*, 1993).

Studies indicate that the colonization of eukaryotes occurred independently many times in evolution, involving bacteria of diverse evolutionary origin. Endosymbionts and free-living strains are, in some cases, closely related, but absolute identity of the rDNA sequences has never been observed. In fact some sequences of endosymbionts represent novel lines of descent. Symbiosis is not restricted to members of the domain *Bacteria*, but includes methanogenic archaeal endosymbionts of anaerobic ciliates, originating from several of the major lineages of methanogens (Embley *et al.*, 1992; Finlay *et al.*, 1993). The endosymbionts of bacterial origin are presently restricted to members of the proteobacteria. One ciliate species has been found to contain a member of the alpha proteobacteria, for which the generic name *Holospora* was introduced (Amann *et al.*, 1991). Symbionts of homopteran insects consist of endosymbiotic bacteria that have their closest free-living relatives among the gamma proteobacteria, for example, the aphids (Unterman *et al.*, 1989; Munson *et al.*, 1991a, b), or the beta proteobacteria, in the case of mealy bugs (Munson *et al.*, 1992). On the other hand, endosymbionts responsible for parthenogenesis or cytoplasmic incompatibility in different insects appear to have originated from a single lineage in the alpha proteobacteria (O'Neil *et al.*, 1992; Stouthamer *et al.*, 1993). Six marine invertebrates which thrive under reduced conditions contain chemoautotrophic symbiotic bacteria from the gamma proteobacteria that oxidize reduced sulphur compounds. The symbiotic bacteria in the gills of ship worms of different genera are also members of this subclass and constitute a novel phylogenetic lineage (Distel et al., 1991).

The Environment of a Continuous Bioleach Bioreactor
(Goebel and Stackebrandt, 1994)

In order to assess the bioleachability of a sulphide ore concentrate under long-term continuous conditions, a laboratory-scale bioreactor containing the ore concentrate was established. This system was initially inoculated with a natural microbial consortium which had been enriched from the natural acidic runoff (pH 2.5) of a chalcocite overburden heap situated at the Mt Isa Mines leasehold, Mt Isa, Queensland, Australia. Over a 12-month period, the continuous bioreactor was repeatedly reinoculated with acidic samples from the batch enrichment culture that contained, as derived from phenotypic and genetic analysis, a diverse collection of *Thiobacillus*, '*Leptospirillum*', and *Acidiphilium* strains. After several months of continuous flow, only two strain types could be recovered from the continuous bioreactor, represented by *Thiobacillus thiooxidans* MIM SH12 and '*Leptospirillum ferrooxidans*' MIM Lf30. Complete and partial 16S rDNA sequences of ten and four isolates, respectively, indicated that the sequences of each of the two strain types were almost identical to sequences published previously (Lane et al., 1992), although strain MIM SH12 was most closely related to *Thiobacillus ferrooxidans* strain LM-2. The finding that strains of *Acidiphilium* and *T. ferrooxidans* could no longer be isolated was a clear indication of the population fluctuation occurring during the transition from a natural environment to the harsh, selective conditions of a continuous bioreactor. A 16S rDNA clone library was generated from a sample taken from the continuous bioreactor. The heterogeneous PCR products amplified from the bulk DNA were blunt-end cloned into the vector pBS (+/-), and subsequently investigated by sequence analysis and a multiprimer PCR assay. Of 57 clones which were shown to contain insert DNA, 43 were either identical or at least highly similar in sequence to *T. thiooxidans* MIM SH12 and seven belonged to '*Leptospirillum ferrooxidans*' MIM Lf30, whereas seven clones of a new sequence type from this environment (represented by clone A70) were closely related to the unnamed facultatively thermophilic iron-oxidizing strains ALV and BC (Lane et al., 1992), originally isolated from coal mine drainage. An organism of the A70 type was recently isolated from the batch and continuous systems when cultivation conditions were changed to match specifically conditions described for the isolation of strains ALV and BC.

Analysis of cultured strains and the clone library from the continuous bioleach operation showed the lowest degree of biodiversity of all environments that have so far been analysed by molecular techniques and accompanying cultivation studies. Both the number of species and the heterogeneity of 16S rDNAs of clones of the same sequence type were small. This can be explained by the highly selective conditions of the bioreactor, such as low pH, elevated temperature, gas composition, dilution rate, and concentration of dissolved metals. Thus, based on knowledge of the composition of bioleaching sites, the traditional isolation techniques have provided a rather accurate picture of the bacterial composition of such environments. The composition of the microbiota may depend on physicochemical parameters, and phylogenetic analysis of mixed populations, using different ore types, should show whether it is possible to design specific consortia with defined bioleach properties suitable for batch or continuous-flow bioreactors.

Forrested Soil
(Liesack and Stackebrandt, 1992a, b; Stackebrandt et al., 1993)

One of the challenges in the investigation of the microbial biodiversity of soil using molecular methods was to solve the problem of possible interference of humic acids with nucleic acids during the process of DNA isolation. A solution to this problem was determined for the environment of an acidic forrested soil from Australia. Due to the initial emphasis of the study, less care was taken to restrict the sampling site to a defined layer. A rather thick layer of 3 cm in depth was investigated. A 16S rDNA library was generated by sticky-end cloning of PCR-amplified partial 16S rDNA, using primers that were specifically designed to target members of the genus *Streptomyces* and related taxa. More than 100 rDNA clones were either sequenced over a stretch of 300 nucleotides (region 5' terminus through position 350), or identified by hybridization with clone-specific probes. In addition, for selected members of individual groups, 1130 nucleotides were analysed, in order to determine phylogenetic position more precisely. In parallel, about one hundred strains with different colony types were isolated from the adjacent sampling site. The study produced some surprising findings: (i) the analysis of clone data indicated the almost complete absence of organisms that were frequently isolated, such as streptomycetes, bacilli or Gram-negative rods; (ii) none of the 16S rDNA sequences of the isolates matched those of the the clone sequences; (iii) as mentioned above, several individual lines of descent were represented by more than one sequence (closely or remotely related); and (iv) certain clone sequences indicated the presence of organisms that would be expected to thrive in this environment (although they could not be isolated), including alpha-1 proteobacteria related to a cluster containing *Rhodopila globiformis* and *Acidiphilium* species (about 2% of the clones), and alpha-2 proteobacteria related to a cluster consisting of

Bradyrhizobium, '*Photorhizobium*', and *Rhodopseudomonas palustris* (about 50% of the total).

About 45% of the clones, however, suggested the presence in this particular environment of organisms that were not anticipated. The majority of these (20% of the total) represent a novel line of descent, comparable to the rank of phylum, branching intermediate to the majority of other bacterial phyla. In certain aspects the sequences shared similarities with those of planctomycetes, but did not appear specifically related.

The other group of sequences, about 6% of the total number, showed the typical idiosyncratic features of planctomycetes, and clustered within the radiation of *Planctomyces*, *Pirellula*, *Isosphaera* and *Gemmata*. The presence of these organisms in a terrestrial habitat was surprising, as their natural habitat has been described previously to be aquatic. Furthermore, only a single strain of *Gemmata obscuriglobus* has been isolated to date, but phylogenetic analysis indicated that at least five different sequences indicative of species rank are present in this single random soil sample.

The third group of clones (about 14% of total) were unexpectedly found to be deep-branching members of the order *Actinomycetales*. Two separate clusters were found, one grouping with the iron-oxidizing strain TH3, the other with the misclassified species *Lactobacillus minutus*. Despite the fact that growth conditions for these two culturable organisms and for the planctomycetes are well known, all attempts to use this information for isolation of similar types of organisms from the Australian soil sample failed.

Acidothermal Soil (F.A. Rainey, unpublished)

Studies of thermal environments have yielded a diverse group of thermophilic microorganisms which fall within both the *Bacteria* and the *Archaea*. Although the majority have been isolated from environments with pH values approximately neutral (Kristjansson and Stetter, 1992) relatively few organisms have been recovered from acidic thermal habitats. In extensive studies selecting for thermophilic, cellulolytic organisms, none were recovered from acidic thermal samples (Hudson *et al.*, 1990). Of the organisms isolated from acidothermal environments, those with pH optima for growth of < 5 and temperature optima of > 60°C have been mainly aerobic archaea (Kristjansson and Stetter, 1992) and acidophilic thermophilic bacilli (Hudson *et al.*, 1989).

In order to determine the degree of microbial diversity in an environment with extremes of both pH and temperature an acidic thermal soil was chosen for a molecular ecology study. The acidothermal soil sample was obtained from Orakei Korako thermal reserve in the central north island of New Zealand. This thermally heated soil was at a temperature of 75°C and had a pH of 3.5. Total genomic DNA was directly extracted from the soil and used as a template for PCR-mediated amplification of the 16S rRNA genes. The primer pair used for PCR was designed to amplify both bacterial and archaeal 16S rDNA. The amplified fragments obtained were cloned and 160 clones were selected for hybridization and sequence analysis.

Hybridization with an archaeal-specific probe demonstrated the presence of archaeal 16S rDNA in 33 of the 160 clones. Sequence analysis of 18 clones allocated them to two main groups, one to a novel group falling between the *Crenarchaeota* and the *Euryarchaeota*, and the other composed of two sequence types closely related to the genus *Sulfolobus*.

Probing with a Gram-positive phylum-specific probe gave a positive signal for a further 24 clones. Those clones for which sequence data were obtained were of three sequence types, falling within the radiation of the genera *Bacillus* and *Alicyclobacillus*, and within the order *Actinomycetales*. Sequence diversity within these three sequence groups was minimal.

Random selection and sequence analysis of a further 17 clones showed all to belong to a new group within the beta-subclass of the *Proteobacteria*. As found in other environmental studies, this novel multimember branch that was well separated from phylogenetic neighbours, showed very little sequence difference among the tips of the branch.

The molecular analysis so far applied to this acidothermal environmental sample has revealed a relatively limited microbial diversity with only six different sequence groups detected. With the exception of the beta *Proteobacteria* group, the other groups detected have previously been isolated from such environments. Further sequence and probe analysis is required to determine the identity of the beta *Proteobacteria* group and the degree of diversity within the group.

When compared to the large microbial diversity found in the Mt Cootha soil, this study has demonstrated at the molecular level the limited diversity of a soil environment with extremes of temperature and pH and supports the findings of isolation studies in similar habitats. As compared to the two kinds of populations determined by isolation and genetic analysis from the thermal Octopus Spring (Ward *et al.*, 1990; see below), the microbial diversity revealed in the acidothermal soil was found to be totally different.

Hot Spring Environments
(Ward *et al.*, 1990, 1992; Weller *et al.*, 1991)

Octopus Spring in Yellowstone National Park attracted the attention of ecologists because the cyanobacterial mats developing at 55°C were believed to contain a lower degree of biodiversity than mats developing at less extreme conditions. A comparison of 16S rDNA sequences obtained from molecular ecological studies, with those from pure cultures isolated from this and similar habitats, was among the first to demonstrate the inability of the cultivation techniques to describe biodiversity in a natural sample. Although *Synecchococcus lividans* was believed to be the sole representative of *Cyanobacteria*, molecular analysis indicated the presence at this site of four different clone sequences that were unrelated not only to *S. lividans*, but also among themselves. Other clone sequences showed specific but distant similarites to those of *Chloroflexus* and related non-phototrophic taxa, and to those of spirochaetes, but several sequences could not be assigned to any known phylum. On the other hand, sequences of organisms that were previously isolated from the same environment could not be detected in the clone library. The importance of this finding will be discussed below.

Marine Environments
(DeLong *et al.*, 1989a; Giovannoni *et al.*, 1990; Britschgi and Giovannoni, 1991; Schmidt *et al.*, 1991; Fuhrman *et al.*, 1992, 1993)

Several studies have been performed in the marine environment to determine the biodiversity of near-surface and subsurface microbial communities. In each of these experiments, cells were collected, nucleic acids extracted and a clone library generated using different strategies. The early results from the Sargasso Sea (SAR clones) and from the North-Central Pacific Ocean (ALO clones) have been summarized and discussed by Ward *et al.* (1993). Although separated by the American continents, and despite different strategies used for the generation of clone libraries, a significant part of the microbial composition was surprisingly homogeneous in both oceans. Two clusters of cyanobacterial sequences of high similarity, resembling that of a *Synechococcus* strain isolated in the Sargasso Sea, were found in both environments. Recently *Prochlorococcus marinus* has been described, the 16S rDNA sequence of which shows very high relationship to sequences of clones SAR 6 and ALO 7 and their related sequences. The second group of clones indicated the presence of *Proteobacteria*. The Pacific clone sequences grouped with those from members of the alpha and gamma subclasses, but each of these appeared to represent novel lines of as yet unknown bacteria. Because studies of Sargasso Sea material favoured the recognition of cyanobacterial sequences, only a few sequences from the alpha subclass were found. These sequences (SAR 1, SAR 11), remotely related to *Caulobacter crescentus* and *Pseudomonas diminuta*, were almost indistinguishable from those recovered from the Pacific clone library (ALO 21, ALO 39).

The extensive studies by Fuhrman *et al.* (1993) compared the microbial population of two different sites, one from the western California Current (two different depths) and one from a 10-m depth in the Atlantic Ocean. The results pointed to the presence of broader microbial assemblages than had been indicated by other studies of similar environments. As in all other molecular environmental studies, the phylogenetic lineages constituted novel and undescribed groups. The most exciting finding was the presence of two different clusters of archaeal sequences from oligotrophic open waters. Sequences that were only remotely related to those of culturable archaeal strains were, on the other hand, closely related to clone sequences found by DeLong (1992) in Atlantic and Pacific US coastal waters. The data also indicated the presence of similar bacterial populations in both oceans, and that some of the sequences were highly related to those found by Giovannoni *et al.* (1990) and De Long *et al.* (1989a). This was the case for cyanobacterial sequences of the *Synechococcus–Prochlorococcus* group from the Pacific and Sargasso Sea, and of sequences that related the SAR 1 and SAR 11 sequences of the alpha proteobacteria to a large cluster of sequences retrieved from material from the Pacific and Bermuda. Novel lineages were found among the gamma proteobacteria, the actinomycetes, the flavobacteria and a group that could not be affiliated to any of the recognized phyla of *Bacteria*.

Holothurian (N. Ward, unpublished)

The holothurians, or sea cucumbers, are marine invertebrates that form a prominent element of the macrofauna of benthic communities. Sea cucumbers feed by the ingestion of marine sediment or by filtration of sea water.

The selection of the intestinal tract of a deposit-feeding holothurian, *Holothuria atra*, as a site for a microbial diversity study was based on the fact that the total microbial diversity associated with this ecological niche had not previously been investigated. The presence of a metabolically active gut flora, however, had been demonstrated in a holothurian species (Deming *et al.*, 1981; Deming and Colwell, 1982), but the total composition of the microbial community was not determined.

Determination of the microbial diversity associated with the holothurian gut was attempted using both enrichment and isolation techniques (followed by identification of the isolates obtained) and 16S rRNA-based molecular ecological methods, in parallel.

An isolation study, using a range of different growth media under aerobic conditions, yielded 44 strains, which were identified and characterized by 16S rDNA sequence analysis and a small number of phenotypic tests. Of these isolates, a large number (24) were phylogenetically related to members of the genus *Vibrio* and neighbouring taxa. Other isolates included members of the genus *Bacillus*, of the alpha and gamma subclasses of the *Proteobacteria*, of the *Cytophaga–Flavobacterium–Bacteroides* line of descent, and some actinomycetes.

The microbial diversity determined by a molecular approach, involving generation of a 16S rDNA clone library from a gut content sample, was substantially different from that revealed by conventional isolation procedures. 16S rDNA sequence analysis, and hybridization of clones with taxon-specific oligonucleotide probes, showed that the majority of clone sequences investigated (67.5%) originated from organisms belonging to the beta subclass of the *Proteobacteria*, a group which was not recovered by isolation. These clone sequences were phylogenetically most closely related to an organism with an anaerobic, photosynthetic phenotype, suggesting that the failure to isolate this organism may have been due to the limitations of the isolation study, which was performed using only aerobic cultivation conditions. Which was the dominant group physiologically, however, is not known, i.e. the activities of the uncultured group obviously is completely unknown at the present time. A large number of sequences originating from members of the family *Pasteurellaceae* was also demonstrated, and the inability to cultivate these organisms may also have been due to the use of an inappropriate incubation atmosphere and/or the failure to supply a required growth factor. Again, their role in the metabolism of the sea cucumber is entirely unknown, i.e., whether they are actively metabolizing or not.

Other clone sequences were found to be phylogenetically related to members of the alpha and gamma *Proteobacteria* subclasses, of the *Cytophaga–Flavobacterium–Bacteroides* line of descent and an actinomycete. Although organisms from these taxa were isolated, their sequences were not even closely related to those of the clones. The large number of *Vibrio* strains obtained by isolation was not reflected in the composition of the 16S rDNA clone library in which no *Vibrio* clone sequences were found. This discrepancy may have been due to the presence of recognition sites for the cloning enzymes within the 16S rDNA sequence of the *Vibrio* isolates. This methodological problem is discussed below, together with recent developments that allow it to be overcome.

The results of this investigation thus provide an example of how two alternative strategies can be used together to reveal an increased proportion of the actual microbial diversity than could be demonstrated using either method alone. A similar approach may be suitable for determination of the diversity of microbial populations associated with other marine invertebrates, which, with a few exceptions, have not been thoroughly investigated to date.

Limitations of the 16S rDNA Approach

Gene Amplification and Clone Libraries

There is little doubt that within a very short time the combined application of PCR technology, cloning, sequence analysis and probing has revolutionized microbial ecology. The measurement of physiological activity alone is acknowledged to be inadequate for the description of the underlying microbial diversity. But it must be stressed that, although the techniques allow recognition of molecular diversity, the corresponding phenotype of the organisms cannot be circumscribed and detailed information about community structure, i.e. the interaction of organisms and the relative proportion of taxa or their metabolism, cannot be obtained. As long as the influence of several parameters involved in the recovery of cells and nucleic acids, in gene amplification and cloning, and in the binding of probes to whole cells is not known, the scope of the molecular approach includes only the phylogenetic information from the clone sequences and detection of individual cells in the natural sample, but as yet not the quantification of taxa based on numbers of similar or identical sequences and fluorescently labelled cells. The following is a short compilation of problems encountered in these studies.

One of the most serious problems is the recovery from the environment of nucleic acids that represent, both qualitatively and quantitatively, the composition of the sample. The data obtained so far from different environments seem to indicate that the majority of bacteria are Gram-negative. This finding may be accurate, but it may also be that lysis procedures favour the selective disintegration of members of *Gracilicutes*, whereas the spores and resting stages of Gram-positive bacteria, including actinomycetes, may resist lysis. This phenomenon was observed in the studies of the acidic Australian soil, where the lack of sequences representing Gram-positive bacteria was in sharp contrast to the isolation of dozens of different strains of the genera *Streptomyces* and *Bacillus*. Subsequent studies using the same isolated DNA as a template for PCR, but a different pair of PCR primers (10–30 and 1390 versus 10–30 and 1224) for the amplification of 16S rDNA resulted in a dramatic increase in the number of Gram-positive sequences (50% versus 3%), as determined by stringent hybridization of the clone library with a probe specific for Gram-positive bacteria. The percentage of members of the alpha-2 subclass of Proteobacteria was reduced from 50% to only 1% (Liesack and Stackebrandt, 1992b; Stackebrandt *et al.*, 1993; Rainey *et al.*, 1994). Obviously, the selection of primers is crucial and the results are somewhat unpredictable. Even the use of so-called 'universal' primers does not guarantee the amplification of all types of rDNA. Similarly, the use of selective primers may result in the amplification of a much wider spectrum of different rDNAs than anticipated (Britschgi and Giovannoni, 1991; Liesack and Stackebrandt, 1992b; Stackebrandt *et al.*, 1993). The position of primers along the primary structure of rDNA, the influence of the nucleotide composition of the 3' terminus of the primer, the presence of targets of similar sequence in the same molecule, and the length of the region to be amplified may have an influence on the formation of amplificates, and therefore on the composition of clone libraries (Reysenbach *et al.*, 1992). Further extensive studies are necessary to determine this kind of influence and to eliminate biased formation of amplificates.

A second important factor is the cloning system. Of the three cloning alternatives, shotgun, sticky-end, and blunt-end cloning, the first alternative is too laborious, and the second one may introduce a significant bias. So it is only the third one that appears to be appropriate. Until recently, sticky-end cloning was the most widely used method. PCR primers with flanking regions containing motifs for the recognition of restriction enzymes allowed the rapid generation of clonable amplificates. However, the use of restriction enzymes to transform the amplificates to allow ligation with a vector molecule may lead to a loss of diversity if the recognition sites are present not only in the overhangs, but also in the amplificates themselves. A survey of 500 sequences for the presence of recognition sites for eight enzymes showed that these sites are widely distributed, sometimes restricted to certain taxa, and often present in multiple copies per amplificate. *Sal*I appeared to have the fewest recognition sites within 16S rDNA sequences and to be most useful, whereas the use of *Bam*HI, *Eco*RI (not cutting in Archaea), *Sma*I, *Nae*I, *Pst*I, *Hin*dIII and *Xho*I would result in a significant loss of biodiversity. The finding that *Taq* polymerases add an additional A-residue at the 3' terminus of the amplificate led to the development of the TA PCR cloning system (Invitrogen) that combines the advantages of both systems: the ease of the sticky-end cloning (without the use of restriction enzyme), and the quantitative insertion of amplificates. On the other hand, for certain applications the use of restriction enzymes may be advantageous. If an environment is dominated by one population, i.e., *Vibrio* species, the full diversity may not be detected by screening a clone library. Knowledge of the presence of restriction sites in *Vibrio* sequences, and the use of these enzymes during cloning, may result in the removal of the dominating amplificates and the percentage increase in minority amplificates in the clone library. This kind of deselection may also be achieved by the use of selective probes that remove target amplificates by means of magnetic beads, antigen–antibody reactions or similar assays.

Whole-cell Probes

Application of fluorescent probes for the identification and enumeration of microorganisms in complex samples is in its infancy (Amann *et al*., 1990, 1992; Tsien *et al*., 1990; Devereux *et al*., 1992; Hicks *et al*., 1992; Manz *et al*., 1992; Poulson *et al*., 1993; Wagner *et al*., 1993). This technology complements molecular diversity studies by identifying the assumed members of a community directly in their natural habitat. Despite rapid progress in methods of probe labelling and detection, probe penetration through physical barriers, and probe binding, considerable method development remains to be made. Soil matrices may cause problems with background fluorescence and microorganisms, occurring as dormant or resting stages in nature, may bind insufficient probe due to the physicochemical composition of cell walls and low ribosome content. The tertiary structure of ribosomal RNA within ribosomes, together with structural changes of the RNA–protein complex as a result of the use of fixatives for permeabilization of cells, may cause probes to miss their *in situ* rRNA targets, although the same probes may work reliably on isolated RNA and rDNA. Experiments are currently being performed with different solvents and enzymes to unmask target regions and to facilitate probe entry (Yu and Gorovsky, 1986; Hahn *et al.*, 1993). In order to compensate for low probe binding efficiency, methods for increasing the signal strength are also under investigation (Zarda *et al.*, 1991; Poulson *et al.*, 1993).

The Challenge

Knowing the diversity of a microbial community, and being able to identify some of its members within the natural sample, are a greater achievement than microbial ecologists were able to accomplish five years ago. Information about the phylogenetic position of members of a community, the realization that the number of uncultured microorganisms is probably significantly higher, more than 99% of the total number in some cases, and the exciting recognition that total biodiversity is not encompassed by previously described species have contributed to a renewed and increased interest in microbial ecology. On the other hand, those working in the field have recognized that this period is only the beginning of an epoch, in which there will be a need to develop collaborative efforts, in order to address questions about short-term and long-term interactions between microorganisms, macroorganisms and the inorganic environment. One of the most burning questions is the identity of the organisms shown to be present by the molecular analysis of an environment. It was hoped that knowledge of the phylogenetic relatedness between a clone sequence and the sequence of its closest culturable relative would provide information about how to grow the previously unculturable (or uncultured) strain. These hopes have faded for two reasons. First, almost none of the clone sequences in the examples provided here were so closely related that they could be assigned to a cultured organism at the species level, except for the bioleach environment. Second, it was realized that genetic or epigenetic characters, metabolic traits, biochemical pathways or morphologies cannot be predicted from the phylogenetic position of a clone sequence. Without this knowledge, the development of selection and enrichment media for a previously uncultured organism is almost as difficult as it would be in the absence of information about phylogenetic position. In some cases, organisms were isolated that were related to clone sequences, e.g. *Synechococcus* and *Prochlorococcus*, *Streptomyces*, *Bacillus* and other organisms that grow easily on complex media. In other cases, attempts failed, even when the clone sequences indicated membership to a taxon of defined metabolic characters, such as planctomycetes, and actinomycetes of the TH3 type.

The molecular studies described above were performed with the objectives of developing methods, and obtaining insights into the composition of microbial communities. Now that all results obtained to date point to the same conclusions, the need to explore novel environments using the established methods appears less urgent than the need to develop means by which the number of members of the community can be quantified, and the unculturables can be cultured. Even determination of which organisms of the community represent the metabolically active portion turned out to be a serious problem. The most common way is by the strength of a probing fluorescent response. It is widely believed that metabolic activity is linked to ribosome content and protein synthesis. Indeed, Stahl *et al.* (1988) could demonstrate the relationship between strength of fluorescence and the addition of monensin, using probes directed against *Fibrobacter* strains. This assumption may be true, but the information available about the precise relationship between probe response and physiological status is limited to a few cell types (DeLong *et al.*, 1989a; Kerkhof and Ward, 1993; Poulson *et al.*, 1993). Even 'inactive' cells may still contain significant amounts of rRNA (Flardh *et al.*, 1992; Kramer and Singleton, 1992). In summary, relationships between ribosome content, probe signal, and metabolic status remain poorly understood. It appears logical to combine the molecular work on probes and clone libraries with the application of microprobes, developed to measure physicochemical parameters in microcosms, such as 1 mm deep layers. The restriction of these studies to microenvironments of such small dimensions will probably limit the detectable biodiversity. The methods that are most likely to reveal a comprehensive picture include (not listed in the order of importance): (i) the search for meaningful chemotaxonomic markers (e.g., lipids; Bååth *et al.*, 1992); (ii) careful measurement of end-products and gases (e.g. O_2, H_2S, CH_4; Revsbech and Jørgensen, 1986; Gundersen *et al.*, 1992); (iii) the application of probes that target not only rRNA but also genes coding for a wide range of properties, e.g. nitrogen fixation, hydrogen production, acetogenesis, methanogenesis, ammonia oxidation, iron and sulphur oxidation, or genes coding for enzymes involved in biodegradation or bioremediation; (iv) more rapid determination of the complexity of a microbial community within an environmental sample and the potential to determine easily the temporal and spatial changes of some of its members by DGGE (Muyzer *et al.*, 1993); and (v) the generation of comprehensive clone libraries that also cover the minor components of an ecosystem. The results of such a multidisciplinary approach may provide the basis for estimation of numbers of organisms and phenotypic circumscription of those organisms for which phylogenetic position and taxonomic neighbours are known. A sixth, equally important, component will be the patience, innovation, and experience of microbiologists ultimately necessary to transform the hidden 'uncultured' microorganisms to culturable isolates.

Acknowledgement

Studies of the bioleaching environment, the acidothermal soil, the holothurian and the acidic forrested soil were performed at The Centre for Microbial Diversity and Identification, Department of Microbiology, The University of Queensland, Brisbane, Australia.

References

Amann, R.I., Krumholz, L. and Stahl, D.A. (1990) Fluorescent-oligonucleotide probing of whole cells for determinative, phylogenetic and environmental studies in microbiology. *Journal of Bacteriology* 172, 762–770.

Amann, R.I., Springer, N., Ludwig, W., Görtz, H.-D. and Schleifer, K.-H. (1991) Identification of *in situ* phylogeny of uncultured bacterial endosymbionts. *Nature* 351, 161–164.

Amann, R.I., Stromley, J., Devereux, R., Key, R. and Stahl, D.A. (1992) Molecular and microscopic identification of sulfate-reducing bacteria in multispecies biofilms. *Applied and Environmental Microbiology* 58, 614–623.

Bååth, E., Frostegård, Å. and Fritze, H. (1992) Soil bacterial biomass, activity, phospholipid fatty acid pattern, and pH tolerance in an area polluted with alkaline dust deposition. *Applied and Environmental Microbiology* 58, 4026–4031.

Britschgi, T. and Giovannoni, S.J. (1991) Phylogenetic analysis of a natural marine bacterioplankton population by rRNA gene cloning and sequencing. *Applied and Environmental Microbiology* 57, 1707–1713.

DeLong, E.F. (1992) Archaea in coastal marine environments. *Proceedings of the National Academy of Sciences of the USA* 89, 5685–5689.

DeLong, E.F., Schmidt, T.M. and Pace, N.R. (1989a) Analysis of single cells and oligotrophic picoplancton populations using 16S rRNA sequences. In: Hattori, T., Ishida, Y., Maruyama, Y., Morita, R.Y. and Uchida, A. (eds) *Recent Advances in Microbial Ecology*. Japanese Science Society Press, Japan, pp. 697–700.

DeLong, E.F., Wickham, G.S. and Pace, N.R. (1989b) Phylogenetic strains. Ribosomal RNA based probes for the detection of single cells. *Science* 243, 1360–1363.

Deming, J.W. and Colwell, R.R. (1982) Barophilic bacteria associated with digestive tracts of abyssal holothurians. *Applied and Environmental Microbiology* 44, 1222–1230.

Deming, J.W., Tabor, P.S. and Colwell, R.R. (1981) Barophilic growth of bacteria

from intestinal tracts of deep sea invertebrates. *Microbial Ecology* 7, 85–94.
Devereux, R., Kane, M.D., Winfrey, J. and Stahl, D.A. (1992) Genus- and group-specific hybridisation probes for determinative and environmental studies of sulfate-reducing bacteria. *Systematic and Applied Microbiology* 15, 601–609.
Distel, D.L., DeLong, E.F. and Waterbury, J.B. (1991) Phylogenetic characterisation and *in situ* localisation of the bacterial symbiont of shipworms (Teredinidae. Bivalvia) by using 16S rRNA sequence analysis and oligodeoxynucleotide probe hybridisation. *Applied and Environmental Microbiology* 57, 2376–2382.
Embley, T.M., Finlay, B.J., Thomas, R.H. and Dyal, P.L. (1992) The use of rRNA sequences and fluorescent probes to investigate the phylogenetic positions of the anaerobic ciliate *Metopus palaeformis* and its archaeobacterial endosymbiont. *Journal of General Microbiology* 138, 1479–1487.
Finlay, B.J., Embley, T.M. and Fenchel, T. (1993) A new polymorphic methanogen, closely related to *Methanocorpusculum parvum*, living in stable symbiosis within the anaerobic ciliate *Trimyema* sp. *Journal of General Microbiology* 139, 371–378.
Flardh, K., Cohen, P.S. and Kjelleberg, S. (1992) Ribosomes exist in a large excess over the apparent demand for protein synthesis during carbon starvation in marine *Vibrio* sp. strain CCUG 15956. *Journal of Bacteriology* 174, 6780–6788.
Fuhrman, J.A., McCallum, K. and Davis, A.A. (1992). Novel major archaebacterial group from marine plankton. *Nature* 356, 148–149.
Fuhrman, J.A., McCallum, K. and Davis, A.A. (1993) Phylogenetic diversity of subsurface marine microbial communities from the Atlantic and Pacific oceans. *Applied and Environmental Microbiology* 59, 1294–1302.
Giovannoni, S.J., Britschgi, T.B., Moyer, C.L. and Field, K.G. (1990) Genetic diversity in Sargasso Sea bacterioplankton. *Nature* 345, 60–63.
Goebel, B.M. and Stackebrandt, E. (1994) Cultural and phylogenetic analysis of mixed microbial populations found in natural and commercial bioleach environments. *Applied and Environmental Microbiology* 60, 1614–1621.
Gundersen, J.K., Jørgensen, B.B., Larsen, E. and Jannasch, H.W. (1992) Mats of giant sulphur bacteria on deep-sea sediments due to fluctuating hydrothermal flow. *Nature* 360, 454–456.
Hahn, D., Amman, R.I. and Zeyer, J. (1993) Whole-cell hybridisation of *Frankia* strains with fluorescence- or digoxigenin-labelled, 16S rRNA-targeted oligonucleotide probes. *Applied and Environmental Microbiology* 59, 1709–1716.
Hicks, R.E., Amman, R.I. and Stahl, D.A. (1992) Dual staining of natural bacterioplankton with 4',6-diamidino-2-phenylindol and fluorescent oligonucleotide probes targeting kingdom-level 16S rRNA sequences. *Applied and Environmental Microbiology* 58, 2158–2163.
Höfle, M.G. (1990) In: Overbeck, J. and Chrost, R.J. (eds) *Biochemical and Molecular Approaches in Aquatic Microbial Ecology*. Science Tech/Springer, New York, pp. 129–159.
Hudson, J.A., Daniel, R.M. and Morgan, H.W. (1989) Acidophilic and thermophilic *Bacillus* strains from geothermally heated Antarctic soil. *FEMS Microbiology Letters* 60, 279–285.
Hudson, J.A., Daniel, R.M. and Morgan, H.W. (1990) A survey of cellulolytic anaerobic thermophiles from hot springs. *Systematic Applied Microbiology* 13, 72–76.

Kerkhof, L. and Ward, B.B. (1993) Comparison of nucleic acid hybridisation and fluorometry for measurement of the relationship between RNA/DNA ratio and growth rate in a marine bacterium. *Applied and Environmental Microbiology* 59, 1303–1309.

Kessler, C. (1992) *Nonradioactive Labeling and Detection of Biomolecules.* Springer-Verlag, Berlin.

Kramer, J.G. and Singleton, F.L. (1992) Variations in rRNA content of marine *Vibrio* spp. during starvation-survival and recovery. *Applied and Environmental Microbiology* 58, 201–207.

Kristjansson, J.K. and Stetter, K. (1992) Thermophilic bacteria. In: Kristjansson, J.K. (ed.) *Thermophilic Bacteria.* CRC Press, Boca Raton, FL, pp. 1–18.

Lane, D.J., Harrison, A.P., Stahl, D.A., Pace, B., Giovannoni, S.J., Olsen, G.J. and Pace, N.R. (1992) Evolutionary relationships among sulfur- and iron-oxidising eubacteria. *Journal of Bacteriology* 174, 269–278.

Lee, S. and Fuhrman, J.A. (1990) DNA hybridization to compare species compositions of natural bacterioplancton assemblages. *Applied and Environmental Microbiology* 56, 739–746.

Liesack, W. and Stackebrandt, E. (1992a) Unculturable microbes detected by molecular sequences and probes. *Biodiversity and Conservation* 1, 250–262.

Liesack, W. and Stackebrandt, E. (1992b) Occurrence of novel groups of the domain Bacteria as revealed by analysis of genetic material isolated from an Australian terrestrial environment. *Journal of Bacteriology* 174, 5072–5078.

Manz, W., Amann, R.I., Ludwig, W., Wagner, M. and Schleifer, K.-H. (1992) Phylogenetic oligodeoxynucleotide probes for the major subclasses of proteobacteria. Problems and solutions. *Systematic and Applied Microbiology* 15, 593–600.

Munson, M.A., Baumann, P., Clark, M.A., Baumann, L., Moran, N.A., Voegtlin, D.J. and Campbell, B.C. (1991a) Aphid-eubacterial symbiosis: evidence for its establishment in an ancestor of four aphid families. *Journal of Bacteriology* 173, 6321–6324.

Munson, M.A., Baumann, P. and Kinsey, M.G. (1991b) *Buchnera* gen. nov. and *Buchnera aphidicola* sp. nov. designation for a phylogenetic taxon consisting of the primary endosymbionts of aphids. *International Journal of Systematic Bacteriology* 41, 566–568.

Munson, M.A., Baumann, P. and Moran, N.A. (1992) Phylogenetic relationships of the endosymbionts of mealybugs (Homoptera. Pseudococcidae) based on 16S rDNA sequences. *Molecular Phylogeny and Evolution* 1, 26–30.

Muyzer, G., De Waal, E. and Uitterlinden, A.G. (1993) Profiling of complex microbial populations by denaturating gradient gel electrophoresis analysis of polymerase chain reaction-amplified genes coding for 16S rRNA. *Applied and Environmental Microbiology* 59, 695–700.

O'Neill, S.L., Giordano, R., Colbert, A.M.E., Karr, T.L. and Robertson, H.M. (1992) 16S rRNA phylogenetic analysis of the bacterial endosymbionts associated with cytoplasmic incompatibility in insects. *Proceedings of the National Academy of Sciences of the USA* 89, 2699–2702.

Pace, N.R., Stahl, D.A., Lane, D.J. and Olsen, G.J. (1985) Analyzing natural microbial populations by rRNA sequences. *American Society for Microbiology News* 51, 4–12.

Pace, N.R., Stahl, D.A., Lane, D.J. and Olsen, G.J. (1986) The analysis of natural microbial populations by ribosomal RNA sequences. *Advances in Microbial Ecology* 9, 1–55.

Poulson, L.K., Ballard, G. and Stahl, D.A. (1993) Use of rRNA fluorescence *in situ* hybridisation for measuring the activity of single cells in young and established biofilms. *Applied and Environmental Microbiology* 59, 1354–1360.

Revsbech, N.P., and Jørgensen, B.B. (1986) Microelectrodes. Their use in microbial ecology. *Advances in Microbial Ecology* 9, 293–352.

Reysenbach, A.-L., Giver, L.J., Wickham, G.S. and Pace, N.R. (1992) Differential amplification of rRNA genes by polymerase chain reaction. *Applied and Environmental Microbiology* 58, 3417–3418.

Ried, T., Baldini, A., Rand, T.C. and Ward, D.C. (1992) Simultaneous visualisation of seven different DNA probes by *in situ* hybridisation using combinatorial fluorescence and digital imaging microscopy. *Proceedings of the National Academy of Sciences of the USA* 89, 1388–1392.

Sayler, G.S. and Layton, A.C. (1990) Environmental application of nucleic acid hybridisation. *Annual Review of Microbiology* 44, 625–648.

Schmidt, T.M., DeLong, E.F. and Pace, N.R. (1991) Analysis of a marine picoplankton community by 16S rRNA gene cloning and sequencing. *Journal of Bacteriology* 173, 4371–4378.

Spring, S., Amman, R.I., Ludwig, W., Schleifer, K.-H. and Peterson, N. (1992) Phylogenetic diversity and identification of nonculturable magnetotactic bacteria. *Systematic and Applied Microbiology* 15, 116–122.

Stackebrandt, E., Liesack, W. and Goebel, B.M. (1993) Bacterial diversity in a soil sample from a subtropical Australian environment as determined by 16S rDNA analysis. *FASEB Journal* 7, 232–236.

Stahl, D.A. and Amann, R.I. (1991) Development and application of nucleic acid probes in bacterial systematics. In: Stackebrandt, E. and Goodfellow, M. (eds) *Nucleic Acid Techniques in Bacterial Systematics.* John Wiley, Chichester, pp. 205–248.

Stahl, D.A., Flesher, B., Mansfield, H.R. and Montgomery, L. (1988) Use of phylogenetically based hybridisation probes for studies of ruminal microbial ecology. *Applied and Environmental Microbiology* 54, 1079–1084.

Stouthamer, R., Breeuwer, J.A.J., Luck, R.F. and Werren, J.H. (1993) Molecular identification of microorganisms associated with parthenogenesis. *Nature* 361, 66–68.

Torsvik, V., Goksøyr, J. and Daae, F.L. (1990) High diversity in DNA of soil bacteria. *Applied and Environmental Microbiology* 56, 782–787.

Tsien, H.C., Bratina, B.J., Tsuji, K. and Hanson, R.S. (1990) Use of oligodeoxynucleotide signature probes for identification of physiological groups of methylotrophic bacteria. *Applied and Environmental Microbiology* 56, 2858–2865.

Unterman, B.M., Bauman, P. and McLean, D.L. (1989) Pea aphid symbiont relationships established by analysis of 16S rRNAs. *Journal of Bacteriology* 171, 2970–2974.

Wagner, M., Amman, R.I., Lemmer, H. and Schleifer, K.-H. (1993) Probing activated sludge with oligonucleotides specific for proteobacteria. Inadequacy of culture dependent methods for describing microbial community structure.

Applied and Environmental Microbiology 59, 1520–1525.

Ward, D.M., Weller, R. and Bateson, M.M. (1990) 16S rRNA sequences reveal numerous uncultured inhabitants in a natural community. *Nature* 345, 63–65.

Ward, D.M., Bateson, M.M., Weller, R. and Ruff-Roberts, A. (1992) Ribosomal analysis of microorganisms as they occur in nature. *Advances in Microbial Ecology* 12, 219–286.

Weller, R., Weller, J.W. and Ward, D.M. (1991) 16S rRNA sequences of uncultivated hot spring cyanobacterial mat inhabitants retrieved as randomly primed cDNA. *Applied Environmental Microbiology* 57, 1146–1151.

Yu, S.-M. and Gorovsky, M.A. (1986) *In situ* dot blots: quantification of mRNA in intact cells. *Nucleic Acids Research* 14, 7597–7615.

Zarda, B., Amann, R., Wallner, G. and Schleifer, K.-H. (1991) Identification of single bacterial cells using digoxigenin-labelled rRNA targeted oligonucleotides. *Journal of General Microbiolology* 137, 2823–2830.

THE IMPACT OF MICROORGANISMS ON GLOBAL ECOLOGY AND NUTRIENT CYCLING III

A Neglected Carbon Sink? 6
Biodegradation of Rocks

W.E. KRUMBEIN

Geomicrobiology, ICBM, Carl von Ossietzky-Universität Oldenburg PO Box 2503, D-26111 Oldenburg, Germany

Introduction

What, if not life, controls the cycling of carbon dioxide, organic carbon and inorganic carbonate through the compartments of planet Earth? Did the individual amounts of carbon being reduced into organic and inorganic compounds, oxidized, dissolved, precipitated, volatilized change with the changing history of Earth? Did residence times and reservoir sizes change? Did the ratios between reduced and oxidized carbon change? How does the amount of carbon being reduced relate to nitrogen being reduced and phosphate being bound in the energized living bond? Is the secular increase of atmospheric carbon dioxide by a factor of 15% in 50 years a first time 'première' on Earth? Why is it only 15% and not 30% or more? If the theoretical increase by anthropogenic fossil fuel burning and deforestation of more than 30% over 50 years is not measured in the atmosphere, why then is not all additional carbon dioxide output immediately and totally buffered by the geophysiological reactions of Earth as a whole? How and where is that part of the annual anthropogenic carbon dioxide 'overproduction' bound and stored, which does not accumulate in the atmosphere? Is it stored in a reservoir or does it vanish into a sink? The modern science of 'Global Change' speaks of carbon or carbon dioxide 'sinks'. So last but not least: What is a sink?

In this chapter we first draw attention to general matters, in order to arrive at the core of the subject. We stress, however, from the beginning that we trust in the coordinated ways of planet Earth as a living or at least life-supporting natural body existing as a whole for 4500 million years and in its

© 1995 CAB INTERNATIONAL. *Microbial Diversity and Ecosystem Function*
(eds D. Allsopp, R.R. Colwell and D.L. Hawksworth)

self-supporting and self-coordinating way in geophysical, astrophysical and geochemical equilibria for about 3700 million years. In 1969 it was claimed (Krumbein, 1969) that biologically accelerated weathering is one factor, while speeding up of mass cycles by weathering, sedimentation, and biogeochemical segregation of crustal materials keeps the system going (Krumbein, 1983).

The question of rock weathering as a global carbon sink needs to be considered from the view of general problems of the carbon cycle on Earth. The atmosphere, hydrosphere, lithosphere and biosphere are not separate boxes or reservoirs with individual chemical or physical regimes. They are interlocked, as are individual organs of a body or organelles of a cell. Hutton (1795) believed that Earth should be regarded as a living entity or superorganism and that physiology is the method to study it. In his time however, physiology was synonymous with physics and not with medicine as Lovelock (1989) stated. Despite this unifying approach we have to deal with 'boxes, sinks, and subtopics', which include:

- carbon dioxide increase in the atmosphere;
- carbon pools and fluxes between the pools;
- balance of carbon dioxide sources and sinks;
- the ocean as a sink and the lysocline;
- calcium carbonate weathering, dissolution and precipitation;
- silicate weathering, dissolution and precipitation;
- general balance between ocean and continents;
- biology of rock degradation and rock formation; and
- geophysiology.

Definitions

Sink

A sink of matter or energy in global dynamics is that part of the total system that receives energy or matter from a known and definite compartment (or reservoir) and keeps it for an unknown period of time out of the flow of energy (or matter). Its dimensions and rates of intake are not known. A sink of energy or matter can, but may not necessarily be at the same time or at a given time a source. If it is a source at the same time as it serves as a sink it could be regarded as a reservoir of unknown storage capacity and fluxes.

Defined parts of the Earth's crust are reservoirs of silica, calcium and other elements, when calculable. They represent sources and sinks depending on the direction of the transferred materials, when the content and fluxes are unknown. All these terms and assumptions, however, make it evident that the modern view of global dynamics or global change must necessarily take a geophysiological direction, in which external sources of energy and matter are structured in a metabolic and anatomical sense.

Geophysiology

A geophysiological approach to the problem of energy and material cycles was published recently by Krumbein (1990) and Krumbein and Schellnhuber (1992). Data supporting the physical theory are provided by microbiologists and palaeomicrobiologists. The energy and materials necessary to maintain biogeomorphological equilibrium are fed into the system mainly from outside. They are stored and released on a geological time scale mainly by microbial activities. These are: (i) the energy and electron channelling processes of microbial photosynthesis, respiration and disproportioning inorganic and organic fermentation; (ii) microbially controlled accumulation and storage of energy and mineral reservoirs; and (iii) biologically controlled transportation, translocation, and release of the latter from mantle and crust material into surface-related geotectonical, biogeomorphogenetic, and biogeochemical cycles. Primordial (non-biological) energies and matter distributions are replaced by living matter driven energy and material cycles.

Rock Weathering or (Bio)degradation

The term weathering is an awkward term in view of the fact that at or near the Earth's surface practically no physical or chemical process operates that is not indirectly or directly under the control of living matter. Weathering on the other hand relates much to physical and chemical changes produced by the physical forces of weather and its long-term average, i.e. climate. The terms (bio)transfer, (bio)corrosion, (bio)deterioration, (bio)abrasion, (bio)erosion are intimately connected and related as indicated by the prefix. Krumbein and Dyer (1985) have therefore suggested the division of the processes of material or rock destruction into physical and chemical transfer reactions. Physical transfer includes biogenic and abiogenic processes through which particles are mechanically removed from a material and transferred into the hydrosphere, atmosphere or pedosphere as particulates, colloids and aerosols. Chemical transfer embraces all chemical reactions, biotic and abiotic, through which materials are removed and transferred into the form of gases, solutes, colloids and particulates (chemical reactions can yield solid particles after a chemical transfer and chemical reaction ultimately yielding precipitates after an intermediate gaseous or liquid state). For descriptions of physical and chemical biotransfer actions, see Krumbein and Dyer (1985).

Carbon Dioxide Increase in the Atmosphere

V.I. Vernadsky in several different publications on biogeochemistry, living matter, and the biosphere (Vernadsky, 1929, 1930, 1944) and later, Lovelock (1989) claimed that the chemical composition of the atmosphere is biologically controlled and practically each atom of carbon that is represented in living matter has passed many times through the atmosphere and back to the same living organism (natural body) during its lifetime, despite the fact that the total amount of carbon in the atmosphere is less than 1% of that represented in living matter. They further claimed that the atmosphere is kept at its present composition, more or less stable against external changes, by the force of living matter. Anderson (1984), Krumbein (1983, 1990) and Krumbein and Schellnhuber (1990, 1992) expanded this view on the status of the Earth's crust and its relations to the Earth's mantle.

Balance of Carbon Dioxide Sources and Sinks

Figure 6.1 shows the 'notorious' Keeling curve of actual atmospheric carbon dioxide concentrations, compared to a theoretical line if all carbon dioxide emissions by industrialized societies remained in the atmosphere. There is enough evidence from the calculated annual increase by fossil fuel burning and other processes (Fig. 6.2) that the actual increase as depicted in Fig. 6.1 is only reflecting about 45–50% of the theoretical increase. The input into the atmosphere therefore must be balanced by some factor(s) and mechanism(s), as follows:

1. shallow and deep ocean by diffusion and eddy mixing and the main marine currents upwelling, sinking of water bodies;
2. increase of oceanic and terrestrial photosynthetic binding of carbon dioxide;
3. increase of oceanic and/or terrestrial calcium carbonate and humus (kerogen) production and trapping in terrestrial or oceanic sinks;
4. increase of terrestrial and/or oceanic weathering (physical and chemical weathering of carbonate and silicate rocks under binding of CO_2).

From the Keeling curve (Fig. 6.1) it can be concluded that the transfer rates between atmosphere and aquatic and terrestrial systems is on the level of an annual exchange of about 20% of the atmospheric pool. This means that within 5–7 years all carbon atoms of the atmosphere are transferred into other reservoirs and that this happens almost exclusively by photosynthesis and respiration. Sinks and fluxes into sinks then should have the same order of magnitude of masses and fluxes.

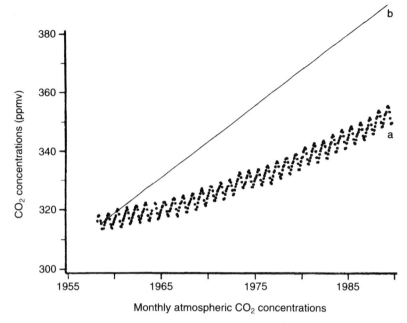

Fig. 6.1. The 'Keeling' curve of measured carbon dioxide concentrations (a) and the theoretical line (b) if all carbon dioxide emissions by industrialized society remain in the atmosphere. (After Boden et al. 1990.)

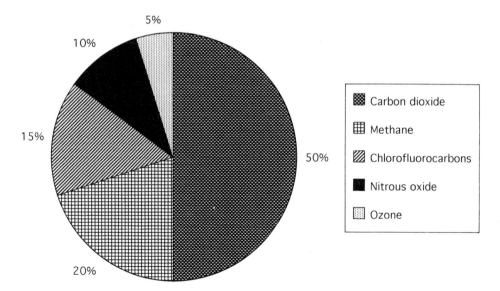

Fig. 6.2. Gas emissions of 'so-called' greenhouse gases (after Pearman, 1992). The values of the individual greenhouse effects are discussed by Andreae and Schimel (1989).

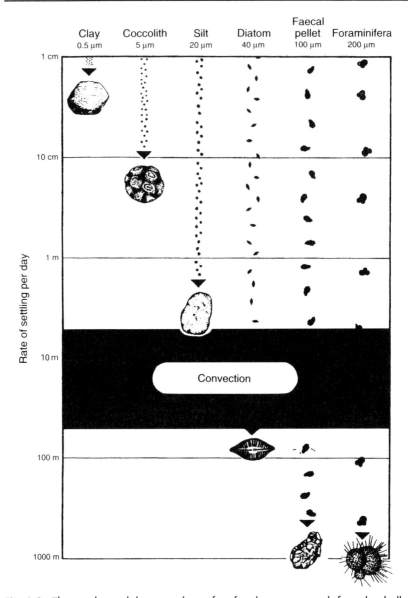

Fig. 6.3. The accelerated downward transfer of carbon compounds from the shallow to the deep ocean (after Degens, 1989).

Diffusion into the oceanic reservoir is far too slow to account for more than a negligible fraction. Photosynthesis, however, in combination with particle formation and calcified tissue production will accelerate the downwards process in the ocean considerably (Moore and Bolin, 1987; Degens, 1989; Tans *et al.*, 1990). Figure 6.3 delineates the transfer velocities of individual carbon-containing compounds from the ocean surface to the deep ocean. Detailed analyses of the oceanic regions, the transfer through warm and cold surfaces, measurements of the carbon dioxide balance in the northern and the southern hemisphere (Moore and Bolin, 1987; Tans *et al.*, 1990) further complicate the situation. Carbonate particles once falling down to the deeper parts of the ocean will be redissolved under the regime of higher solubility of carbonate in cold water under higher pressure (lysocline effect). If photosynthesis-driven calcification is a factor in the elimination of carbon dioxide from the atmosphere then the lysocline should change its position downwards. This, however, is difficult to detect with present day methodologies, although sediment traps have been extremely useful tools in this (Degens, 1989). Further it has become more and more evident that the major part of the carbon dioxide release and increase is registered in the northern hemisphere and that the two carbon pools of the southern and northern hemisphere act as separate boxes. Since the distribution of land and water masses is opposite it is practically impossible that the ocean functions as a primary sink for atmospheric surplus fluxes.

Another approach is the question of fossil fuel burning related to terrestrial photosynthesis, terrestrial organic matter sinks (forests, humus, peat, tundra organic materials). Houghton (1991), Keller *et al.* (1991), and Pearman (1992) as well as several modelling groups (e.g. Karpe *et al.*, 1990; Kohlmaier, 1990) have considered these possibilities. They come to the overall conclusion that industrialized agriculture, deforestation, land erosion through excessive agriculture, and other factors result in no increase of productivity and humus reservoirs but rather an additional contribution to carbon dioxide in the atmosphere (see Fig. 6.2). Thus the terrestrial surface biota has no sink but a source function. There remains, however, the possibility that tundra, peat, and swamps of boreal regions may have a trapping and sink function. Degens (1989) and others (Lerman *et al.*, 1977) hypothesized that despite the release of large amounts of carbon dioxide through deforestation an increase in terrestrial photosynthesis and/or weathering should be measurable through higher organic and inorganic carbon load in the major world rivers. Thus the SCOPE carbon centre in Hamburg undertook to study these fluxes and the possibilities of an oceanic sink for terrestrial carbon trapping from the atmosphere (for references see Degens, 1989). In the large number of publications emerging from this study, however, no conclusive data have been drawn about this channelling from a hypothetical photosynthetic trap into the ocean as the ultimate sink. Furthermore, neither our techniques of measuring and computing global photosynthesis rates, nor those of measuring increases in aquatic dissolved organic carbon and particulate organic carbon (DOC, POC) are precise enough to measure the potential annual change in the order of magnitude needed to pinpoint the annual loss of 50% of the total (anthropogenic and natural) additional release into the atmosphere.

Thus from the above-mentioned four possible sources and sinks for atmospheric excess carbon dioxide perhaps only the question of potentially accelerated weathering of rocks and soils remains.

Rock Weathering as a Potential Carbon Dioxide Sink

Microbial influences on chemical and physical weathering may be perceived, assuming surface areas interacting with atmospheric carbon dioxide and humidity to be about one million or even ten million times larger than the reactive oceanic interface with its neuston and plankton biota near the surface.

Krumbein et al. (1991) and Gehrmann et al. (1988) have determined that the size scale of microbial physical and chemical attack on rock surfaces exists in and creates surface areas and topographies of scales between $20\ \mu m^2$ and $20{,}000\ \mu m^2$ with depth extensions or vertical morphologies as steep as the Alps of a dimension of $10{,}000\ \mu m$ and more. From this we derive that the terrestrial interactive topography or biogeomorphogenetically created surface is several if not many magnitudes higher than that of the aquatic environment. Exact fractal dimensions are difficult to calculate at present but we can assume that the area of some parts of the calcareous Alps or the Mediterranean and Namibian limestone hills, e.g. in which biogenic weathering is prevailing, is to be assumed as a minimum of $10^8\ km^2$ reactive surface instead of the topographically derived $10^4\ km^2$ in a simple Mercator projection.

Although some authors are of the opinion that rock weathering may be an additional source of carbon dioxide (lime burning was perhaps in their minds, see Fig. 6.2) it is generally agreed that rock weathering is a potential sink of carbon dioxide (Schwartzman and Volk, 1990, 1991a,b; Walker, 1993). The mechanisms of the distribution of rocks and reactive compartments of the Earth in terms of weathering and weardown as well as the creation of new reservoirs of rocks have been known since Greek, Hebrew and Roman times (references can be found in the Bible (see Krumbein and Jens, 1981), Herodotus, and Lucretius). Keller (1957) gave an excellent overview of the chemical equations and the possible biological factors. Krumbein and Dyer (1985), Krumbein and Urzì (1993) and many others have pointed to biological processes as the main accelerating factors.

Conclusions

As a general conclusion it can be stated that:

1. rock weathering involving the binding of carbon dioxide is perhaps really the missing carbon dioxide sink in the global carbon flux equations;
2. carbonate and silicate rocks are binding carbon dioxide from the atmosphere into the aquatic and sedimentary cycle at a ratio of 2:1 or at least 1:1 for each mole of carbon dioxide bound for one mole of rock dissolved;
3. the fractal calculus of reactive surface areas makes the terrestrial surface the optimal sink of this kind with response possibilities as viewed through fractal surface calculations that by far exceed all other potential sinks and certainly also the present day dimensions of additional carbon dioxide production. The reactivity is on the level of microbial physiology, i.e. faster than the physiological speed of macroorganisms (Vernadsky, 1930).
4. it seems that both rock formation and weathering have been to a certain extent under rate-limiting biological control since the onset of life on Earth.
5. Schwartzman and Volk (1990) give reasonable arguments that biotic soil formation and rock weathering may have been important cooling mechanisms in the Precambrian.

Acknowledgements

Many discussions of the topic with B.D. Dyer, L. Margulis, C. Urzì and colleagues of the COSY group of ICBM (some of their suggestions were not followed as they were more cautious with the applications of the fractal calculus) are acknowledged as well as the help of Elaine Johnston and C. Urzì in preparing some of the figures. Parts of our work on biodeterioration of rocks was supported by grants of the BMFT, Volkswagen Foundation and the CEC commission XII, Science and Technology.

References

Anderson, D.L. (1984) The Earth as a planet: paradigms and paradoxes. *Science* 223, 347–354.

Andreae, M. and Schimel, D.S. (1989) *Exchange of Trace Gases Between Terrestrial Ecosystems and the Atmosphere*. Wiley, Chichester, 347 pp.

Boden, T.A., Kanciruk, P. and Farrel, M.P. (1990) *Trends '90 – A Compendium of Data on Global Change*. Oak Ridge, Tennessee, 257 pp.

Degens, E. (1989) *Perspectives on Biogeochemistry*. Springer, Berlin, 423 pp.

Gehrmann, C., Krumbein, W.E. and Petersen, K. (1988) Lichen weathering activities on mineral and rock surfaces. *Studia Geobotanica* 8, 33–45.

Houghton, R.A. (1991) Tropical deforestation and atmospheric carbon dioxide. *Climatic Change* 19, 99–118.
Hutton, J. (1795) *Theory of the Earth*, 2 vols. Edinburgh.
Karpe, H.-J., Otten, D. and Trinidade, S.C. (1990) *Climate and Development*. Springer, Berlin.
Keller, M., Jacob, D.J., Wofsy, S.C. and Harris, R.C. (1991) Effects of tropical deforestation on global and regional atmospheric chemistry. *Climatic Change* 19, 139–158.
Keller, W.D. (1957) *The Principles of Chemical Weathering*. Lucas Brothers, Columbia, Missouri, 111 pp.
Kohlmaier, G.H. (1990) Contributions to atmospheric CO_2 increase by changes in the land biosphere: Analysis of the past and present, including possible future developments. In: Karpe, H.-J., Otten, D. and Trinidade, S.C. *Climate and Development*. Springer, Berlin, pp. 219–254.
Krumbein, W.E. (1969) Über den Einfluss der Mikroflora auf die exogene Dynamik (Verwitterung und Krustenbildung). *Geologische Rundschau* 58, 333–363.
Krumbein, W.E. (1983) Introduction. In: Krumbein, W.E. (ed.) *Microbial Geochemistry*. Blackwell Scientific Publications, Oxford, pp. 1–4.
Krumbein, W.E. (1990) Der Atem Cäsars. *Mitteilungen Geologisch-Paläontologisches Institut Hamburg* 69, 267–301.
Krumbein, W.E. and Dyer, B. (1985) This planet is alive – Weathering and biology, a multi-faceted problem. *Chemistry of Weathering* 149, 143–160.
Krumbein, W.E. and Jens, K. (1981) Biogenic rock varnishes of the Negev Desert (Israel): an ecological study of iron and manganese transformation by cyanobacteria and fungi. *Oecologia* 50, 25–38.
Krumbein, W.E. and Schellnhuber, H.J. (1990) Geophysiology of carbonates as a function of bioplanets. In: Ittekott, A.V., Kempe, S. Michaelis, W. and Spitzy, A. (eds) *Facets of Modern Biogeochemistry*. Springer, Berlin, pp. 5–22.
Krumbein, W.E. and Schellnhuber, H.-J. (1992) Geophysiology of mineral deposits – a model for a biological driving force of global changes through Earth history. *Terra Nova* 4, 351–362.
Krumbein, W.E. and Urzì, C. (1993) Biodeterioration processes of monuments as a part of (man-made?) global climate change. In: Thiel M.-J. (ed.) *Conservation of Stone and Other Materials*. Chapman & Hall, London, pp. 558–564.
Krumbein, W.E., Urzì, C.E. and Gehrmann, C. (1991) On the biocorrosion and biodeterioration of antique and mediaeval glass. *Geomicrobiology Journal* 9, 139–160.
Lerman, A., Bernhard, M., Bolin, B., Delwiche, C.C., Ehalt, D.H., Gessel, S.P., Kester, D.R., Krumbein, W.E., Likens, G.E., Mackenzie, F.T., Reiners, W.A., Stumm, W., Woodwell, G.M. and Zinke, P.J. (1977) Fossil fuel burning: its effects on the biosphere and biogeochemical cycles. Group Report. In: Stumm, W. (ed.) *Global Chemical Cycles and their Alterations by Man*. Dahlem Konferenzen, Berlin, pp. 275–289.
Lovelock, J. (1989) *The Ages of Gaia*. Oxford University Press, Oxford, 252 pp.
Moore III, B. and Bolin, B. (1987) The oceans, carbon dioxide and global climate change. *Oceanus* 29, 9–15.
Pearman, G.I. (1992) *Limiting Greenhouse Effects Controlling Carbon Dioxide Emissions*. Wiley, Chichester, 631 pp.

Schwartzman, D.W. and Volk, T. (1990) Biotic enhancement of weathering and the habitability of earth. *Nature* 340, 457–460.

Schwartzman, D.W. and Volk, T. (1991a) Biotic enhancement of weathering and surface temperatures on earth since the origin of life. *Palaeoecology* 90, 357–371.

Schwartzman, D.W. and Volk, T. (1991b) When soil cooled the world. *New Scientist* 51, 33–36.

Tans, P.P., Fung, I.Y. and Takahashi, T. (1990) Observational constraints on the global atmospheric CO_2 budget. *Science* 247, 1431–1438.

Vernadsky, V.I. (1929) *La biosphère*. Alcan, Paris, 232 pp.

Vernadsky, V.I. (1930) *Geochemie in ausgewählten Kapiteln*. Akad. Verlagsgesellsch., Leipzig, 370 pp.

Vernadsky, V.I. (1944) Problems of biogeochemistry II. *Transactions of the Connecticut Academy of Arts and Science* 35, 483–517.

Walker, J.C.G. (1993) Biogeochemical cycles of carbon on a hierarchy of time scales. In: Oremland, R.S. (ed.) *Biogeochemistry of Global Change*. Chapman & Hall, New York, pp. 3–28.

Lichens in Southern Hemisphere Temperate Rainforest and Their Role in Maintenance of Biodiversity

7

D.J. GALLOWAY

Department of Botany, The Natural History Museum, Cromwell Road, London SW7 5BD, UK.

Introduction

Arguably one of the greatest environmental debates of our time focuses on the loss of tropical rainforests, an environment which is rich in lichens and bryophytes (Gradstein, 1992) as it is in vascular plants (Raven, 1987). Appalling though the current and continuing loss of tropical rainforest is, it has tended to divert public attention from the equally appalling fate of much of the world's temperate rainforests among which the rainforests of the Southern Hemisphere are in the front rank in terms of habitat and species biodiversity. A resource indeed which we are coming almost too late to recognize the high value of; a resource which needs careful management if its rich biodiversity is to be maintained for the use and enjoyment of future generations.

The distribution and nature of temperate forests is determined by the asymmetric distribution of land and sea between Northern and Southern Hemispheres, with temperate needle-leaved coniferous forests and deciduous broad-leaved forests characteristic of the continental, winter-cold climates of North America and Eurasia, whereas evergreen broad-leaved forests dominate in the southern temperate region where winters are milder (Ovington, 1983).

To a lichenologist, and especially one who sees the world from a largely European perspective, the abundance and profusion of lichens in Southern Hemisphere rainforests in general, and in those of New Zealand and southern Chile in particular, are an overwhelming sight, not easily forgotten. The lichens seem to be almost too prolific, their luxuriance and size

© 1995 CAB INTERNATIONAL. *Microbial Diversity and Ecosystem Function*
(eds D. Allsopp, R.R. Colwell and D.L. Hawksworth)

a rebuke to the widely held view that lichens are among the slowest-growing of terrestrial organisms (which of course they are but in other, more extreme environments). In Southern Hemisphere temperate rainforests the astonishing diversity and biomass of lichens is a given fact which makes one rethink accepted views on a group of fungi which, until recently, ecologists have tended to neglect as being of little more than academic interest in discussions on forest ecosystem functioning and biodiversity. On the contrary, rainforest lichens have evolved (and are evolving still) with the rainforest itself and play a dynamic role in the rainforest ecosystem contributing large amounts of fixed carbon and fixed nitrogen to the forest's nutrient budget.

What are lichens, and why is it that they have been so neglected in most contemporary discussions of southern temperate rainforest? (Ovington and Pryor, 1983; Veblen *et al.*, 1983; Wardle *et al.*, 1983; Beard, 1990; Axelrod *et al.*, 1991; Schmalzt, 1991). Lichens are composite organisms consisting of a usually dominant fungal partner and one or more photosynthetic partners growing together in a symbiotic association, often cited as a good example of mutualism (Honegger, 1991), or regarded as a case of controlled parasitism (Ahmadjian, 1982). The fungal partner or mycobiont is normally an ascomycete, rarely a basidiomycete, whereas the autotrophic partner may be a green alga or a cyanobacterium.

Currently estimated at some 20,000 taxa worldwide (Galloway, 1992c), lichens are an extremely catholic group of organisms in terms of habitat colonization, occurring from the poles to equatorial deserts and from sea level to nival habitats. The peculiar biology of the lichen symbiosis makes them especially sensitive to habitat and environmental disturbance and for this reason lichens are widely used as biomonitors. The potential importance of lichens in monitoring changes in both tropical and temperate forests, and of their role in nutrient cycling in a variety of ecosystems is increasingly recognized (Galloway, 1993a).

Rainforest as a Habitat for Lichens

Beard (1990) states 'The much higher proportion of sea to land in the Southern Hemisphere creates conditions favouring temperate rain forest on west-facing coasts, so that forests dominated by *Nothofagus* spp. and southern conifers in *Araucariaceae* and *Podocarpaceae* have survived since Cretaceous times ...' Temperate rainforest exists in all of the major landmasses of the Southern Hemisphere with those of South America (Veblen *et al.* 1983; Schmaltz, 1991), New Zealand (Wardle *et al.*, 1983; Wardle, 1991) and Tasmania (Jarman *et al.*, 1991) being major centres of diversity, with strong affinities between them all.

These austral affinities are particularly apparent in the epiphytic lichen floras (Galloway, 1987, 1988a, 1991a) which are characterized by both extensive biomass and high species diversity. Dominant epiphytic and ground lichens are taxa from the following genera: *Cladina* (Ahti and Kashiwadani, 1984; Archer, 1992a), *Cladonia* (Ahti and Kashiwadani, 1984; Archer, 1992b), *Coccocarpia* (Arvidsson, 1983), *Collema* (Degelius, 1974), *Degelia* (Arvidsson and Galloway, 1981; Jørgensen and James, 1990; Jørgensen and Galloway, 1992), *Erioderma* (Jørgensen and Galloway, 1992), *Fuscoderma* (Jørgensen and Galloway, 1992), *Homothecium* (Henssen, 1979), *Hypogymnia* (Elix, 1980, 1992), *Leioderma* (Galloway and Jørgensen, 1987; Jørgensen and Galloway, 1992), *Lepolichen* (Galloway and Watson-Gandy, 1992), *Leptogium* (Verdon, 1992) *Menegazzia* (Santesson, 1942; James and Galloway, 1992), *Metus* (Galloway and James, 1987), *Nephroma* (White and James, 1988), *Pannaria* (Jørgensen and Galloway, 1992), *Pannoparmelia* (Calvelo and Alder, 1992), *Parmelia* (Hale, 1987), *Parmeliella* (Jørgensen and Galloway, 1992), *Phlyctis* (Galloway and Guzmàn, 1988), *Protousnea* (Krog, 1976), *Pseudocyphellaria* (Galloway, 1988b, 1992b), *Psoroma* (Galloway, 1985), *Roccellinastrum* (Henssen et al., 1982), *Sagenidium, Sphaerophorus* (Tibell, 1987), *Sticta* (Galloway, 1985, 1994; Galloway and Pickering, 1990) and *Teloschistes* (Almborn, 1992).

Many of the lichens endemic to Southern Hemisphere temperate rainforest are associated with species of *Nothofagus* (Hill, 1992) or with *Araucaria* and other southern gymnosperms and it appears that the evolution of certain austral lichen groups is closely associated with the evolution of their major phorophytes. The work of Kantvilas and his colleagues in Tasmania (Kantvilas et al., 1985; Kantvilas and James, 1987, 1991; Kantvilas, 1988; Kantvilas and Minchin, 1989; Jarman et al., 1991; Kantvilas and Jarman, 1993) with its emphasis on lichen communities stands as the most detailed source of information on Southern Hemisphere rainforest lichens and provides a model for similar investigations in New Zealand and South America. Similarly, the pioneering work of Green and his colleagues on the ecophysiology of rainforest lichens in New Zealand (see references in Green and Lange, 1991) is a benchmark for corresponding work on rainforest lichens in Tasmania and South America.

The rainforest environment in southern temperate forests extends from sea level to a treeline which varies with latitude from 1800 m down to nearly sea level in southern South America. A dominating feature of the southern rainforests is their evergreen canopy which produces low light levels at or close to the forest floor, in the range of 1–5% of incident light at the canopy. Canopy lichens receive much higher photosynthetically active radiation (PAR) values, and the forest interior is generally buffered from the extremes of strong radiation fluxes produced by clear skies which canopy lichens experience (Green and Lange, 1991). The constancy of the radiation environment together with high, well-spread rainfall (2000–5000 mm per annum) leads to high relative humidities.

The general picture of southern temperate rainforest is of a dim, moist environment with ample rainfall and low evaporation within the forest. In southern South America deciduous species of *Nothofagus* (*N. antarctica, N. dombeyi, N. pumilio*) often form natural treelines on the Andean cordillera, and on leaf-fall in the winter, the canopy and trunk lichens are exposed to high insolation (including reflection from snow) and drying winds, and these conditions impose particular stresses on the exposed lichen vegetation.

Lichen biomass from the forest floor to 2 m in height in a beech forest (*Nothofagus menziesii, N. fusca*) in North Island, New Zealand was estimated at 100 kg dry weight ha^{-1} with 80% of this being the single taxon *Pseudocyphellaria homoeophylla* (Green *et al.*, 1980). Growth rates in southern temperate forest are also high with species of *Pseudocyphellaria* and *Sticta*, again in New Zealand, achieving radial growth of up to 27 mm per year (Green and Lange, 1991). Morphological and physiological adaptations of rainforest lichens are discussed by Green and Lange (1991) and include: (i) changes in thallus morphology to allow increased water storage; (ii) adaptations to deep shade; (iii) desiccation sensitivity; (iv) humidity sensitivity; and (v) photobiont versatility.

Lichens and the Maintenance of Biodiversity in Southern Temperate Rainforest

The combination of high biomass and high biodiversity of the lichen vegetation in Southern Hemisphere rainforests invites speculation on how the lichens themselves might influence biodiversity generally in these habitats and also on how the observed lichen biodiversity in these apparently extremely favourable habitats might be explained. In partial answer to these questions three main themes will be discussed, all of which are interrelated. These are: (i) chemically mediated responses of lichens to light levels; (ii) photobiont versatility; and (iii) nutrient cycling.

Chemical Responses to Light Levels

The success of the lichen symbiosis is undoubtedly the harnessing of photosynthetic metabolites produced by the autotrophic photobiont, for growth of the heterotrophic mycobiont which is generally the dominant partner in the symbiosis. However, the success of this symbiosis in a wide variety of ecosystems is strictly dependent on protection of the autotroph's photosynthetic membranes, and lichens have a number of chemical strategies for coping with stresses to the photosynthetic apparatus which may be summarized as follows.

The xanthophyll cycle – a response to high light stress

Light stress results from an excess of absorbed light over and above that utilized in photosynthesis. Green plants, including the photobionts of lichens, have evolved a major photoprotective process, the xanthophyll cycle, to cope with the dissipation of excess energy absorbed by chlorophll pigments during photosynthesis (Demmig-Adams and Adams, 1992). This involves participation of carotenoids and especially of zeaxanthin in the xanthophyll cycle which allows dissipation of excessive excitation energy in photosynthetic membranes. This xanthophyll cycle is present in the thylakoid membranes of all flowering plants, ferns, mosses, algae and lichens (Demmig-Adams and Adams, 1992). It has been shown that lichens with green photobionts have the xanthophyll cycle and produce zeaxanthin readily, and those with cyanobionts do not have the xanthophyll cycle and produce zeaxanthin only slowly (Demmig-Adams *et al.*, 1990a, b).

Light screening compounds

Lichen cortical pigments such as usnic acid and vulpinic acid are widely assumed to be effective light filters, screening the delicate photosynthetic membranes from damaging radiation, with amounts of pigment deposited in the cortex varying with the degree of exposure to light (Lawrey, 1984, 1986). The often massive canopy burdens of *Protousnea* and *Usnea* seen in southern hemisphere rainforests, especially close to the treeline, owe their distinctive colour, which is frequently visible at long distances, to such screening compounds.

UV-B protection compounds

Secondary metabolites that act as chromophores for UV-B are well-known in plants. Flavonoids produced via the shikimic acid pathway are now widely implicated in UV-B protection and photosystems that absorb specifically in the UV-B waveband can induce production of shikimic acid pathway metabolites. Markham *et al.* (1990) have shown that absolute levels of photoprotective flavonoids in mosses from the Ross Sea region of Antarctica collected between 1957 and 1989 correlated directly with measured levels of ozone over the period 1965–1989.

Workers in Chile have shown that a number of widely distributed lichen metabolites such as atranorin, chloroatranorin, pannarin, gyrophoric acid, divaricatic acid, physciosporin, usnic acid, psoromic acid, norstictic acid, etc., absorb short wavelength radiation and re-emit this as fluorescence at longer wavelengths which can then be transferred to photosynthetic pigments (see Quilhot *et al.*, 1992). The chromophoric unit effecting this UV-B absorption and re-emission is characterized by a carbonyl group

(CHO) and an ortho-hydroxyl group in the same benzene ring, a structure occurring in all the compounds mentioned above. These results have great significance, and confirm and extend the pioneering observations of Rao and LeBlanc (1965) who first reported this photochemical behaviour of atarnorin in lichens. An analogous situation obtains in cyanobacteria where scytonemin, a yellow–brown pigment of cyanobacterial extracellular sheaths, is produced as an adaptive strategy of photoprotection against short wavelength solar radiation (Garcia-Pichel and Castenholz, 1991). The environmental importance of lichen secondary compounds is discussed further in Galloway (1993b).

Photobiont Versatility

A notable characteristic of the epiphytic lichen vegetation of southern temperate rainforest is the high number of cyanobacterial lichens (see above) with cyanobacteria present either as the primary photobiont (especially in the families *Lobariaceae*, *Nephromataceae*, *Pannariaceae* and *Stictaceae*), as cephalodia (James and Henssen, 1976) or as photosymbiodemes (Renner and Galloway, 1982).

The ability of cyanobacterial lichens to photosynthesize at low light intensities (Demmig-Adams *et al.*, 1990a, b; Green and Lange, 1991), to be physiologically active when moistened with liquid water rather than with water vapour (Lange *et al.*, 1986) and to show morphological and biochemical diversity in the formation of photosymbiodemes (Renner and Galloway, 1982; Lange *et al.*, 1988; Demmig-Adams *et al.*, 1990a, b; Galloway, 1992b) allows them to exploit a wide variety of habitat conditions in temperate rainforests and particularly to grow at low light intensities.

The creation of gaps in the forest canopy by either natural or man-made disturbance results in many changes in the forest understorey microenvironment, the most obvious being in the quality and quantity of light (McDonald and Norton, 1992) which leads to significant differences also in local lichen vegetation. The exploitation of changing light levels in rainforest by fast-growing species of *Pseudocyphellaria* is explained in part by the fact that apparently cyanobacterial species contain a significant green algal co-primary photobiont (Green and Lange, 1991; Galloway, 1992b) which can contribute up to 30% of net photosynthesis. Cyanobacterial lichens also appear to operate a CO_2-concentrating mechanism (CCM) which is capable of considerable elevation of internal CO_2 and is similar to that reported for free-living cyanobacteria. The CCM of green algal lichens accumulates much less CO_2 and is probably less effective than that which operates in cyanobacterial lichens (Badger *et al.*, 1993).

Nutrient Cycling

Lichens undoubtedly contribute to the nutrient budgets of rainforests through cycling of minerals (Brown and Brown, 1991) and enrichment of carbon and more especially of nitrogen budgets through either leakage in rainwater, or through decomposition of dislodged lichens in the litter (Forman, 1975; Guzmán *et al.*, 1990). It is becoming clear that the ability of lichens of contribute to the nitrogen budget of forest ecosystems on nitrogen-limited soils is a major ecological role for these organisms.

Lichens having cyanobacteria either as primary photobionts or as cephalodia can fix substantial amounts of atmospheric nitrogen into combined organic nitrogen using heterocyst nitrogenase (Haselkorn, 1986; Rai, 1990). Green *et al.* (1980) have reported a possible nitrogen contribution of between 1 and 10 kg N ha^{-1} year^{-1} from large foliose lichens (*Pseudocyphellaria* and *Sticta*) in temperate rainforest in New Zealand. In temperate rainforests of South America where the proportion of species of *Nephroma*, *Pseudocyphellaria* and *Sticta* having cyanobacterial photobionts is higher than it is in New Zealand, one could confidently expect a similar or indeed a higher level of nitrogen enrichment (White and James, 1988; Galloway, 1992a, 1994). The contribution of other cyanobacterial lichens such as *Coccocarpia*, *Coccotrema*, *Collema*, *Degelia*, *Dictyonema*, *Erioderma*, *Fuscoderma*, *Homothecium*, *Lepolichen*, *Leptogium*, *Pannaria*, *Parmeliella*, *Peltigera*, *Placopsis*, *Polychidium*, *Psoroma*, *Psoromidium*, *Ramalodium*, *Siphulastrum* and *Stereocaulon*, to the nitrogen budgets of a variety of ecosystems is undoubtedly substantial, since many of these genera have rapid growth, produce a large biomass and appear to be strongly competitive.

The contribution to nutrient budgets of southern temperate forests, and especially in the cycling of nitrogen, by taxa from the order *Peltigerales* and the family *Lobariaceae* (Galloway, 1991a, b, 1992a, b, 1994) is undoubtedly a contributory factor, and perhaps a major one, in the maintenance of biodiversity in these forests which are often on nitrogen-limited soils. In most forest ecosystems nutrients limiting growth are nitrogen and phosphorus. The contribution that cyanobacterial lichens make to rainforest nitrogen budgets must therefore be assumed to be substantial. It is a field for exciting future research.

References

Ahmadjian, V. (1982) The nature of lichens. *Natural History* 91, 31–36.
Ahti, T. and Kashiwadani, H. (1984) The lichen genera *Cladia*, *Cladina* and *Cladonia* in southern Chile. In: Inoue, H. (ed.) *Studies on Cryptogams in Southern Chile*. Kenseisha Ltd, Tokyo, pp. 125–151.
Almborn (1992) Some overlooked or misidentified species of *Teloschistes* from South

America and a key to the South-American species. *Nordic Journal of Botany* 12, 361–364.
Archer, A.W. (1992a) *Cladina*. *Flora of Australia* 54, 108–111.
Archer, A.W. (1992b) *Cladonia*. *Flora of Australia* 54, 111–143.
Arvidsson, L. (1983) A monograph of the genus *Coccocarpia*. *Opera Botanica* 67, 1–96.
Arvidsson, L. and Galloway, D.J. (1981) *Degelia*, a new lichen genus in the Pannariaceae. *Lichenologist* 13, 27–50.
Axelrod, D.I., Arroyo, M.T.K. and Raven, P.H. (1991) Historical development of temperate vegetation in the Americas. *Revista Chilena de Historia Natural* 64, 413–446.
Badger, M.R., Pfanz, H., Büdel, B., Heber, U. and Lange, O.L. (1993) Evidence for the functioning of photosynthetic CO_2-concentrating mechanisms in lichens containing green algal and cyanobacterial photobionts. *Planta* 191, 57–70.
Beard, J.S. (1990) Temperate forests of the southern hemisphere. *Vegetatio* 89, 7–10.
Brown, D.H. and Brown R.M. (1991) Mineral cycling and lichens: the physiological basis. *Lichenologist* 23, 293–307.
Calvelo, S. and Adler, M. (1992) *Pannoparmelia anzioides*, a taxonomic synonym of *Pannoparmelia angustata* (Parmeliaceae, Lichenes). *Mycotaxon* 43, 487–498.
Degelius, G. (1974) The lichen genus *Collema* with special reference to the extra-European species. *Symbolae Botanicae Upsalienses* 20(2), 1–215.
Demmig-Adams, B. and Adams, W.W. (1992) Photoprotection and other responses of plants to high light stress. *Annual Reviews of Plant Physiology and Plant Molecular Biology* 43, 599–626.
Demmig-Adams, B., Adams III, W.W., Green, T.G.A., Czygan, F.-C. and Lange, O.L. (1990a) Differences in the susceptibility to light stress in two lichens forming a phycosymbiodeme, one partner possessing and one lacking the xanthophyll cycle. *Oecologia* 84, 451–456.
Demmig-Adams, B., Maguas, C., Adams III, W.W., Meyer, A., Kilian, E. and Lange, O.L. (1990b) Effect of high light on the efficiency of photochemical energy conversion in a variety of lichens with green and blue–green phycobionts. *Planta* 180, 400–409.
Elix, J.A. (1980) ['1979'] A taxonomic revision of the lichen genus *Hypogymnia* in Australasia. *Brunonia* 2, 175–245.
Elix, J.A. (1992) *Hypogymnia*. *Flora of Australia* 54, 201–213.
Forman, R.T.T. (1975) Canopy lichens with blue–green algae: a nitrogen source in a Colombian rain forest. *Ecology* 56, 1176–1184.
Galloway, D.J. (1985) *Flora of New Zealand: Lichens*. P.D. Hasselberg, N.Z. Government Printer, Wellington.
Galloway, D.J. (1987) Austral lichen genera: some biogeographical problems. *Bibliotheca Lichenologica* 25, 385–399.
Galloway, D.J. (1988a) Plate tectonics and the distribution of cool temperate Southern Hemisphere macrolichens. *Botanical Journal of the Linnean Society* 96, 45–55.
Galloway, D.J. (1988b) Studies in *Pseudocyphellaria* (lichens) I. The New Zealand species. *Bulletin of the British Museum (Natural History), Botany* 17, 1–267.
Galloway, D.J. (1991a) Phytogeography of Southern Hemisphere lichens. In: Nimis,

P.L. and Crovello, T.J. (eds) *Quantitative Approaches to Phytogeography*. Kluwer, Dordrecht, pp. 233–262.

Galloway, D.J. (1991b) Chemical evolution in the order Peltigerales: triterpenoids. *Symbiosis* 11, 327–344.

Galloway, D.J. (1992a) Studies in *Pseudocyphellaria* (lichens) III. The South American species. *Bibliotheca Lichenologica* 46, 1–275.

Galloway, D.J. (1992b) Lichens of Laguna San Rafael, Parque Nacional 'Laguna San Rafael', southern Chile: indicators of environmental change. *Global Ecology and Biogeography Letters* 2, 37–45.

Galloway, D.J. (1992c) Biodiversity: a lichenological perspective. *Biodiversity and Conservation* 1, 312–323.

Galloway, D.J. (1993a) Lichens as indicators of environmental change: a world view. In: Symoens, J.-J., Rammeloo, J., Devos, P. and Verstraeten, C. (eds) *Biological Indicators of Global Change*. Royal Academy of Overseas Sciences, Brussels, pp. 223–232.

Galloway, D.J. (1993b) Global environmental change: lichens and chemistry. *Bibliotheca Lichenologica* 53, 87–95.

Galloway, D.J. (1994) Studies on the genus *Sticta* (Schreber) Ach.: I. Southern South American species. *Lichenologist* 26(3), 223–282.

Galloway, D.J. and Guzmán, G. (1988) A new species of *Phlyctis* from Chile. *Lichenologist* 20, 393–397.

Galloway, D.J. and James, P.W. (1987) *Metus*, a new austral lichen genus and notes on an Australasian species of *Pycnothelia*. *Notes from the Royal Botanic Gardens Edinburgh* 44, 561–579.

Galloway, D.J. and Jørgensen (1987) Studies in the family Pannariaceae II. The genus *Leioderma* Nyl. *Lichenologist* 19, 345–400.

Galloway, D.J. and Pickering, J. (1990) *Sticta ainoae*, a new species from cool temperate South America. *Bibliotheca Lichenologica* 38, 91–97.

Galloway, D.J. and Watson-Gandy, L.A. (1992) *Lepolichen coccophorus* (lichenized Ascomycotina, Coccotremataceae) in South America. *Bryologist* 95, 227–232.

Garcia-Pichel, F. and Castenholz, R.W. (1991) Characterization and biological implications of scytonemin, a cyanobacterial sheath pigment. *Journal of Phycology* 27, 395–409.

Gradstein, S.R. (1992) The vanishing tropical rain forest as an environment for bryophytes and lichens. In: Bates, J.W. and Farmer, A.M. (eds) *Bryophytes and Lichens in a Changing Environment*. Clarendon Press, Oxford, pp. 234–258.

Green, T.G.A. and Lange, O.L. (1991) Ecophysiological adaptations of the lichen genera *Pseudocyphellaria* and *Sticta* to south temperate rainforests. *Lichenologist* 23, 267–282.

Green, T.G.A., Horstmann, J., Bonnett, H., Wilkins, A.L. and Silvester, W.B. (1980) Nitrogen fixation by members of the Stictaceae (Lichenes) of New Zealand. *New Phytologist* 84, 339–348.

Guzmán, G., Quilhot, W. and Galloway, D.J. (1990) Decomposition of species of *Pseudocyphellaria* and *Sticta* in a southern Chilean forest. *Lichenologist* 22, 325–331.

Hale, M.E. (1987) A monograph of the lichen genus *Parmelia* sensu stricto (Ascomycotina, Parmeliaceae). *Smithsonian Contributions to Botany* 66, 1–55.

Haselkorn, R. (1986) Cyanobacterial nitrogen fixation. In: Broughton, W.J. and

Pühler, A. (eds) *Nitrogen Fixation. Vol. 4. Molecular Biology.* Clarendon Press, Oxford, pp. 168–193.

Henssen, A. (1979) New species of *Homothecium* and *Ramalodium* from S America. *Botaniska Notiser* 132, 257–282.

Henssen, A., Vobis, G. and Renner, B. (1982) New species of *Roccellinastrum* with an emendation of the genus. *Nordic Journal of Botany* 2, 587–599.

Hill, R.S. (1992) *Nothofagus*: evolution from a southern perspective. *Trends in Ecology and Evolution* 7, 190–194.

Honegger, R. (1991) Functional aspects of the lichen symbiosis. *Annual Reviews of Plant Physiology and Plant Molecular Biology* 42, 553–578.

James, P.W. and Galloway, D.J. (1992) *Menegazzia. Flora of Australia* 54, 213–246.

James, P.W. and Henssen, A. (1976) The morphological and taxonomic significance of cephalodia. In: Brown, D.H., Hawksworth, D.L. and Bailey, R.H. (eds) *Lichenology: Progress and Problems.* Academic Press, London, pp. 22–77.

Jarman, S.J., Kantvilas, G. and Brown, M.J. (1991) Floristic and ecological studies in Tasmanian rainforest. *Tasmanian Rainforest Conservation Program Report* 3, 1–67.

Jørgensen, P.M. and Galloway, D.J. (1992) Pannariaceae. *Flora of Australia* 54, 246–293.

Jørgensen, P.M. and James, P.W. (1990) Studies in the family Pannariaceae IV: The genus *Degelia. Bibliotheca Lichenologica* 38, 253–276.

Kantvilas, G. (1988) Tasmanian rainforest lichen communities: a preliminary classification. *Phytocoenologia* 16, 391–428.

Kantvilas, G. and James, P.W. (1987) The macrolichens of Tasmanian rainforest. Key and notes. *Lichenologist* 19, 1–28.

Kantvilas, G. and James, P.W. (1991) New and noteworthy records of crustose lichens from Tasmanian rainforest. *Mycotaxon* 41, 271–286.

Kantvilas, G. and Jarman, S.J. (1993) The cryptogamic flora of an isolated rainforest fragment in Tasmania. *Botanical Journal of the Linnean Society* 111, 211–228.

Kantvilas, G. and Minchin, P. (1989) An analysis of epiphytic lichen communities in Tasmanian cool temperate rainforest. *Vegetatio* 84, 99–112.

Kantvilas, G., James, P.W. and Jarman, S.J. (1985) Macrolichens in Tasmanian rainforest. *Lichenologist* 17, 76–83.

Krog, H. (1976) *Lethariella* and *Protousnea*, two new lichen genera in Parmeliaceae. *Norwegian Journal of Botany* 23, 83–106.

Lange, O.L., Kilian, E. and Ziegler, H. (1986) Water vapour uptake and photosynthesis of lichens: performance differences in species with green and blue–green algae as phycobionts. *Oecologia* 71, 104–110.

Lange, O.L., Green, T.G.A. and Ziegler, H. (1988) Water status related photosynthesis and carbon isotope discrimination in species of the lichen genus *Pseudocyphellaria* with green or blue–green photobionts and in photosymbiodemes. *Oecologia* 75, 494–501.

Lawrey, J.D. (1984) *Biology of Lichenized Fungi.* Praeger, New York.

Lawrey, J.D. (1986) Biological role of lichen substances. *Bryologist* 89, 111–122.

Markham, K.R., Franke, A., Given, D.R. and Brownsey, P. (1990) Historical ozone level trends from herbarium specimen flavonoids. *Bulletin de Liaison-Groupe Polyphénols* 15, 230–235.

McDonald, D. and Norton, D.A. (1992) Light environments in temperate New

Zealand podocarp rainforests. *New Zealand Journal of Ecology* 16, 15–22.
Ovington, J.D. (1983) Introduction. In: Ovington, J.D. (ed.) *Ecosystems of the World* 10. *Temperate Broad-leaved Evergreen Forests.* Elsevier, Amsterdam, pp. 1–4.
Ovington, J.D. and Pryor, L.D. (1983) Temperate broad-leaved evergreen forests of Australia. In: Ovington, J.D. (ed.) *Ecosystems of the World* 10. *Temperate Broad-leaved Evergreen Forests.* Elsevier, Amsterdam, pp. 73–101.
Quilhot, W., Hidalgo, M.E., Fernandez, E., Peña, W. and Flores, E. (1992) Posible rol biologico de metabolitos secundarios en liquenes antarticos. *Serie Científica Instituto Antarctico Chileno* 42, 53–59.
Rai, A.N. (1990) Cyanobacterial–fungal symbioses: the cyanolichens. In: Rai, A.N. (ed.) *CRC Handbook of Symbiotic Cyanobacteria.* CRC Press, Boca Raton, pp. 9–41.
Rao, D.N. and LeBlanc, F. (1965) A possible role of atranorin in the lichen thallus. *Bryologist* 68, 284–289.
Raven, P.H. (1987) *The Global Ecosystem in Crisis.* J.D. and C.T. MacArthur Foundation, Chicago.
Renner, B. and Galloway, D.J. (1982) Phycosymbiodemes in *Pseudocyphellaria* in New Zealand. *Mycotaxon* 16, 197–231.
Santesson, R. (1942) South American *Menegazziae*. *Arkiv för Botanik* 30A(11), 1–35.
Schmaltz, J. (1991) Deciduous forests of southern South America. In: Röhrig, E. and Ulrich, B. (eds) *Ecosystems of the World* 7. *Temperate Deciduous Forests.* Elsevier, Amsterdam, pp. 557–578.
Tibell, L. (1987) Australasian Caliciales. *Symbolae Botanicae Upsalienses* 27(2), 1–279.
Veblen, T.T., Schlegel, F. and Oltremari, J.V. (1983) Temperate broad-leaved forests of South America. In: Ovington, J.D. (ed.) *Ecosystems of the World* 10. *Temperate Broad-leaved Evergreen Forests.* Elsevier, Amsterdam, pp. 5–31.
Verdon, D. (1992) *Leptogium. Flora of Australia* 54, 173–192.
Wardle, P. (1991) *Vegetation of New Zealand.* Cambridge University Press, Cambridge.
Wardle, P., Bulfin, M.J.A. and Dugdale, J.E. (1983) Temperate broad-leaved evergreen forests of New Zealand. In: Ovington, J.D. (ed.) *Ecosystems of the World* 10. *Temperate Broad-leaved Evergreen Forests.* Elsevier, Amsterdam, pp. 33–71.
White, F.J. and James, P.W. (1988) Studies on the genus *Nephroma* II. The southern temperate species. *Lichenologist* 20, 103–166.

Mineral Cycling by Microorganisms: Iron Bacteria 8

D.B. JOHNSON

School of Biological Sciences, University of Wales, Bangor, Gwynedd LL57 2UW, UK.

Iron is a major element in the lithosphere; only oxygen, silicon and aluminium are more abundant. It occurs in numerous minerals, including ferromagnesium silicates, oxides, carbonates and sulphides (Lundgren and Dean, 1979). Iron is essential for the growth of living organisms (with the possible exception of *Lactobacillus* spp.) though in most cases it is required only in relatively low ('trace') concentrations. Assimilation of iron into microbial biomass (with the possible exception of magnetotactic bacteria) has only minor significance in terms of global scale biogeochemical cycling of the metal. In contrast, dissimilatory transformations of iron, in which oxidoreduction of the metal is linked to microbial energy metabolism, involve large-scale turnover of the metal. The term 'iron bacteria' has no unique definition, and has been used to refer to various groups of microorganisms by different researchers. Winogradsky (1888) defined 'iron bacteria' exclusively as those capable of autotrophic growth using ferrous iron as energy source. Pringsheim (1949) described the group as including all bacteria that were involved in the dissimilatory oxidation and/or deposition of iron, whether or not such bacteria were capable of chemolithotrophic growth on ferrous iron; a similar approach was used by Jones (1986). Lundgren and Dean (1979) extended the definition of 'iron bacteria' to include bacteria which cause the reduction of ferric iron. This latter approach will be adopted in this review.

Besides its pure metallic form (Fe^0) which is highly reactive, iron exists in two ionic states: ferrous (Fe^{2+}) and ferric (Fe^{3+}). Because of their very different stabilities and solubilities, the relative abundance of these two species in a particular environment depends on the prevailing physicochemical characteristics. Ferric iron is the dominant form in most aerated soils and waters; chemical oxidation of ferrous iron proceeds rapidly in most aerated environments, its rate depending on pH, oxygen concentration, temperature and ionic strength of the solution (Nealson, 1983). For example, the half-life of ferrous iron added to sea water is 1–3 min, and in fresh water even less, due to the lower ionic strength of the aqueous phase (Nealson, 1983). In all but the most extremely acidic environments, ferric iron tends to hydrolyse, producing a variety of amorphous and crystalline forms, so that concentrations of soluble, non-complexed ferric iron are very low (in the order of 10^{-17} M) in most environments. However, in the presence of suitable chelating agents, concentrations of soluble iron will exceed those predicted from calculations based on purely inorganic systems. Iron (both ferrous and ferric) may be complexed by a wide range of organic molecules, such as oxalate, citrate, humic colloids, haem and siderophores (Nealson, 1983). Phase stability diagrams based on pH and Eh parameters (e.g. Aristovskaya and Zavarzin, 1971) are useful in predicting the ionic state and solid phase compounds of the metal that are most stable in different environments, and have been quoted widely.

The E'_0 of the Fe^{2+}/Fe^{3+} redox couple (+770 mV) is such that microorganisms may use the couple both as an electron sink and as a supply of electrons, depending on environmental constraints. However, the energy available from ferrous iron oxidation ($\Delta F = -30$ kJ mol^{-1}, at pH 2.0) is small, even in comparison with some other chemolithotrophic reactions (Kelly, 1978). This is because the E'_0 value of the Fe^{2+}/Fe^{3+} couple is close to that of $\frac{1}{2}O_2/H_2O$; oxygen is the only feasible electron acceptor for iron-oxidizing autotrophs. However, the fact that bacteria are associated with iron deposits does not necessarily imply that they are responsible, either directly or indirectly, for their formation. Although indigenous bacteria may have been involved in iron deposition, either by oxidizing ferrous iron or by accumulating ferric iron compounds, they may merely be associated fortuitously with the deposit. The exact situation would need to be resolved, following isolation of pure cultures, in the laboratory, using carefully designed controls. A further consideration is that microorganisms may transform iron (oxidation or reduction) either directly (e.g. enzymatically, or by binding to specific proteins) or indirectly (e.g. by changing the Eh and/or pH characteristics in the environment which promote abiotic iron oxidation or reduction; Nealson, 1983).

Two distinct mechanisms have been described for the formation of iron minerals by biological systems (Frankel and Blakemore, 1990). Biologically controlled mineralization (BCM) involves deposition of minerals in or on preformed organic matrices or vesicles formed by the organism; biologically induced mineralization (BIM) involves extracellular mineral formation as a result of cellular export of metabolic products. Iron minerals formed as a result of BCM reactions tend to have narrow size distributions, in contrast to those deriving from relatively uncontrolled BIM reactions which often have a large size distribution and no unique morphology. Examples of BCM processes are the formation of magnetite in magnetotactic bacteria, and the formation of ferritin, which involves a protein vesicle that controls the deposition of ferrihydrite (Frankel and Blakemore, 1990).

Bacteria involved in iron transformations are a highly diverse group, often with little or no phylogenetic relationship to each other. They include eubacteria and archaea, mesophiles and thermophiles, aerophiles and aerophobes. An approach that has been used frequently to subdivide iron bacteria (other than whether they catalyse iron oxidation or reduction) is on the basis of their pH optima and range, generally identifying two groups: 'neutrophilic' (pH optima of *ca.* 7) and 'acidophilic' (pH optima 2–3) microorganisms. However, it should be noted that pH ranges of some bacteria in these groups overlap, allowing them to inhabit the same environmental niche, provided that the pH does not become too extreme either way. Furthermore, although most 'iron bacteria' are exclusively involved in either oxidation or reduction of the metal, there are others (particularly among acidophilic isolates) that have the capacity either to oxidize or reduce iron, depending on oxygen availability.

Iron Transformations in Circum-neutral pH Environments

Iron Oxidation

There is a prima facie problem with differentiating between microbially induced and abiotic iron oxidation in aerobic neutral pH environments, due to the rapid chemical oxidation of ferrous iron under such conditions. For those bacteria which do actively oxidize iron, it does not necessarily follow that they utilize the energy released during iron oxidation, either as their sole or supplementary energy source. Regardless of the mechanism of ferrous iron oxidation, deposits of ferric oxyhydroxide compounds accumulate in neutral pH environments, sometimes extensively as with 'ochre' deposits in field drains. One mechanism for microbial iron deposition under these conditions is ligand destruction, i.e. the break-up of soluble ferric iron complexes resulting in the rapid precipitation of released iron. Ferrous iron complexes may also be microbially degraded, iron oxidation (chemical or biological) proceeding rapidly after ligand destruction in aerated, neutral environments. Bacteria which cause ferric iron compounds to precipitate in such environments have been grouped together as 'iron depositing' bacteria, a term which avoids the contentious issue of the mechanism of iron accumulation (Jones, 1986). Considerable numbers of different bacterial genera, representing a range of morphological types, have been reported to

Table 8.1. Neutrophilic iron-oxidizing/depositing bacteria.

(a) Bacteria for which there is evidence of direct ferrous iron oxidation.

Genus/species (original description)	Nutritional metabolisms(s)	pH range	G+C (mol %)	Notes
Gallionella ferruginea (Ehrenberg, 1836)	Autotroph/mixotroph	6.0–7.6	54.6	Energy acquisition from Fe^{2+} oxidation confirmed; Mn^{2+} not oxidized
Leptothrix spp, (Kutzing, 1843)	Heterotrophs	6.0–7.5	69.5–71	No evidence for energy conserved from Fe^{2+} oxidation; Mn^{2+} also oxidized
Sphaerotilus natans (Kutzing, 1833)	Heterotroph	6.5–7.5	70	No evidence for energy conserved from Fe^{2+} oxidation; Mn^{2+} not oxidized

(b) Bacteria which deposit ferric iron, but for which firm evidence of ferrous iron oxidation is lacking.

Morphological type	Genera/species	Notes
Budding/prosthecate	*Pedomicrobium ferrugineum*, *Hyphomicrobium*	pH range 3.5–10. Fe^{3+} deposition variable, depending on growth conditions
Budding/non-prosthecate	*Blastocaulis/Planctomyces* group	Heterogeneous collection of bacteria; taxonomy unclear
Non-budding, non-prosthecate, appendaged bacteria	*Metallogenium*	The existence/nature of this genus has been much debated; oxidation of Mn^{2+} reported.
	Seliberia	Accumulation of Fe^{3+} compounds characteristic of mixed cultures, or in pure cultures with ~1% CO_2
Encapsulated bacteria	*Siderocapsa, Siderococcus, Ochrobium, Naumanniella*	Taxonomic uncertainty surrounds all four genera
Sheathed bacteria	*Crenothrix polyspora, Lieskeela, Clonothrix*	Taxonomic position unclear
Gliding bacteria	*Toxothrix*	Produces long filaments (often U-shaped) on which Fe^{3+} oxyhydroxides may be deposited

deposit ferric iron compounds. However, not all of these have been isolated in pure culture, and the taxonomic position of some genera/species remains unclear. In recent years, firm evidence for direct iron oxidation by some neutrophilic bacteria has accumulated. These bacteria are described below, and listed in Table 8.1, together with those bacteria that deposit ferric iron compounds but for which there is no firm evidence for ferrous iron oxidation.

Stalked bacteria

The genus *Gallionella* currently contains only one member species, *G. ferruginea*, though over five species have been described in past classifications (Hanert, 1989). It is one of the easiest of all bacteria to identify; the apical, bean-shaped cells synthesizing characteristic, spirally twisted stalk structures which consist of a bundle of fibres of inorganic (ferric oxyhydroxide) origin. Liberated swarmer cells are motile. *G. ferruginea* is an obligate aerobe, though it grows best under reduced oxygen tensions. Characteristically, in its favoured environments redox potentials are fairly low (typically +200 to +320 mV). Under such conditions, ferrous ions are relatively stable even at circum-neutral pH values, which allows the bacteria to exploit the energy released from enzymatic iron oxidation. *G. ferruginea* is most abundant in relatively pure (low organic matter content), iron-rich waters, such as ferruginous mineral springs and wells, though it is tolerant of salinity and of moderate temperatures (to 47°C). It grows slowly in the laboratory, reaching a cell density of only about 2×10^6 cells ml^{-1} when grown in mineral media, though cell yield can be increased by the addition of organic substrates to growth media (Hallbeck and Pederson, 1991). The carbon metabolism of this bacterium has only recently been resolved. Hanert (1989) reported that ribulose-1,5-bisphosphate carboxylase was present in *G. ferruginea*, but actual fixation of carbon dioxide was first demonstrated some time later (Hallbeck and Pederson, 1991). Hallbeck and Pederson (1991) also showed that *G. ferruginea* could grow either autotrophically or mixotrophically, in the absence or presence of sugars such as glucose. The relative contributions of carbon dioxide and organic carbon to the overall cell budgets varied with medium concentrations of the latter; at glucose concentrations of 10 μM or greater, all of the cell carbon was estimated to have derived from the sugar. Similar traits have been found in some acidophilic iron-oxidizing bacteria, though not in other neutrophilic iron oxidizing/depositing bacteria.

Sheathed bacteria

Sheathed bacteria form filamentous growths comprising rod-shaped cells encased in a hollow, tubular envelope or sheath. In some situations the sheaths are associated with iron (or manganese) deposits, though this is dependent on genus/species as well as environmental conditions. The most well-known bacteria in this group are members of the genera *Sphaerotilus* and *Leptothrix*. Manganese (II) was shown to be enzymatically oxidized by *Leptothrix* spp. (but not by *Sphaerotilus* spp.; Ghiorse, 1984) but, in contrast, the precipitation of iron oxyhydroxides on bacterial sheaths had been considered to be a non-specific process, in which ferrous ions bound to anionic groups on the sheath surface were abiotically oxidized to ferric. However, an iron-oxidizing factor (considered to be a protein) has recently been identified in spent cultures of a non-sheath-forming strain (SS-1) of *Leptothrix discophora* (Corstjens et al., 1992). This macromolecule could be separated from a second, which was able to oxidize manganese (II), by SDS-PAGE. In the same work, it was found that cell lysates of *Sphaerotilus natans* could oxidize ferrous iron but not manganese (II), a result which parallels observations with viable cultures. *Sphaerotilus* spp. are obligate heterotrophs, in which ferrous iron oxidation appears to be a peripheral activity. However, there has been considerable debate over the question of whether *Leptothrix* spp. are capable of autotrophic or mixotrophic metabolism. Most experimental work in this area has involved manganese rather than ferrous iron oxidation; the current consensus is that *Leptothrix* spp. are also obligately heterotrophic, and do not utilize the energy from iron and manganese oxidation. The environmental niches occupied by *Leptothrix* spp. and *Sphaerotilus* spp. differ somewhat; *S. natans* thrives in polluted waters whereas *Leptothrix* spp. are most abundant in slow-running, unpolluted water (often at aerobic/anaerobic interface zones) though *L. cholodnii* is also found in activated sludge (Mulder, 1989).

Other bacteria

Sheathed bacteria other than *Leptothrix/Sphaerotilus* may also be associated with iron deposits (Table 8.1). However, considerably less is known about most of these organisms, and some are yet to be isolated in pure culture. Extracelluar accumulation of ferric oxyhydroxides has also been reported for some encapsulated, budding, appendaged (non-budding) and gliding bacteria (Table 8.1). Most species of the 'genus' *Metallogenium* are neutrophiles associated with manganese deposits rather than iron, although a moderately acidophilic (pH optimum 4.1), iron-depositing isolate was described (Walsh and Mitchell, 1972). However, considerable controversy surrounds *Metallogenium*, since there is no evidence that the stellate structures formed by the 'bacteria' are associated with living material, and some inorganic precipitates may form similar morphological features (Ehrlich, 1990).

There have been reports of ferrous iron oxidation by some photosynthetic bacteria which resemble *Rhodomicrobium* spp. (Widdel *et al.*, 1993). Iron oxidation by these anoxygenic bacteria requires only photosystem I, and allows the intriguing possibility that large-scale oxidation of iron occurred in Archaean times before molecular oxygen became widespread as an oxidant.

Iron Reduction

Ferric iron is a powerful oxidizing agent, though, for reasons discussed earlier, its extremely low solubility means that it is present almost exclusively in solid (amorphous and crystalline) forms in circum-neutral pH environments. In the past there has been considerable debate on whether or not dissimilatory ferric iron reduction is enzymatically mediated, but current consensus is that microorganisms may cause the reduction of ferric to

Table 8.2. Neutrophilic iron-reducing bacteria.

Fermentative bacteria
   Achromobacter
   Aerobacter
   Arthrobacter
   Bacillus
   Bacteroides
   Clostridium
   Corynebacterium
   Escherichia
   Micrococcus
   Paracoccus
   Proteus
   Pseudomonas
   Staphylococcus
   Vibrio

Non-fermentative bacteria
   Geobacter metallireducens
   Pseudomonas sp.
   Shewanella putrefaciens

Sulphate-reducing bacteria
   Desulfotomaculum spp.
   Desulfovibrio spp.

ferrous by either direct or indirect mechanisms. Munch and Ottow (1982) found that direct contact between bacteria and haematite was necessary for iron reduction to occur, indicating an enzymatic mechanism. In contrast, spent, cell-free media from a glucose-fermenting *Vibrio* sp. was found to induce substantial ferric iron reduction (Jones, 1986), implying an indirect mechanism. Organic compounds such as phenolic materials and some organic acids may chemically reduce ferric iron, though the significance of such a mechanism has been considered to be minor compared with biological processes (Lovley, 1987). Another proposed indirect mechanism is that, by lowering the redox potential of an environment, bacteria may induce spontaneous ferric iron reduction. This view has been challenged by Lovley (1991) on the basis that, in non-sulphidic environments, the measured redox potential is itself a reflection of the Fe^{2+}/Fe^{3+} equilibrium, so that it is more accurate to reason that ferric iron reduction is the causative agent of decreased redox potentials in terrestrial and aquatic environments, than to use the converse argument.

Microbiological reduction of ferric iron has been the subject of recent reviews by Lovley (1991) and by Nealson and Myers (1992). Although some fungi seem to be able to reduce ferric iron (Ehrlich, 1990), bacteria are more well known and more important in this respect. Iron-reducing neutrophilic bacteria may be conveniently divided into three groups (Table 8.2): fermentative, non-fermentative, and sulphate-reducing bacteria.

Fermentative bacteria

Most bacteria that have been reported to reduce ferric iron do so during fermentative metabolism; further details of these may be found in reviews by Jones (1986) and Lovley (1987). Glucose, amino acids and malate may serve as fermentable substrates. However, electron flow to ferric iron has been shown to be a minor (though variable) pathway during fermentation. Jones *et al.* (1984) found that *Vibrio* spp. fermenting glucose or malate transferred only 0.13% and 0.03% of reducing equivalents, respectively, to ferric iron; Lovley (1991) quoted <5% as a general figure for this type of metabolism. The metabolic strategy of using ferric iron as a minor electron sink during fermentation has been calculated to be thermodynamically more efficient than fermentation alone (Lovley, 1991). Although the amount of energy released by oxidizing fermentable substrates, such as glucose, to carbon dioxide using ferric iron as sole electron acceptor would be greater, in theory, than by fermentation, no such bacteria have been isolated. However, it has been argued that the primary thermodynamic consideration in anoxic environments is the amount of energy released per mole of substrate metabolized, and on that basis glucose fermentation is more energetically favourable than oxidation using ferric iron as the exclusive electron acceptor (Lovley, 1991).

Fermentative bacteria may use ferric iron via an electron transport system, or as a hydrogen sink associated with substrate level phosphorylation or regeneration of NAD (Jones, 1986). The early opinion that nitrate reductase acted universally as the iron-reducing enzyme was later countered by findings that not all bacteria possessing nitrate reductase were able to reduce ferric iron, and, conversely, that some mutant strains lacking the enzyme still continued to reduce iron (Jones, 1986). A second enzyme system, 'ferrireductase', has been implicated in such cases. Evidence for ferric iron reduction being linked to substrate level phosphorylation was provided by a malate-fermenting *Vibrio* sp., which, in the presence of ferric iron, was able to change the balance of its fermentation end products (less ethanol and more acetate) to be more energetically favourable (Jones *et al.*, 1984).

Non-fermentative bacteria

Bacteria which are able to couple the oxidation of organic substrates or hydrogen to the reduction of ferric iron have been isolated and characterized only recently. A *Pseudomonas* sp. capable of coupling hydrogen oxidation with ferric iron reduction has been described (Balashova and Zavarzin, 1980) but, at present, only two bacterial species are known that can completely oxidize carbon compounds using ferric iron exclusively as the electron sink. These are *Shewanella putrefaciens* (Nealson and Myers, 1992) and *Geobacter metallireducens* (Lovley *et al.*, 1993). *S. putrefaciens* is a facultative anaerobe, capable of using a variety of electron acceptors, though it is less versatile in its potential sources of energy. Formate, acetate, lactate, ethanol and hydrogen may be used as electron donors; rates of growth and of iron reduction may be increased by addition of some carbohydrates, amino acids and other organic compounds to laboratory media (Nealson and Myers, 1992). In contrast, *G. metallireducens* is an obligate anaerobe which is highly versatile in respect of energy source, and capable of utilizing a number of aliphatic (e.g. acetate), aromatic (e.g. benzoate) or inorganic (hydrogen) substrates (Lovley *et al.*, 1993). In addition to ferric iron, manganese (IV), uranium (VI) and nitrate (which is reduced to ammonium) are used as electron acceptors by *G. metallireducens*.

Sulphate-reducing bacteria

The obligately anaerobic, sulphate-reducing bacteria, *Desulfovibrio desulfuricans* and *Desulfotomaculum nigrificans*, can reduce ferric iron to ferrous (Jones, 1986). Sulphide produced via a dissimilatory metabolism can reduce ferric iron chemically (Lovley, 1991).

Iron Transformations in Low pH Environments

Microorganisms that populate extremely acidic environments (generally regarded as those with pH values of <3) are mostly obligately acidophilic rather than acid-tolerant, and are therefore distinct from the microflora of neutral pH environments. This holds true for microorganisms involved in iron transformations. Ferrous iron oxidation at low pH has been much more thoroughly researched than ferric iron reduction. Indeed, it is only recently that the diversity of acidophiles capable of iron reduction and their environmental significance have been realized. One reason for this is that iron-oxidizing acidophiles are able to oxidize pyrite and many other sulphidic minerals, a process at the heart of 'microbial leaching', the use of metal-mobilizing bacteria to extract or concentrate metals from low-grade ores. The same bacteria are responsible for the genesis of 'acid mine drainage', a serious form of pollution associated with effluents from old mines and mine spoils. Most reviews of acidophilic bacteria tend, therefore, to concentrate on aspects of bacterial iron and mineral oxidation, though both oxidation and reduction of iron by acidophilic bacteria have been reviewed by Pronk and Johnson (1993).

Iron Oxidation

Chemical oxidation of ferrous iron proceeds much more slowly in highly acidic conditions, so that distinguishing biologically mediated from abiotic

Table 8.3. Iron-oxidizing, acidophilic bacteria.

Species	G+C (mol %)	Nutritional metabolism
(a) *Mesophiles* (Gram-negative, non-spore-forming eubacteria)		
Thiobacillus ferrooxidans	58	Autotrophic
'Thiobacillus ferrooxidans' m-1	65	Autotrophic
Thiobacillus prosperus	64	Autotrophic
Leptospirillum ferrooxidans	51–55	Autotrophic
(b) *Moderate thermophiles* (Gram-positive, spore-forming eubacteria)		
Sulfobacillus thermosulfidooxidans	45–49	Autotrophic/heterotrophic/ mixotrophic/chemolitho- heterotrophic
(c) *Extreme thermophiles* (Gram-negative archaebacteria)		
Acidianus brierleyi	31	Autotrophic/heterotrophic

oxidation does not involve the problems described for neutral pH environments. All acidophilic iron-oxidizers described to date have been bacteria, and these can be conveniently divided into three subgroups according to their response to temperature, viz. mesophiles, moderate thermophiles and extreme thermophiles (Table 8.3). Further information about these iron-oxidizers can be found in Norris (1990).

Mesophilic acidophiles

By far the most thoroughly researched of this group, and indeed of all iron-oxidizing microorganisms, is *Thiobacillus ferrooxidans*. This acidophile, first isolated and described over 40 years ago, is capable of oxidizing ferrous iron and a number of reduced sulphur compounds and may effectively catalyse, directly or indirectly, the dissolution of many sulphidic minerals. Although many bacteria have been classified as strains of *T. ferrooxidans* on the basis of their cellular morphologies (single or paired straight rods), acidophily and ability to oxidize ferrous iron, individual isolates may have little or no genomic relationship with each other (e.g. on the basis of DNA homologies; Harrison, 1986). A major reason why *T. ferrooxidans* has generally been reported to be ubiquitous in iron-rich, acidic waters (which it probably is) and has frequently been implied to be the dominant ferrous-oxidizing acidophile in such environments (which is more open to question) relates to the common practice of isolating iron-oxidizing bacteria using enrichment cultures of ferrous sulphate liquid media. Strains of *T. ferrooxidans* tend to grow more rapidly on ferrous iron (mean generation time (t_d) ~5–12 h) than other iron-oxidizing acidophiles (t_d often = 10–15 h) thus causing them to dominate mixed populations of iron-oxidizers (Pronk and Johnson, 1993). Liquid media have often been used preferentially because of the frequently reported problems of growing *T. ferrooxidans* (and other iron-oxidizing acidophiles) satisfactorily on solid media. Recent improvements in media design, and, in particular, the use of overlaid, agarose-gelled media which gave very high (>90%) plating efficiency (e.g. Johnson and McGinness, 1991a) have facilitated the direct isolation of iron-oxidizing acidophiles from environmental samples. Variation in morphologies of the ferric iron-encrusted colonies is a useful starting point to separate potentially different species and strains, which may be extended following examination of the cellular morphologies and physiological characteristics of isolates.

One distinctive '*T. ferrooxidans*-like' isolate which was reported to be unable to oxidize sulphur (which should disqualify it from being classified as a *Thiobacillus* sp.) is strain m-1 (Harrison, 1986). Although it now appears that strain m-1 can oxidize elemental sulphur on prolonged incubation, comparison of 16S rRNA sequences has shown that strain m-1 is not closely related to 'mainstream' strains of *T. ferrooxidans* and, in fact, occurs in a different subdivision of the *Proteobacteria* group (Lane *et al.*,

1992). *Thiobacillus prosperus*, on the other hand, has many physiological characteristics in common with *T. ferrooxidans*, with the major exception that it is halotolerant (up to 6% NaCl; Huber and Stetter, 1989). In contrast, *Leptospirillum ferrooxidans* differs from *T. ferrooxidans* in its morphology (vibrioid, highly motile cells), its inability to grow on reduced sulphur compounds, and its greater sensitivity to many metals (Norris, 1990). *L. ferrooxidans* is more acidophilic and tolerant of ferric iron than *T. ferrooxidans*, and may bring about more complete oxidation of sulphidic minerals such as pyrite. Analysis of 16S rRNA sequences has shown that *L. ferrooxidans* is not closely related to *T. ferrooxidans* (or isolate m-1) or to any other bacteria of known sequencing data (Lane *et al.*, 1992). Although frequently overlooked in the past, *L. ferrooxidans* is probably as ubiquitous as *T. ferrooxidans*; classification of iron-oxidizers isolated directly from three mines as '*T. ferrooxidans*-like' and '*L. ferrooxidans*-like' isolates has indicated that *L. ferrooxidans* was dominant at two of the sites (D.B. Johnson, unpublished data). Iron oxidation by autotrophic acidophiles is a somewhat unusual 'primary production' system, though it has been shown, using laboratory-based ecosystems, to form the basis of an intricate food web (Fig. 8.1; Johnson, 1991).

All of the acidophilic iron-oxidizers described above are considered to be obligate chemoautotrophs, though it is known that *T. ferrooxidans* can incorporate trace quantities of some organic compounds (e.g. amino acids and nucleic acid bases; Oliver and Van Slyke, 1988) and may also grow

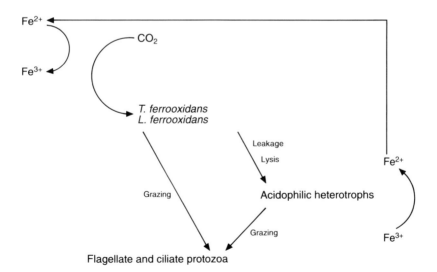

Fig. 8.1. Food web based on the oxidation of ferrous iron by acidophilic, autotrophic bacteria (after Johnson, 1991).

anaerobically using formic acid as electron donor (Pronk et al., 1991). However, there have been reports of iron-oxidizing acidophiles that are obligately heterotrophic, i.e., they do not display an ability to fix carbon dioxide and require a source of organic carbon (such as yeast extract) for growth. One of these, isolate CCH7 (Johnson et al., 1992), was isolated from a sulphur mine containing copious growths of 'acid streamers'. Morphologically, CCH7 was distinct from other iron-oxidizers in being filamentous, and there was tentative evidence of sheath formation. It formed gelatinous, streamer-like growths up to 5 cm or so long *in vitro*, on which ferric oxyhydroxides were deposited at pH >2.5. The isolate has not yet been classified. Another novel iron-oxidizing heterotrophic acidophile isolated from the same mine site has recently been described. This bacterium (designated T-21) occurs as small (1–1.5 µm long) motile rods which, unlike CCH7, appear to utilize the energy available from ferrous iron oxidation (in 'lean' organic media, cell yields correlate with initial ferrous iron concentrations) but, like CCH7, has been found unable to fix carbon dioxide. Isolates similar to T-21 have been isolated from a cobalt mine in Idaho, USA and Parys mountain copper mine in North Wales (D.B. Johnson, unpublished data). Initial experiments incorporating T-21 in mixed cultures (together with *T. ferrooxidans*, *L. ferrooxidans*, heterotrophic

Table 8.4. Novel, unclassified, iron-oxidizing acidophilic bacteria.

Isolate	Morphology	Nutritional metabolism
'Iron-oxidizing bacterium' (Cameron et al., 1981)	Long filaments; individual cells, short rods	Autotrophic
'OP14' (Johnson et al., 1989)	Medium-long rods; filamentous	Autotrophic
'Iron bacterium' (Emtiazi et al., 1989)	Filamentous, 'sheathed', Gram-positive, spore-forming	Heterotrophic
'CCH7' (Johnson et al., 1992)	Long filaments; individual cells, motile rods	Heterotrophic
'T-21' (Pronk and Johnson, 1993)	Single rods, motile	Heterotrophic
'T3.2' (de Siloniz et al., 1993)	Single/paired motile rods	Autotrophic[a]/ mixotrophic

[a]No autotrophic growth on Fe^{2+} alone; yeast extract, glucose or $S_2O_3^{2-}$ required.

acidophiles and moderate thermophiles) in shake flasks containing pyrite have indicated that they compete successfully with other acidophiles in such circumstances (Johnson et al., 1993a). Novel, though so far unclassified, acidophilic iron-oxidizing bacteria are listed in Table 8.4.

Mineral oxidation may be catalysed by acidophilic bacteria that do not oxidize ferrous iron, as evidenced by *Thiobacillus cuprinus* (Huber and Stetter, 1990). A recent *Thiobacillus*-like isolate described by de Siloniz et al. (1993) oxidized pyrite generating ferric iron, but would only oxidize ferrous iron if media were supplemented with reduced sulphur or organic (yeast extract or glucose) compounds. The sulphur-oxidizing acidophile *Thiobacillus thiooxidans* does not oxidize sulphidic minerals (Norris, 1990).

Moderately thermophilic acidophiles

Iron-oxidizing acidophilic bacteria with temperature optima of about 50°C have been noted to be dissimilar to mesophilic isolates in a number of fundamental respects (Norris, 1990; Ghauri and Johnson, 1991). First, they are Gram-positive, rather than Gram-negative eubacteria, and most isolates have been observed to form endospores. Second, most strains can grow autotrophically on ferrous iron only when supplied with a source of reduced sulphur. Third, they have highly versatile nutritional metabolisms, and may grow mixotrophically (e.g. in ferrous iron/glucose medium, where both carbon dioxide and glucose are used as carbon sources), chemolithoheterotrophically (e.g. in ferrous iron/yeast extract medium, in which iron oxidation is used as energy source and yeast extract as carbon source), heterotrophically (e.g. in ferric iron/yeast extract medium), or autotrophically (e.g. on pyrite). Most moderate thermophilic iron-oxidizers appear to have a requirement for enhanced concentrations of carbon dioxide for maximum growth and mineral oxidation rates when growing autotrophically (Norris, 1990), a fact that may greatly limit their potential in commercial mineral extraction.

Moderately thermophilic iron-oxidizing bacteria have been isolated from geothermal springs (e.g. in Iceland and Yellowstone National Park, Wyoming), self-heating coal spoils and leach dumps (Norris, 1990; Ghauri and Johnson, 1991). All isolates described to date have been rods of varying sizes, and with differing propensities to grow as filaments. Classification of these bacteria is still at an early stage, even though the first isolate of this type was recorded as far back as 1977 (Le Roux et al., 1977). Karavaiko et al. (1988) have proposed the genus *Sulfobacillus*, and species *S. thermosulfidooxidans* for their isolates from eastern Europe. However, differences in (G+C) contents and data from 16S rRNA analysis (Lane et al., 1992) of three other moderate thermophiles have indicated that these may belong to more than one genus.

Extremely thermophilic acidophiles

Iron-oxidizing acidophilic 'bacteria' with pH optima at *ca.* 70–75°C have also been isolated from geothermal waters and self-heating coal spoils. Isolates belong to the archaean domain. Unevenly lobe-shaped archaea belonging to the genus *Sulfolobus* are aerobes that may grow either autotrophically or heterotrophically (Segerer *et al.*, 1986), whereas *Acidianus* spp. are facultative aerobes. *A. brierleyi* (originally classified as *S. brierleyi*) can grow chemolithotrophically on ferrous iron (Segerer *et al.*, 1986), though yeast extract has been reported to be required (Norris, 1990). As with the moderate thermophiles, the mineral-oxidizing activity of a *Sulfolobus* sp. was found to be stimulated by enhanced concentrations of carbon dioxide (Norris, 1990). A more recent extremely thermophilic isolate, *Metallosphaera sedula*, oxidizes sulphidic minerals but was reported not to oxidize ferrous iron (Huber *et al.*, 1989).

Iron Reduction

The increased solubility of ferric iron in low pH environments would suggest that it may have more widespread use as a direct and indirect electron sink under microaerophilic and anoxic conditions. However, relatively little is known about this area of iron biogeochemistry, although interest has grown in recent years. Both autotrophic and heterotrophic acidophiles have been observed to reduce ferric iron *in vitro* (Table 8.5).

Table 8.5. Iron-reducing, acidophilic bacteria.

Organism	Nutritional metabolism
(a) Mesophiles	
Thiobacillus ferrooxidans	Autotrophic (S^0, electron donor); heterotrophic (formic acid, electron donor)
Thiobacillus thiooxidans	Autotrophic (S^0, electron donor)
Acidiphilium spp., *Thiobacillus acidophilus*	Heterotrophic (glycerol, electron donor)
(b) Moderate thermophiles	
Sulfobacillus-like isolates TH1, THWX, YTF1, HPTH	Heterotrophic (glycerol, electron donor)
(c) Extreme thermophiles	
Sulfolobus acidocaldarius	Autotrophic (S^0, electron donor)

Autotrophic acidophiles

Brock and Gustafson (1976) noted that ferric iron was reduced in cultures of *T. thiooxidans* growing on elemental sulphur under aerobic conditions, and also showed that both *T. thiooxidans* and *T. ferrooxidans* coupled the anaerobic oxidation of sulphur to the reduction of ferric iron. In most situations, any ferric iron reduced by *T. ferrooxidans* would, in the presence of oxygen, be rapidly reoxidized, though at very low pH (<1.3) ferrous iron does accumulate in aerobic cultures (Sand, 1989). Pronk *et al.* (1991) showed that ferric iron could act as electron acceptor in anaerobic cultures of *T. ferrooxidans* in which either elemental sulphur or formic acid acted as electron donor. In anaerobic sulphur cultures, biomass production was proportional to the amount of ferric iron reduced (Pronk *et al.*, 1992), confirming that *T. ferrooxidans* is a facultative anaerobe.

Ferric iron reduction has also been reported for thermophilic acidophiles. Some moderate thermophiles can reduce ferric iron when growing heterotrophically (Ghauri and Johnson, 1991), and iron cycling by pure cultures of these bacteria has been demonstrated by changing dissolved oxygen concentrations (Johnson *et al.*, 1993b). Brock and Gustafson (1976) showed that the extreme thermophile *Sulfolobus acidocaldarius* could reduce ferric iron when oxidizing sulphur.

Heterotrophic acidophiles

Obligately heterotrophic acidophilic bacteria have been isolated from cultures of *T. ferrooxidans*, acid mine drainage and mineral leachates (e.g. Harrison, 1984). Many appear to belong to the genus *Acidiphilium*, and were initially considered to be obligate aerobes. However, many isolates are now known to be facultative anaerobes, and can use ferric iron as electron acceptor in the absence of oxygen (Johnson and McGinness, 1991b). Both soluble and insoluble ferric iron compounds may be reduced, though at different rates; soluble ferric iron is most readily reduced, followed by amorphous ferric hydroxides, and finally crystalline (e.g. jarosites, haematite) forms (Pronk and Johnson, 1993). The electron donor is an organic substrate, such as glycerol or glucose. Strictly anoxic conditions are not required; indeed, ferric iron reduction by these bacteria proceeds most rapidly under microaerophilic conditions, and may be demonstrated using aerobically incubated Petri plate cultures (Johnson and McGinness, 1991b). Mixed cultures of iron-oxidizing and iron-reducing acidophilic bacteria may demonstrate rapid cycling of iron between the ferrous and ferric states in laboratory cultures (Fig. 8.2), the extent of cycling generally being limited by the availability of organic substrates (Johnson *et al.*, 1993b). Interestingly, although ferric iron reduction has been noted in cultures of moderately thermophilic iron-oxidizing bacteria (described above), no

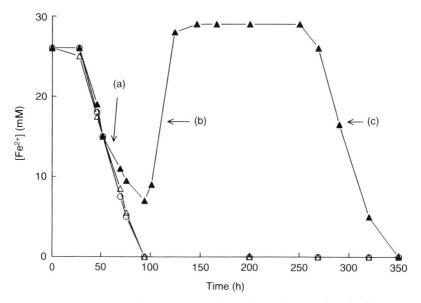

Fig. 8.2. Transformations of iron in pure and mixed populations of acidophilic bacteria, grown in 25 mM ferrous iron (+/− 10 mM glycerol) in unshaken flasks at 30°C. ○, *T. ferrooxidans* (− glycerol); △, *T. ferrooxidans/Acidiphilium* SJH (− glycerol); ▲, *T. ferrooxidans/Acidiphilium* SJH (+ glycerol). (a) First phase of net iron oxidation; (b) phase of net iron reduction; (c) second phase of net iron oxidation.

reduction was observed in cultures of five obligately heterotrophic thermophiles isolated from Yellowstone National Park (Johnson and McGinness, 1991b).

Very little is known about the capacity of iron-oxidizing heterotrophic bacteria to reduce ferric iron; limited ferric iron reduction has been reported for the single-celled isolate T-21, but no reduction has been observed in cultures of the filamentous isolate CCH7 (Pronk and Johnson, 1993).

Overview and Future Prospects of Microbial Oxidoreduction of Iron

Microbially mediated geochemical cycling of iron in the environment is an on-going process. On a local scale, cycling may be rapid. Soils, for example, are known to be complex ecosystems, composed of myriads of micro-environments which may vary tremendously in physicochemical characteristics within a single profile, and which are subject to rapid and dramatic changes, e.g. in pH, Eh and oxygen tensions, all of which would influence

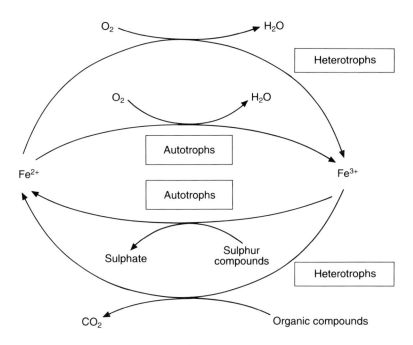

Fig. 8.3. Microbial oxidoreduction of iron in extremely acidic environments (after Pronk and Johnson, 1993).

microbial oxidoreduction of iron. In extremely acidic environments, oxidation and reduction of iron may be brought about by both autotrophic and heterotrophic acidophiles, as outlined in Fig. 8.3. On a global scale, cycling of iron is a much slower process, though it has been accelerated by human activities, for example, by exposing pyritic minerals to moisture and aeration and thereby facilitating microbial iron oxidation.

It has been speculated that the reduction of ferric iron may have been the first globally important mechanism for the microbial oxidation of organic matter to carbon dioxide (Lovley, 1991). In many present day sedimentary environments ferric iron continues to be an important electron sink, even though there is a wide range of electron acceptors available for organic matter oxidation. Iron reduction in sediments may inhibit the production of methane (thereby controlling global methane fluxes) and the reduction of sulphate (thereby limiting the production of phytotoxic sulphides) in waterlogged soils and sediments. On the other hand, microbial oxidation/deposition of iron in waters draining hydromorphic soils can result in the accumulation of ochreous precipitates, and failure of land drainage systems. Acid mine drainage, a serious form of pollution that effects many hundreds of kilometres of streams and rivers throughout the world, is generated primarily by iron-oxidizing acidophiles. Conversely, remediation of acid mine drainage in constructed wetlands depends upon the activities of indigenous, non-defined populations of iron- and sulphate-reducing bacteria.

Iron bacteria may, however, be exploited in industrial operations. Microbial iron reduction has been proposed as a mechanism to solubilize and extract iron selectively from iron ores, and may also be used to remove iron impurities from kaolins (Lovley, 1991). However, it is the acidophilic iron-oxidizing bacteria and the established biotechnology of mineral leaching that have attracted most attention. Although *T. ferrooxidans* has been the major bacterium highlighted in leaching studies, there is no doubt that other bacteria, many possibly yet to be isolated and characterized, will prove to be of at least equal significance in future developments.

It is something of a paradox with 'iron bacteria' that, although some of these microorganisms were among the earliest to be described, only relatively recently has it become apparent that characterized isolates represent only a small percentage of those present in the environment. Non-fermentative, ferric iron-reducing isolates, for example, are a subgroup of 'iron bacteria' that were mostly unknown several years ago. There is clearly a need to extend our knowledge of the biodiversity of this important group of bacteria, and the need to develop more efficient methodologies has been stressed on several occasions. Although genetic probes are very useful in evaluating the abundance of known strains of an organism in mixed communities, they have often indicated that cultured strains are numerically insignificant in the environment. The objectives of improving techniques for isolating the more diffident 'iron bacteria', and of maintaining pure and mixed cultures in the laboratory are currently being addressed by some groups. Such fundamental research should advance our understanding of these bacteria in forthcoming years.

References

Aristovskaya, T.V. and Zavarzin, G.A. (1971) Biochemistry of iron in soil. In: McLaren, A.D. and Skujins, J.J. (eds) *Soil Biochemistry*, Vol. 2. Marcel Dekker, New York, pp. 385–408.

Balashova, V.V. and Zavarzin, G.A. (1980) Anaerobic reduction of ferric iron by hydrogen bacteria. *Microbiology* 48, 635–639.

Brock, T.D. and Gustafson, J. (1976) Ferric iron reduction by sulfur- and iron-oxidising bacteria. *Applied and Environmental Microbiology* 32, 567–571.

Cameron, F.J., Edwards, C. and Jones, M.V. (1981) Isolation and preliminary characterization of an iron-oxidizing bacterium from an ochre-polluted stream. *Journal of General Microbiology* 124, 213–217.

Corstjens, P.L.A.M., de Vrind, J.P.M., Westbroek, P. and de Vrind-de Jong, E.W. (1992) Enzymatic iron oxidation by *Leptothrix discophora*: identification of an

iron-oxidizing protein. *Applied and Environmental Microbiology* 58, 450–454.
de Siloniz, M.I., Lorenzo, P., Murua, M. and Perera, J. (1993) Characterization of a new metal-mobilising *Thiobacillus* isolate. *Archives of Microbiology* 159, 237–243.
Ehrlich, H.L. (1990) *Geomicrobiology*. Marcel Dekker, New York.
Emtiazi, G., Habibi, M.H. and Setareh, M. (1989) Novel filamentous sporeforming iron bacteria causes bulking in activated sludge. *Journal of Applied Bacteriology* 67, 99–108.
Frankel, R.B. and Blakemore, R.P. (1990) *Iron Biominerals*. Plenum, New York.
Ghauri, M.A. and Johnson, D.B. (1991) Physiological diversity amongst some moderately thermophilic iron-oxidising bacteria. *FEMS Microbiology Ecology* 85, 327–334.
Ghiorse, W.C. (1984) Biology of iron- and manganese-depositing bacteria. *Annual Review of Microbiology* 38, 515–550.
Hallbeck, L. and Pederson, K. (1991) Autotrophic and mixotrophic growth of *Gallionella ferruginea*. *Journal of General Microbiology* 137, 2657–2661.
Hanert, H.H. (1989) Non-budding, stalked bacteria. Genus *Gallionella*. In: Staley, J.T. (ed.) *Bergey's Manual of Systematic Bacteriology*, Vol. 3. Williams and Wilkins, Baltimore, pp. 1974–1979.
Harrison, A.P. Jr (1984) The acidiphilic thiobacilli and other acidophilic bacteria that share their habitat. *Annual Review of Microbiology* 38, 265–292.
Harrison, A.P. Jr (1986) Characteristics of *Thiobacillus ferrooxidans* and other iron-oxidising bacteria, with emphasis on nucleic acid analysis. *Biotechnology and Applied Biochemistry* 8, 249–257.
Huber, H. and Stetter, K.O. (1989) *Thiobacillus prosperus* sp. nov. represents a new group of halotolerant metal-mobilizing bacteria isolated from a marine geothermal field. *Archives of Microbiology* 151, 479–485.
Huber, H. and Stetter, K.O. (1990) *Thiobacillus cuprinus* sp. nov., a novel facultatively organotrophic metal-mobilizing bacterium. *Applied and Environmental Microbiology* 56, 315–322.
Huber, H., Spinnler, C., Gambacorta, A. and Stetter, K.O. (1989) *Metallosphaera sedula* gen. and sp. nov. represents a new genus of aerobic, metal-mobilizing, thermoacidophilic archaebacteria. *Systematic and Applied Microbiology* 12, 38–47.
Johnson, D.B. (1991) Diversity of microbial life in highly acidic, mesophilic environments. In: Berthelin, J. (ed.) *Diversity of Environmental Biogeochemistry*. Elsevier, Amsterdam, pp. 225–235.
Johnson, D.B. and McGinness, S. (1991a) A highly efficient and universal solid medium for growing mesophilic and moderately thermophilic, iron-oxidizing, acidophilic bacteria. *Journal of Microbiological Methods* 13, 113–122.
Johnson, D.B. and McGinness, S. (1991b) Ferric iron reduction by acidophilic heterotrophic bacteria. *Applied and Environmental Microbiology* 57, 207–211.
Johnson, D.B., Said, M.F., Ghauri, M.A. and McGinness, S. (1989) Isolation of novel acidophiles and their potential use in bioleaching operations. In: Salley, J., McCready, R.G.L. and Wichlacz, P.L. (eds) *Biohydrometallurgy 1989; International Symposium Proceedings*. CANMET, Ontario, pp. 403–414.
Johnson, D.B., Ghauri, M.A. and Said, M.F. (1992) Isolation and characterization of an acidophilic, heterotrophic bacterium capable of oxidizing ferrous iron.

Applied and Environmental Microbiology 58, 1423–1428.
Johnson, D.B., Nicolau, P. and Bridge, T.A.M. (1993a) Manipulation and monitoring of acidophilic bacterial populations. In: Torma, A.E., Wey, J.E. and Lakshmanan, V.I. (eds) *Biohydrometallurgical Technologies*, Vol. 1. TMS, Warrendale, Pennsylvania, pp. 673–683.
Johnson, D.B., Ghauri, M.A. and McGinness, S. (1993b) Biogeochemical cycling of iron and sulphur in leaching environments. *FEMS Microbiology Reviews* 11, 63–70.
Jones, J.G. (1986) Iron transformations by freshwater bacteria. *Advances in Microbial Ecology* 9, 149–185.
Jones, J.G., Gardener, S. and Simon, B.M. (1984) Reduction of ferric iron by heterotrophic bacteria in lake sediments. *Journal of General Microbiology* 130, 45–51.
Karavaiko, G.I., Golovacheva, R.S., Pivovarova, T.A., Tzaplina, I.A. and Vartanjan, N.S. (1988) Thermophilic bacteria of the genus *Sulfobacillus*. In: Norris, P.R. and Kelly, D.P. (eds) *Biohydrometallurgy: Proceedings of the International Symposium, Warwick 1987*. Science and Technology Letters, Kew, UK, pp. 29–42.
Kelly, D.P. (1978) Bioenergetics of chemolithotrophic bacteria. In: Bull, A.T. and Meadow, P.M. (eds) *Companion to Microbiology: Selected Topics for Further Discussion*. Longman, London, pp. 363–386.
Lane, D.J., Harrison, A.P. Jr, Stahl, D., Pace, B., Giovannoni, S.J., Olsen, G.J. and Pace, N.R. (1992) Evolutionary relationships among sulfur- and iron-oxidising eubacteria. *Journal of Bacteriology* 174, 269–278.
Le Roux, N.W., Wackerley, D.S. and Hunt, S.D. (1977) Thermophilic thiobacillus-like bacteria from Icelandic thermal areas. *Journal of General Microbiology* 100, 197–201.
Lovley, D.R. (1987) Organic matter mineralization with the reduction of ferric iron: a review. *Geomicrobiology Journal* 5, 375–399.
Lovley, D.R. (1991) Dissimilatory Fe(III) and Mn(IV) reduction. *Microbiological Reviews* 55, 259–287.
Lovley, D.R., Giovannoni, S.J., White, D.C., Champine, J.E., Phillips, E.J.P., Gorby, Y.A. and Goodwin, S. (1993) *Geobacter metallireducens* gen. nov. sp. nov., a microorganism capable of coupling the complete oxidation of organic compounds to the reduction of iron and other metals. *Archives of Microbiology* 159, 336–344.
Lundgren, D.G. and Dean, W. (1979) Biogeochemistry of iron. In: Trudinger, P.A. and Swaine, D.J. (eds) *Biogeochemical Cycling of Mineral Forming Elements*. Elsevier, Amsterdam, pp. 211–251.
Mulder, E.G. (1989) Sheathed bacteria. In: Staley, J.H. (ed.) *Bergey's Manual of Systematic Bacteriology*, Vol. 3. Williams & Wilkins, Baltimore, pp. 1994–2003.
Munch, J.C. and Ottow, J.C.G. (1982) Effect of cell contact and iron (III) oxide form on bacterial iron reduction. *Zeitschrift für Pflanzenernaehrung, Duengung und Bodenkunde* 140, 549–562.
Nealson, K.H. (1983) The microbial iron cycle. In: Krumbein, W.E. (ed.) *Microbial Geochemistry*. Blackwell Scientific Publications, Oxford, pp. 159–190.
Nealson, K.H. and Myers, C.R. (1992) Microbial reduction of manganese and iron: new approaches to carbon cycling. *Applied and Environmental Microbiology* 58, 439–443.

Norris, P.R. (1990) Acidophilic bacteria and their activity in mineral sulphide oxidation. In: Ehrlich, H.L. and Brierley, C.L. (eds) *Microbial Mineral Recovery*. McGraw-Hill, New York, pp. 3–27.

Oliver, D.J. and Van Slyke, J.K. (1988) Sulfur-dependent inhibition of protein and RNA synthesis by iron-grown *Thiobacillus ferrooxidans*. *Archives of Biochemistry and Biophysics* 263, 369–377.

Pringsheim, E.G. (1949) Iron bacteria. *Biological Reviews* 24, 200–250.

Pronk, J.T. and Johnson, D.B. (1993) Oxidation and reduction of iron by acidophilic bacteria. *Geomicrobiology Journal* 10, 153–171.

Pronk, J.T., Liem, K., Bos, P. and Kuenen, J.G. (1991) Energy transduction by anaerobic ferric iron reduction in *Thiobacillus ferrooxidans*. *Applied and Environmental Microbiology* 57, 2063–2068.

Pronk, J.T., de Bruyn, J.C., Bos, P. and Kuenen, J.G. (1992) Anaerobic growth of *Thiobacillus ferrooxidans*. *Applied and Environmental Microbiology* 58, 2227–2230.

Sand, W. (1989) Ferric iron reduction by *Thiobacillus ferrooxidans* at extremely low pH values. *Biogeochemistry* 7, 195–201.

Segerer, A., Neuner, A., Kristjansson, J.K. and Stetter, K.O. (1986) *Acidianus infernus* gen. nov., sp. nov., and *Acidianus brierleyi* comb. nov: facultatively aerobic, extremely acidophilic thermophilic sulfur-metabolizing archaebacteria. *International Journal of Systematic Bacteriology* 36, 559–564.

Walsh, F. and Mitchell, R. (1972) An acid-tolerant iron-oxidizing *Metallogenium*. *Journal of General Microbiology* 72, 369–376.

Widdel, F., Schnell, S., Heising, S., Ehrenreich, A., Assmus, B. and Schink, B. (1993) Ferrous iron oxidation by anoxygenic phototrophic bacteria. *Nature* 362, 834–836.

Winogradsky, S.N. (1888) Uber Eisenbacterin. *Botanisches Zeitung* 46, 262–270.

The Potential Importance of Biodiversity in Environmental Biotechnology Applications: Bioremediation of PAH-contaminated Soils and Sediments

9

P.H. PRITCHARD[1], J.G. MUELLER[2], S.E. LANTZ[2], AND D.L. SANTAVY[1]

[1] *US Environmental Protection Agency, Environmental Research Laboratory, Gulf Breeze, Florida 32561;* [2] *SBP Technologies, Gulf Breeze, Florida 32561, USA.*

Introduction

The biodegradation of hydrocarbons by bacteria and fungi has been actively studied for 20–30 years (Atlas, 1984; Leahy and Colwell, 1990). From these studies, we recognize that a great diversity of hydrocarbon-degrading microorganisms are maintained in most aquatic and terrestrial environments. It can be argued that if we understand how this biodiversity is developed and maintained in natural microbial communities, then we may be able to modify and control the biodiversity as a basis for more effective environmental biotechnology applications, particularly those involving the bioremediation of environmental materials contaminated with hydrocarbons (Atlas, 1984; Bartha, 1986; Lee and Levy, 1989; Morgan and Watkinson, 1989). Success in respect of bioremediation will depend, in part, on understanding the environmental factors responsible for the maintenance of this biodiversity and the ability to manipulate the diversity to improve or stimulate hydrocarbon degradation rates and extents in the environment.

Diversity of Hydrocarbon-degrading Microorganisms

The importance of biodiversity to hydrocarbon degradation can be appreciated by knowledge of the microorganisms involved. The ability to metabolize hydrocarbons is ubiquitous to the microbial world. Numerous genera of bacteria and fungi have been isolated based on their hydrocarbon-degrading capabilities (for general reviews, see Atlas, 1984; Leahy and Colwell, 1990). The majority of these isolations involved hydrocarbons associated with crude oil and petroleum products. Biodegradation of hydrocarbons produced from plants (i.e. terpenes such as a-pinene and oxygenated compounds such as camphor and diterpenic acids), which represent a major source of natural hydrocarbons in the environment, has been less investigated (Button, 1984).

A list of the major hydrocarbonoclastic bacteria and fungi isolated from marine and terrestrial environments is shown in Table 9.1. Most of the organisms in Table 9.1 will grow on a single hydrocarbon or mixture of hydrocarbons. The diversity appears to be quite large. There is even a report of one algal species that degrades straight- and branched-chain alkanes (Walker et al., 1975). The reason so many bacterial species are able to degrade hydrocarbons is probably related to the universal environmental availability of hydrocarbons to microbial communities resulting from the decomposition of plant material and release of petroleum from geological deposits.

The major groups of hydrocarbons that are utilized by microorganisms include the alkanes (linear and branched), alkenes or olefins, cycloalkanes and their alkyl derivatives, aromatics (single and condensed ring, alkylated and non-alkylated), and the nitrogen-, sulphur-, and oxygen-containing heterocyclic chemicals. Included in the alkane and alkene groups are gaseous hydrocarbons, such as methane, ethane, propane, butane, ethylene, propylene, and acetylene. These hydrocarbons are produced biogenically from the decay of organic material and, thus, have origins from sources other than crude oil and petroleum deposits (Schoell, 1980). Many of the bacteria (and some fungi) that oxidize these short-chain hydrocarbons do not oxidize the longer-chain alkanes and alkenes. This is particularly true of the bacteria that oxidize methane, a diverse group of bacteria called the methanotrophs (Higgins et al., 1984).

Aromatic hydrocarbons have been the most intensively studied of the hydrocarbon classes, probably because of their worldwide distribution (they originate from a variety of sources both natural and anthropogenic), extensive use in industry, and potential toxicity (Kusk, 1981; Black et al., 1983; Jacob et al., 1986; Nisbet and LaCoy, 1992). Consequently, many different aromatic hydrocarbon-degrading bacteria and fungi have been isolated and characterized. A list of these organisms is provided in Table 9.2. Cyanobacteria (Cerniglia, 1980; Narro et al., 1992), algae (Cerniglia, 1982;

Table 9.1. Examples of different genera of hydrocarbon-degrading bacteria and fungi from marine, estuarine and terrestrial environments (compiled in part from Atlas, 1984 and Floodgate, 1984).

Bacteria	Fungi
Achromobacter	*Acremonium*
Acinetobacter	*Allescheria*
Actinomyces	*Aureobasidium*
Aeromonas	*Beauveria*
Alcaligenes	*Botrytia*
Arthrobacter	*Candida*
Bacillus	*Chrysosporium*
Beneckea	*Cladosporium*
Brevibacterium	*Cochliobolus*
Chromobacterium	*Cunninghamella*
Corynebacterium	*Cylindrocarpon*
Cytophaga	*Debaryomyces*
Erwinia	*Fusarium*
Flavobacterium	*Geotrichum*
Klebsiella	*Gonytrichum*
Lactobacillus	*Hansenula*
Leucothrix	*Helminthosporium*
Micrococcus	*Humicola*
Moraxella	*Monilia*
Mycobacterium	*Mucor*
Nocardia	*Odiodendron*
Proteus	*Paecilomyces*
Pseudomonas	*Penicillium*
Sarcina	*Phialophora*
Spherotilus	*Rhodosporidium*
Spirillum	*Rhodotorula*
Streptomyces	*Saccharomyces*
Vibrio	*Scolecobasidium*
Xanthomonas	*Torulopsis*
	Trichoderma
	Trichosporon

Warshawasky *et al.*, 1988), and yeasts (MacGillvray and Shiaris, 1993) are reported to transform selected aromatic hydrocarbons, but generally they are unable to utilize these compounds as sources of both carbon and energy.

Table 9.2. Examples of genera and species of bacteria and fungi that are involved in aromatic hydrocarbon metabolism (compiled in part from Cerniglia 1984, 1993).[a]

Bacteria	Fungi
Achromobacter sp.	Absidia glauca
Acinetobacter claocaceticus	Absidia ramosa
Aeromonas sp.	Aspergillus niger
Alcaligenes denitrificans	Aspergillus ochraceus
Alcaligenes eutrophus	Basidiobolus ranarum
Alcaligenes faecalis	Candida maltosa
Beijerinckia sp.	Candida tropicalis
Corynebacterium renale	Candida utilis
Desulfomonile tiedje	Choanephora conjuncta
Flavobacterium sp.	Cokeromyces poitrasii
Klebsiella pneumoniae	Conidiobolus gonimodes
Methylosinus trichosporium	Cunninghamella blakesleeana
Moraxella sp.	Cunninghamella echinulata
Nitrosomonas europaea	Cunninghamella elegans
Nocardia sp.	Gilbertella persicaria
Pseudomonas aeruginosa	Mortierella verrucosa
Pseudomonas cepacia	Mucor hiemalis
Pseudomonas fluorescens	Neurospora crassa
Pseudomonas mendocina	Panaeolus subbalteatus
Pseudomonas pickettii	Penicillium notatum
Pseudomonas pseudoalcaligenes	Penicillium ochro-chloron
Pseudomonas putida	Phanerochaete chrysosporium
Pseudomonas rhodochrous	Phytophthora cinnamomi
Pseudomonas vesicularis	Psilocybe strictipes
Rhodococcus chlorophenolicus	Rhizophlyctis horderi
Rhodopseudomonas palustris	Rhizopus stolonifer
Sphingomonas paucimobilis	Saccharomyces cerevisiae
Xanthobacter sp.	Saprolegnia parasitica
	Smittium culicis
	Syncephalastrum racemosum

[a]Genus and species names are taken from the literature and may differ from those currently accepted.

Basis for Extensive Hydrocarbon Degradation Capabilities in Natural Microbial Communities

Biodiversity in the microbial world is determined, in part, by the organic materials to which the microorganisms are exposed as they attempt to live in the natural environment. In the case of hydrocarbons, these chemicals have been produced in the environment throughout geological time. They are of diverse structure and widely distributed in the biosphere, predominantly as components of surface waxes of leaves, plant oils, cuticles of insects and the lipids of microorganisms (Millero and Sohn, 1991). Civilization has increasingly affected both hydrocarbon concentration and distribution in the environment, largely as a result of wood burning, oil spillage, industrial activities and discharges, and automotive exhaust, among others. Thus, the three major sources of hydrocarbons are biogenic, geochemical and anthropogenic.

In general, many of the alkanes and alkenes are of biogenic origin, being produced from a variety of terrestrial plants and aquatic algae (Millero and Sohn, 1991). In marine systems, good correlations have been documented between the presence of certain straight- and branched-chained alkanes and blooms of primary producers (Gordon et al., 1978). Straight-chain alkanes with carbon number maxima in the C_{17}–C_{21} range are typical of aquatic algae, whereas terrestrial plant sources typically produce alkanes with C_{25}–C_{33} maxima (Millero and Sohn, 1991). The alkanes from terrestrial sources tend to show a predominance of odd-numbered carbon backbones, whereas aquatic systems tend to produce equivalent amounts of odd- and even-numbered homologues (Grimalt et al., 1985; Nishimura and Baker, 1986; Cripps, 1992). Thus, alkanes appear to be a relatively common natural source of carbon in the environment and, certainly, many microorganisms have evolved to use hydrocarbons as growth substrates.

In many other environments, however, alkanes have been detected in significant concentrations as a result of pollution from crude oil and petroleum products (Gordon et al., 1978; Kostecki and Calabrese, 1989). This input is steadily increasing on a worldwide scale and is thereby becoming a prominent, albeit episodic, carbon source for bacteria in the environment.

Among the aromatic hydrocarbons are both mononuclear and polynuclear compounds. The structures of several of these hydrocarbons are shown in Fig. 9.1. They are derivatives of the benzene ring which is thermodynamically stable due to the possession of a large negative resonance energy. Many aromatic hydrocarbons are metabolized by microorganisms (Gibson and Subramanian, 1984). The mononuclear aromatic compounds have been the focus of considerable research because of their prominence as industrial chemicals. Petroleum and coal provide the largest source of these chemicals, but they are also commonly found in the environment as products of biosynthesis in the plant kingdom. Included among these are aromatic carotenoids, lignin, alkaloids, terpenes and flavonoids (Hopper, 1978). Alkyl-substituted benzenes, varying in substitution from single methyl groups to long side chains (often olefinic), are common from these sources.

The polyaromatic hydrocarbons (PAHs) are largely of geochemical

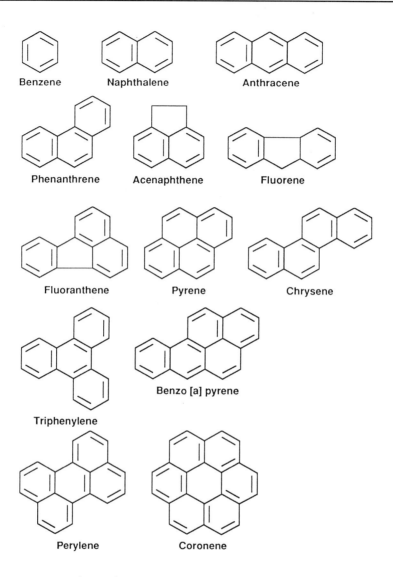

Fig. 9.1. Structures of several common polyaromatic hydrocarbons.

origin, both in terrestrial and marine environments. These compounds consist of two or more fused benzene rings, containing only carbon and hydrogen. The PAHs are formed by a natural pyrolytic process involving alkylated benzene rings (Blumer, 1976). The alkyl groups can be of sufficient length to allow cyclization and then, eventually, these cyclized moieties become aromatized. The temperature at which this process occurs determines the degree of alkyl substitution on the PAH; the higher the temperature, the less substituted the resulting PAHs become (Blumer, 1976). Thus microbial communities have been exposed to PAHs, particularly the two- and three-ring PAHs, for millions of years.

Diversity of Hydrocarbon Degradation Processes

Microorganisms have evolved specialized metabolic processes to degrade hydrocarbons and this represents a part of their biodiversity (Gibson, 1984). The strategy used by bacteria and fungi to degrade hydrocarbons involves the insertion of molecular oxygen into these carbon-rich structures. The process is catalysed by oxygenase enzymes. The enzymes have been well studied in many different bacterial taxa (and some fungi), but predominantly in *Pseudomonas* species, and it appears that they impart uniqueness to a bacterial species or strain, based on the specificity of the enzymes for certain types of hydrocarbons. General metabolic pathways for alkanes, cycloalkanes, monoaromatic and polycyclic aromatic hydrocarbons are shown in Fig. 9.2. Alkanes and alkenes are initially attacked (terminally or subterminally) by a hydroxylase (mixed-function oxidase) to produce the corresponding n-alcohol. The alcohol is further oxidized to the corresponding monocarboxylic acid and metabolized by β-oxidation to provide acetate units for the cell's intermediary metabolism.

Branched alkanes are utilized by fewer organisms and with greater difficulty. The assumption is that the methyl branches either impede uptake into the cell or make the hydrocarbon less compatible with enzymes involved in β-oxidation.

The cyclic alkanes are also initially attacked by a hydroxylase enzyme producing a cycloalkanone and leading to direct insertion of another oxygen molecule into the ring. The resulting lactone is eventually cleaved to produce a dicarboxylic acid.

Aromatic hydrocarbons are attacked by both mono- and dioxygenases (requiring molecular oxygen) producing an initial dihydroxy compound that can subsequently be attacked by a second ring-opening oxygenase. The resulting product, often a dicarboxylic acid or a semialdehyde, is further metabolized to produce intermediates that can be routed into the intermediary metabolism of the cell.

Polycyclic Aromatic Hydrocarbon Degradation and Biodiversity

Although biodiversity is generally not investigated from the standpoint of the compounds metabolized by microorganisms, there are examples where

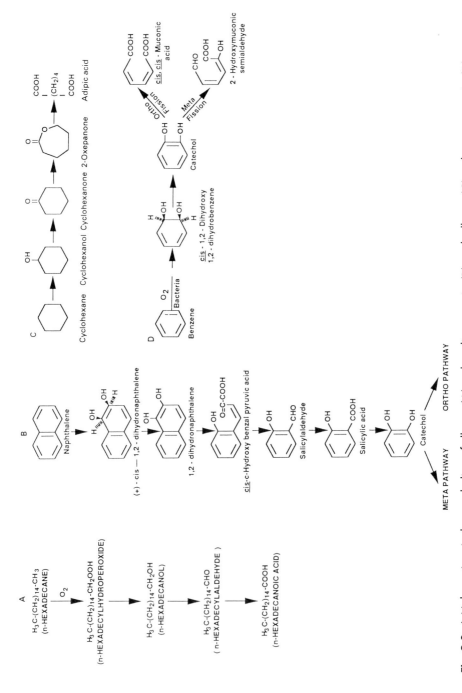

Fig. 9.2. Initial reactions in the metabolism of alkanes (A), polycyclic aromatics (B), cycloalkanes (C) and monoaromatics (D).

investigations of this nature have provided a greater understanding of microbial biodiversity and its phylogenetic basis (West et al., 1984; Takeuchi et al., 1993). The biodegradation of hydrocarbons must overcome problems that are not associated with the degradation of other more oxidized substrates (lipids, carbohydrates, proteins, etc.). Thus a unique diversity of hydrocarbon-degrading microorganisms has evolved. Admittedly, this diversity may be primarily defined at the enzymatic level (for example, the same taxonomic group of bacteria may contain oxygenases of differing specificity for aromatic hydrocarbons), yet there are other cellular characteristics that an organism must possess to degrade certain compounds.

This is particularly true of the polycyclic aromatic hydrocarbons. They are hydrophobic and very insoluble. For bacteria to attack these compounds, they must be able to respond to very low aqueous-phase concentrations. Additionally, because of their hydrophobicity and insolubility, PAHs in the environment are likely to be found sorbed to many kinds of surfaces. Thus, bacteria that use the PAHs as sources of carbon and energy, may have evolved a variety of mechanisms for accommodating the problems associated with sorption and bioavailability. Examples would be the production of emulsifiers or biosurfactants, the possible use of very efficient membrane transport systems, and the recruitment of specialized catabolic plasmids. It may be these mechanisms, if they are organism-specific, that form a basis for further biodiversity studies.

In other cases, we know that PAHs can be cometabolized, that is, the PAHs are transformed by bacteria without necessarily serving as a source of carbon and energy, often by cells grown on a different PAH that is used for energy and growth (Bauer and Capone, 1988; Mueller et al., 1990; Cerniglia, 1993). Cometabolism (as defined here) could be an important mechanism used by microbial communities to degrade mixtures of PAHs. Thus, the attributes of the individuals that make a microbial community could define a unique biodiversity pattern.

Since even the high-molecular-weight PAHs are incorporated into biogeochemical cycles in the environment, bacteria likely exist that attack these compounds where they are found. In fact, we are beginning to realize that there are many microorganisms in soil and sediments, particularly those that are contaminated with PAHs, that are capable of growing on high-molecular-weight PAHs (Foght and Westlake, 1988; Mueller et al., 1990; Weissenfels et al., 1990; Kelly and Ceringlia, 1991; Walter et al., 1991), suggesting that they may possess unique characteristics beyond their enzymatic mechanisms for the initial oxidation of these chemicals. The genes for these enzymes may be located on plasmids (Zylstra and Gibson, 1991).

Importance of Microbial Biodiversity in the Alaskan Bioremediation Experience

The role of biodegradation in the cleanup of oil-contaminated beaches in Prince William Sound, Alaska, appears to have been important (Chianelli et al., 1991; Lindstrom et al., 1991; Pritchard and Costa, 1991; Pritchard et al., 1991, 1993). Although oil biodegradation rates could be enhanced by the addition of fertilizers, they had to compete with high rates of 'natural' biodegradation in control areas, i.e., those that did not receive fertilizer. Chemical analysis of the residual oil from these control plots showed a substantial reduction in the concentrations of PAHs, particularly the alkyl-substituted PAHs. The alkylated PAHs are generally slower to degrade than the unsubstituted parent compound. An example of these results is shown in Fig. 9.3. It is reasonable to assume that the observed reductions in the

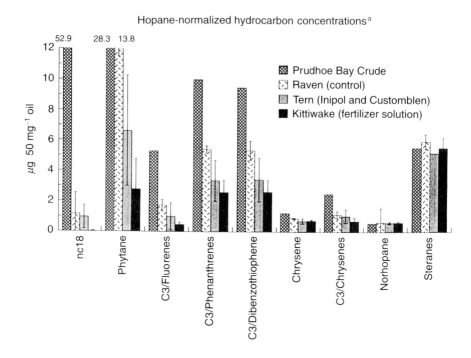

Fig. 9.3. Concentrations of selected PAHs in oil samples from control and fertilizer-treated (TN and KW) beaches in Prince William Sound, Alaska, taken during the summer of 1989. The TN beach was treated with a combination of fertilizer granules and an oleophilic fertilizer. The KW beach was treated with a fertilizer solution. Hopane, a cyclic alkane that is resistant to biodegradation, was used as an internal reference marker (Pritchard et al., 1991). [a] $n = 8$, randomly selected from 20 August 1989.

PAHs, particularly those of higher molecular weight, were due to biodegradation. If this is true, the microbial communities associated with these PAH degradation activities appear to be diverse, given the extent of removal observed. In fact, the PAH results in Alaska potentially represent some of the most extensive biodegradation observed in aquatic environments (Shiaris, 1989).

What is the basis for this biodiversity? Clearly, it is related to the background types of aromatic hydrocarbons (or other types of chemicals) that the microbial communities normally utilize as sources of carbon and energy. Natural concentrations of PAHs are unlikely to be high enough to elicit this type of microbial diversity in Prince William Sound. But it is possible that the degradation of plant-derived hydrocarbons, or similar chemical products, is the basis for this diversity. Button (1984) has proposed that terpenes might be the basis of the microbial food chain in Alaskan waters, and given the large quantities of terpenes and other related chemicals released from the surrounding coniferous forest in Prince William Sound, this is a reasonable proposal. However, there has been little work to show that bacteria capable of degrading these plant products also degrade PAHs. The relatively rapid response of the natural microbial communities to the degradation of PAHs in the *Exxon Valdez* oil spill would tend to suggest a possible connection.

Further information on the biodiversity of natural microbial populations relative to PAH degradation in Alaskan waters can be obtained from work by Braddock *et al.* (1990). Figure 9.4 shows the naphthalene mineralization rates of 44 water and sediment samples collected in Prince William Sound during the summer (1989) of the oil spill. Since the results are for 2-day incubations, any significant mineralization indicates that the natural microbial communities are acclimated for active naphthalene mineralization. Most of the sites that had high numbers of oil degraders (indicating exposure to *Exxon Valdez* oil) had high naphthalene mineralization rates, as would be expected. But several sites (Amakdedori Beach, Chignik Bay), with undetectable numbers of oil degraders, also showed some naphthalene mineralization. There could be, in these cases, exposure to natural sources of aromatic hydrocarbons that is compatible with naphthalene degradation. However, it is noteworthy that when these samples were incubated for a total of 10 days (Fig. 9.4b), many more of the samples with no prior oil exposure (a total of 86% of the sites) showed naphthalene mineralization potential, and many with no apparent exposure to oil. Thus, the microbial diversity of the microbial communities in these waters includes the potential to metabolize naphthalene as a common attribute. This adaptation response is, of course, critical for any biotechnology applications that might be considered for oil-contaminated materials.

Similar studies were conducted by Lindstrom (1991) in summer 1990.

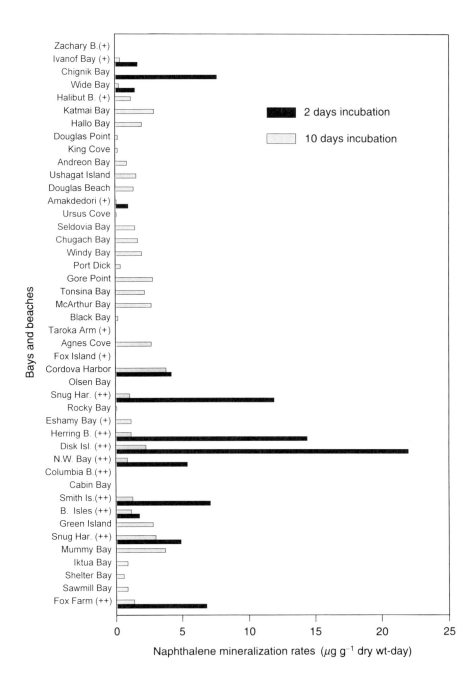

Fig. 9.4. (opposite) Naphthalene oxidation rates in samples taken from Prince William Sound, Alaska, during summer 1989 that were incubated in laboratory test systems for (a) 2 days and (b) 10 days. Beaches with high numbers of oil degraders (>10^5 cells g^{-1} of beach material) are designated with two plus signs. Beaches with low numbers of oil degraders (<10^1 cells g^{-1} of beach material) are designated with a plus sign. From Braddock *et al.* (1990).

He compared the ability of natural microbial communities to mineralize phenanthrene as a function of their previous exposure to oil and/or fertilizers and their exposure to nitrogen and phosphorus nutrients in the laboratory. Some of the results are shown in Fig. 9.5. At the KN401 site, which had oil exposure but no fertilizer added to the beaches, background phenanthrene mineralization was relatively high and the response to the

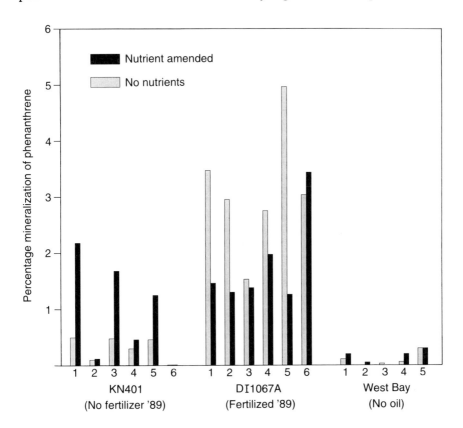

Fig. 9.5. Mineralization of phenanthrene in samples from two oil-exposed beaches and one unexposed beach in Prince William Sound, Alaska. Samples were amended with sea water or sea water with nutrients (nitrogen and phosphorus) in the laboratory prior to the addition of radiolabelled phenanthrene. From Lindstrom (1991).

addition of nutrients in the laboratory was variable and small. Samples from the KN135 site, which had previously been exposed to both oil and fertilizer, showed about the same background mineralization, but a greater response to the addition of nutrients. The unexposed site (West Bay) showed very low phenanthrene mineralization capabilities and did not respond (in the short term) to the addition of nutrients.

The most interesting site was DI1067A (oil- and fertilizer-exposed) which showed very high background phenanthrene mineralization and what appears to be an inhibition of mineralization from the addition of nutrients. A possible explanation for these results is that site DI1067A was previously exposed to plant organic matter which promoted the biodegradation of aromatic compounds and directly or indirectly the metabolism of phenanthrene. In fact, this site has a large amount of peat deposited below the cobble and mixed sand and gravel surface, a characteristic absent at the other sites. Addition of nutrients appears to have induced the bacterial communities to metabolize other organic material rather than the added phenanthrene. Thus the microbial diversity of these sites relative to aromatic hydrocarbon degradation is significant and it is clear that, for biotechnology applications, great care must be taken to understand the role of biodiversity in natural biodegradation responses.

Cometabolism of PAHs and Biodiversity

The mechanisms by which natural microbial communities degrade mixtures of hydrophobic PAHs are both complicated and unclear. There may be a unique biodiversity that is required. Studies on pure culture studies indicate that many high-molecular-weight PAHs are transformed by cometabolism as mentioned above. This is illustrated in Table 9.3 by the bacterial strain, *Pseudomonas paucimobilis* EPA 505 (recently reclassified as *Sphingomonas paucimobilis*), that was isolated by its ability to grow on fluoranthene as a sole source of carbon and energy. If fluoranthene-grown cells are exposed to other PAHs individually, many of these, particularly the high-molecular-weight PAHs, appear to be transformed (cometabolized). We do not know the products produced from these transformations but it is likely that the oxygenase enzyme(s) induced for the initial attack on the fluoranthene also hydroxylates and possibly opens an aromatic ring on the other PAHs. The effectiveness of this cometabolic capability in bioremediation has been demonstrated using inoculation into bioreactors (Mueller *et al.*, 1993). The presence of bacteria in the microbial community with this broad ability to transform PAHs is a critical biodiversity component that is important for biotechnology applications.

In addition, we have isolated 13 bacterial strains from a microbial community exposed to creosote-contaminated soil (American Creosote

Table 9.3. Biotransformation of PAHs by washed, resting cells of EPA505 grown in different media.[a]

Compound	Transformation[b] after growth in		
	Fluoranthene minimal salts medium	Complex medium	Growth substrates[c]
Tween 80® (control)	–	–	–
Naphthalene	+	–	–
2-Methylnaphthalene	–	–	–
1-Methylnaphthalene	–	–	–
Biphenyl	+	+	–
2,6-Dimethylnaphthalene	–	–	–
2,3-Dimethylnaphthalene	+	–	++
Acenaphthene	–	–	–
Fluorene	+	–	–
Phenanthrene	+	–	++
Anthracene	+	–	+
2-Methylanthracene	–	–	–
Anthraquinone	+	+	–
Fluoranthene	+	–	++
Pyrene	+	+	–
Benzo(b) fluorene	+	+	+
Chrysene	+	+	–
Benzo(a)pyrene	–	–	–

[a]From Mueller et al. (1990).
[b]Incubation time was 48 h. Symbols: + indicates transformation, as monitored by change in medium coloration, significant change in UV absorption spectrum, or disappearance of parent compound; – indicates no change.
[c]Measured as the increase in bacterial protein in liquid culture after 10 days of incubation (30°C) in the dark with shaking (200 cycles min^{-1}). Symbols: (–) no increase (<2-fold); (+) >2-fold increase; (++) >20-fold increase.

Works, Pensacola, FL) to determine the breadth of this cometabolic capability. The isolates were selected based on their predominance during a study in which creosote degradation was enhanced (Mueller et al., 1989). Each isolate was grown in a complex medium and then exposed to the neutral fraction of creosote (primarily PAHs). Disappearance of PAHs, relative to an uninoculated control, was monitored. Preliminary results are shown in Table 9.4.

Table 9.4. Percentage reduction (compared to uninoculated controls) in individual PAHs and heterocyclic compounds in creosote as a result of incubation with CRE and EPA bacterial strains in the presence of minimal salts medium, 250 ppm creosote, and 0.03% Triton X-100 for 10 days at 30°C.

Compounds	CRE 8	CRE 9	CRE 10	CRE 11	CRE 12	CRE 13	EPA 505
Naphthalene	99	99	98	85	99	43	98
Thianaphthene	27	31	30	83	21	43	52
2-Methylnaphthalene	99	99	99	91	99	54	99
1-Methylnaphthalene	96	96	92	91	73	53	97
Biphenyl	79	78	62	86	51	28	97
2,6-Dimethylnaphthalene	75	75	58	88	49	37	95
2,3-Dimethylnaphthalene	81	79	67	84	14	15	83
Acenaphthylene	71	71	57	71	57	7	77
Acenaphthene	64	64	42	75	34	10	90
Dibenzofuran	89	87	68	69	36	2	99
Fluorene	43	42	13	60	10	0	93
Dibenzothiophene	43	48	13	54	8	0	92
Phenanthrene	34	34	6	50	8	0	99
Anthracene	33	23	0	46	6	0	64
Carbazole	94	25	0	51	10	0	87
2-Methylanthracene	25	33	0	53	4	0	47
Anthraquinone	23	23	0	48	5	0	53
Fluoranthene	24	24	0	47	5	0	51
Pyrene	24	24	0	48	4	0	23
Benzo(b)fluorene	28	24	3	47	19	0	29
Benz(a)anthracene	28	27	5	50	9	0	28
Chrysene	0	27	6	37	9	0	26
Benzo(b&k)fluoranthene	15	23	0	50	6	0	21
Benzo(a)pyrene	17	35	7	33	2	0	17

Clearly, several members of the community demonstrated metabolic capabilities similar to EPA 505. In other cases, the activities were limited to the lower-molecular-weight PAHs. As a rough generalization, it appears that the ability extensively to metabolize phenanthrene correlates with the ability to metabolize other high-molecular-weight PAHs. It is, therefore, not uncommon to find organisms in creosote-contaminated soil with the ability to transform high-molecular-weight PAHs. It is important to retain their presence if efforts to enhance the biodegradation of creosote-contaminated soils are to be successful.

Ecological Aspects of High-molecular-weight PAH-degraders

The existence of bacteria from different environments with metabolic capabilities similar to *P. paucimobilis* strain EPA 505 is important from the standpoint of ecological diversity and biotechnological risk assessment. Ecologically, it is of interest to know if the ability to grow on fluoranthene and cometabolize other high-molecular-weight PAHs is widespread throughout many bacterial genera in different geographical areas or if it is confined to a small number of genera. From a risk consideration, we can ask if inoculation of soils from different sources worldwide with EPA 505 represents the introduction of an exotic species. To answer these questions, enrichments for fluoranthene and phenanthrene degraders from creosote/

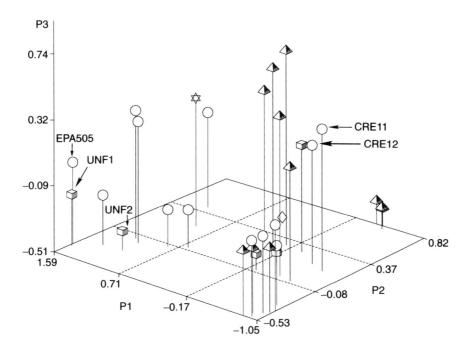

Fig. 9.6. Principal components analysis of data generated from BIOLOG assay and aromatic hydrocarbon degradation studies (19 substrates) for bacterial isolates, employing Jaccard similarity coefficient and unweighted group mean averages. BIOLOG substrates used in the analysis were those common to both Gram-negative and Gram-positive microplates. Clusters of related strains contained microorganisms isolated from different geographical locations. Strains isolated from the given location are represented on the graph by the corresponding symbols: Alaska = cylinder, Florida = circle, Germany = box, Norway = pyramid, and Texas = star.

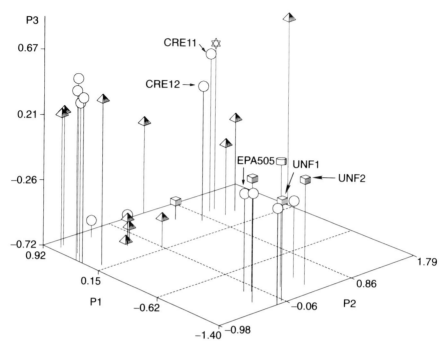

Fig. 9.7. Principal components analysis of data generated from phospholipid analysis of cell membranes (GC-FAME) for bacterial isolates, employing Sorenson similarity coefficient and unweighted group mean averages. Clusters of related strains contained microorganisms isolated from different geographical locations. Strains isolated from the given locations are represented on the graph by the corresponding symbols: Alaska = cylinder, Florida = circle, Germany = box, Norway = pyramid, and Texas = star.

PAH-contaminated soils sampled in Germany, Norway and the US were conducted (Mueller et al., 1994). Enrichments were also done using soils that had no previous exposure to PAHs. Bacteria capable of growing on fluoranthene as sole sources of carbon and energy were isolated from all soils except those with no exposure to PAHs. The isolates were then compared using GC-FAME and BIOLOG characteristics. These are shown in Figs 9.6 and 9.7. A *Mycobacterium* species (Pyr-1 obtained from C. Cerniglia), able to cometabolize pyrene, was also included in the analyses.

The data suggest that fluoranthene degradation capability was dispersed among several taxonomic clusters. There also appear to be isolates similar to EPA 505 (same cluster) that can be found in soils from Germany. In other cases, clusters of related organisms contained isolates from different countries, that is, there seemed to be little tendency for a sampling environment (site) to be unique to any one isolate. Thus, the ability to degrade fluoranthene is concluded to be common to all creosote-contaminated sites examined to date, suggesting that these PAHs may be responsible for part of the biodiversity associated with these sites.

Summary

The biodiversity of hydrocarbon degraders is indeed extensive. This is due, in part, to the ubiquitous nature of hydrocarbons, from both natural and anthropogenic sources. It is proposed that this biodiversity be investigated further because (a) it will improve our understanding of the role that diversity plays in maintaining hydrocarbon mineralization within carbon cycling processes in the environment, and (b) it can provide a strong scientific basis upon which the effectiveness and environmental safety of environmental biotechnologies can be implemented.

This chapter has focused on the aromatic hydrocarbon degraders, because of their current attention as widespread environmental contaminants. Relatively little is known about the ecology and diversity of microorganisms that degrade the polycyclic aromatic hydrocarbons (PAHs), especially those of four or more condensed rings. The importance of the PAHs as worldwide contaminants dictates that further biodiversity characterization of microbial populations in soil be carried out as a means to predict more precisely the fate and ecological effects of the PAHs. If the degradation of the high-molecular-weight PAHs involves a unique group of microorganisms because of the special characteristics associated with the chemistry and environmental distribution of PAHs, then further research may reveal new elements of biodiversity that can be useful as a measure of ecosystem health.

References

Atlas, R.M. (ed.) (1984) *Petroleum Microbiology*. Macmillan, New York, 692 pp.
Bartha, R. (1986) Biotechnology of petroleum pollutant biodegradation. *Microbiology and Ecology* 12, 155–172.
Bauer, J.E. and Capone, D.G. (1988) Effects of co-occurring aromatic hydrocarbons on the degradation of individual polycyclic aromatic hydrocarbons in marine sediment slurries. *Applied and Environmental Microbiology* 54, 1649–1655.
Black, J.A., Birge, W.J., Westerman, A.G. and Francis, P.C. (1983) Comparative aquatic toxicology of aromatic hydrocarbons. *Fundamental and Applied Toxicology* 3, 353–358.

Blumer, M. (1976) Polycyclic aromatic compounds in nature. *Scientific American* 234, 35–45.

Braddock, J.F., Brockman, M.R. and Brown, E.J. (1990) *Microbial hydrocarbon degradation in sediments impacted by the Exxon Valdez oil spill. Final report for Scientific Applications International Corporation.* Everett, WA, Task No. 55-DSNC-9-00016.2911-1/2, 32 pp.

Button, D.K. (1984) Evidence for a terpene-based food chain in the Gulf of Alaska. *Applied and Environmental Microbiology* 48, 1004–1011.

Cerniglia, C.E. (1980) Oxidation of naphthalene by cyanobacteria and microalgae. *Journal of General Microbiology* 116, 495–500.

Cerniglia, C.E. (1982) Naphthalene metabolism by diatoms isolated from the Kachemak Bay region of Alaska. *Journal of General Microbiology* 128, 987–990.

Cerniglia, C.E. (1984) Microbial transformation of aromatic hydrocarbons. In: Atlas, R.M. (ed.) *Petroleum Microbiology.* Macmillan, New York, pp. 99–128.

Cerniglia, C.E. (1993) Biodegradation of polycyclic aromatic hydrocarbons. *Biodegradation* 3, 351–368.

Chianelli, R.R., Aczel, T., Bare, R.E., George, G.N., Genowitz, M.W., Grossman, M.J., Haith, C.E., Kaiser, F.J., Lessard, R.R., Liotta, R., Mastracchio, R.L., Minal-Bernero, V., Prince, R.C., Robbins, W.K., Stiefel, E.I., Wilkinson, J.B., Hinton, S.M., Bragg, J.R., McMillem, S.J. and Atlas, R.M. (1991) Bioremediation technology development and application to the Alaskan spill. In: *Proceedings 1991 Oil Spill Conference.* American Petroleum Institute, Washington, DC, pp. 545–555.

Cripps, G.C. (1992) Baseline levels of hydrocarbons in seawater of the Southern Ocean: natural variability and regional patterns. *Marine Pollution Bulletin* 24, 109–114.

Floodgate, G. (1984) The fate of petroleum in marine ecosystems. In: Atlas, R.M. (ed.) *Petroleum Microbiology.* Macmillan, New York, pp. 335–398.

Foght, J.M. and Westlake, D.W.S. (1988) Degradation of polycyclic aromatic hydrocarbons abd aromatic heterocycles by a *Pseudomonas* species. *Canadian Journal of Microbiology* 34, 1135–1141.

Gibson, D.T. (ed.) (1984) *Microbial Degradation of Organic Compounds.* Marcel Dekker, New York, 535 pp.

Gibson, D.T. and Subramanian, V. (1984) Microbial degradation of aromatic hydrocarbons. In: Gibson, D.T. (ed.) *Microbial Degradation of Organic Compounds.* Marcel Dekker, New York, pp. 181–252.

Gordon, D.C., Keizer, P.D. and Dale, J. (1978) Temporal variations and probable origins of hydrocarbons in the water column of Bedford Basin, Nova Scotia. *Estuarine and Coastal Marine Science* 7, 243–256.

Grimalt, J., Albaiges, J., Al Saad, H.T. and Douabul, A.A. (1985) N-alkane distributions in surface sediments from the Arabian Gulf. *Naturwissenschaften* 72, 35–37.

Higgins, I.J., Scott, D. and Hammond, R.C. (1984) Transformation of C_1 compounds by microorganisms. In: Gibson, D.T. (ed.) *Microbial Degradation of Organic Compounds.* Marcel Dekker, New York, pp. 43–87.

Hopper, D.J. (1978) Microbial degradation of aromatic hydrocarbons. In: Watkinson, R.J. (ed.) *Developments in Biodegradation of Hydrocarbons – 1.* Applied Science Publishers, London, pp. 85–112.

Jacob, J., Karcher, W., Belliardo, J.J. and Wagstaffe, P.J. (1986) Polycyclic aromatic hydrocarbons of environmental and occupational importance. *Fresenius Zeitschrift für Analytische Chemie* 323, 1–10.

Kelly, I. and Cerniglia, C.E. (1991) The metabolism of fluoranthene by a species of *Mycobacterium*. *Journal of Industrial Microbiology* 7, 19–26.

Kostecki, P.T. and Calabrese, E.J. (eds) (1989) *Petroleum Contaminated Soils*. Lewis Publishers, Chelsea, Michigan, 357 pp.

Kusk, K.O. (1981) Comparison of the effects of aromatic hydrocarbons on a laboratory alga and natural phytoplankton. *Botanica Marina* 24, 611–613.

Leahy, J.G. and Colwell, R.R. (1990) Microbial degradation of hydrocarbons in the environment. *Microbiology Reviews* 54, 305–315.

Lee, K. and Levy, E.M. (1989) Biodegradation of petroleum in the marine environment and its enhancement. In: Nriagu, J.A. (ed.) *Aquatic Toxicology and Water Quality Management*. John Wiley and Sons, New York, pp. 217–243.

Lindstrom, J.E., Prince, R.C., Clark, J.C., Grossman, M.J., Yeager, T.R. and Brown, E.J. (1991) Microbial populations and hydrocarbon biodegradation potential in fertilized shoreline sediments affected by the T/V Exxon Valdez oil spill. *Applied and Environmental Microbiology* 57, 2514–2522.

Lindstrom, J.L. (1991) *Microbiology of oil-fouled beaches in Prince William Sound. Report to Alaskan Department of Environmental Conservation*. Oil Spill Response Center, Anchorage, AK, 10 pp.

MacGillvray, A.R. and Shiaris, M.P. (1993) Biotransformation of polycyclic aromatic hydrocarbons by yeasts isolated from coastal sediments. *Applied and Environmental Microbiology* 5, 1613–1618.

Millero, F.J. and Sohn, M.L. (1991) *Chemical Oceanography*. CRC Press, Boca Raton, FL, 531 pp.

Morgan, P. and Watkinson, R.J. (1989) Hydrocarbon degradation in soils and methods for soil biotreatment. *CRC Critical Reviews in Biotechnology* 8, 305–333.

Mueller, J.G., Chapman, P.J., Blattmann, B.O. and Pritchard, P.H. (1989) Action of a fluoranthene-degrading bacterial community on polycyclic aromatic hydrocarbon components of creosote. *Applied and Environmental Microbiology* 55, 3085–3090.

Mueller, J.G., Chapman, P.J., Blattmann, B.O. and Pritchard, P.H. (1990) Isolation and characterization of a fluoranthene-utilizing strain of *Pseudomonas pacuimoblis*. *Applied and Environmental Microbiology* 56, 1079–1086.

Mueller, J.G., Lantz, S.E., Ross, D., Colvin, R.J., Middaugh, D.P. and Pritchard, P.H. (1993) Strategy using bioreactors and specifically selected microorganisms for bioremediation of groundwater contaminated with creosote and pentachlorophenol. *Environmental Science and Technology* 27, 691–698.

Mueller, J.G., Lantz, S.E., Devereux, R., Berg, J.D. and Pritchard, P.H. (1994) Studies on the microbial ecology of polycyclic aromatic hydrocarbon degradation. In: Hinchel, R.E., Leeson, A., Semprini, L. and Ong, S.K. (eds) *Bioremediation of Chlorinated and Polycyclic Aromatic Hydrocarbons*. Lewis Publishers, Boca Raton, FL, pp. 218–230.

Narro, M.L., Cerniglia, C.E., Van Baalen, C. and Gibson, D.T. (1992) Metabolism of phenanthrene by the marine cyanobacterium *Agmenellum quadruplicatum* PR-6. *Applied and Environmental Microbiology* 58, 1351–1359.

Nisbet, I.C. and LaCoy, P.K. (1992) Toxic equivalency (FEFs) for polycyclic aromatic hydrocarbons (PAHs). *Regulatory Toxicology and Pharmacology* 16, 290–300.

Nishimura, M. and Baker, E.W. (1986) Possible origin of n-alkanes with a remarkable even to odd predominance in recent marine sediments. *Geochimica Cosmochimica Acta* 50, 299–305.

Pritchard, P.H. and Costa, C.F. (1991) EPA's Alaskan oil spill bioremediation project. *Environmental Science and Technology* 25, 372–379.

Pritchard, P.H., Costa, C.F. and Suit, L. (1991) *Alaska oil spill bioremediation project*. US EPA, Office of Research and Development Report. EPA/600/9-91/046a, 522 pp.

Pritchard, P.H., Mueller, J.G., Rogers, J.C., Kremer, F.V. and Glaser, J.A. (1993) Oil spill bioremediation: experiences, lessons and results from the Exxon Valdez oil spill in Alaska. *Biodegradation* 3, 315–335.

Schoell, M. (1980) The hydrogen and carbon isotopic composition of methane from natural gases of various origins. *Geochimica Cosmochimica Acta* 44, 649–661.

Shiaris, M.P. (1989) Seasonal biotransformation of naphthalene, phenanthrene, and benzo[a]pyrene in surficial estuarine sediments. *Applied and Environmental Microbiology* 55, 1391–1399.

Takeuchi, M., Kawai, F., Shimada, Y. and Yokota, A. (1993) Taxonomic study of polyethylene glycol-utilizing bacteria: emended description of the genus *Sphingomonas* and new descriptions of *Sphingomonas macrogoltabidus* sp. nov., *Sphingomonas sanguis* sp. nov. and *Sphingomonas terre* sp. nov. *Systematic and Applied Microbiology* 16, 227–238.

Walker, J.D., Colwell, R.R., Vaituzis, Z. and Meyer, S.A. (1975) Petroleum-degrading achlorophyllous alga *Prototheca zopfi*. *Nature* 254, 423–424.

Walter, U., Beyer, M., Klein, J. and Rehm, H.J. (1991) Degradation of pyrene by *Rhodococcus* sp. UW1. *Applied Microbiology and Biotechnology* 34, 671–676.

Warshawasky, D., Radike, M., Jayasimhulu, K. and Cody, T. (1988) Metabolism of benzo[a]pyrene by a dioxygenase enzyme system of the freshwater green alga *Selenstrum capricornutum*. *Biochemical Biophysical Research Communications* 152, 540–544.

Weissenfels, W.D., Beyer, M. and Klein, J. (1990) Degradation of phenanthrene, fluorene, and fluoranthene by pure bacterial cultures. *Applied Microbiology and Biotechnology* 34, 528–535.

West, P.A., Okpokwasili, G.C., Brayton, P.R., Grimes, D.J. and Colwell, R.R. (1984) Numerical taxonomy of phenanthrene-degrading bacteria isolated from the Chesapeake Bay. *Applied and Environmental Microbiology* 48, 988–993.

Zylstra, G.J. and Gibson, D.T. (1991) Aromatic hydrocarbon degradation: a molecular approach. In: J.K. Setlow (ed.) *Genetic Engineering: Principles and Methods*, vol. 13. Plenum Press, New York, pp. 183–203.

MICROORGANISMS AND ECOSYSTEM MAINTENANCE

IV

Bacterial Diversity and Ecosystem Maintenance: An Overview

10

A. BIANCHI AND M. BIANCHI

Microbiologie Marine CNRS UPR 223, Campus de Luminy, case 907 – 13288 Marseille cedex 9, France.

What good are all those species in nature that man can't eat or sell?

(Odum, 1971)

As many microorganisms are ubiquitous and physiologically highly versatile, we can question if species diversity really corresponds to a need for ecosystem maintenance or if diversity of microbial communities is simply due to ubiquity of the microorganisms. In other words: is structural and functional redundancy in a microbial community a disadvantage, extravagant precaution, or necessity?

Genetic diversity is treated elsewhere in this volume, therefore, we consider only the role of phenotypic diversity in an ecosystem, that is, diversity of physiological adaptation to environmental conditions and the effects of resulting diverse biochemical processes on ecosystem characteristics.

Functional Relationship Between Microorganisms

A biotic community is an assemblage of populations living in a defined area or physical habitat. It is an organized unit, to the extent that it possesses characteristics in addition to individual and population components and it functions as a unit via coupled metabolic transformations (Odum, 1971).

From this, it also follows that the scale of time and space of communities depends on the group of organisms under study. Diversity can be appreciated at different levels, for example at the scale of ecosystem, habitat, size of the organisms, i.e., millimetre and, more precisely, micrometre scale, with respect to bacteria (Bianchi and Bianchi, 1991).

In a biotic community, populations of different organisms interact in many ways (Hurst, 1991). For example, in two populations both can be unaffected by interaction (neutralism); one population may benefit, while the second population is unaffected (commensalism) or both populations may benefit (mutualism, synergy) or one population may benefit while the other is negatively affected (parasitism, predation); both populations may be negatively affected (competition).

These interactions will be profoundly influenced by growth conditions to which the organisms are exposed, by their spatial distribution and by a wide variety of environmental factors, such as nutrient availability, pH and redox conditions, water activity, metabolic inhibitors, clay minerals, etc.

Neutralism is the rule in environments where both populations are inactive (not multiplying), such as in the atmosphere or frozen environments. However, environments with very low population concentrations, such as oligotrophic lakes or open ocean far from shore, favour neutralism because low population numbers, ranging between 10^5 and 10^6 cells ml^{-1}, allow for different spatial and temporal niches.

Commensalism, where one population benefits and the other is unaffected, is quite common in microbial systems. Commensalism and mutualism favour diversity in microbial communities. An example often observed in eutrophic waters, soils, and sediments is the interrelationship between aerobic and anaerobic microbes.

At the opposite extreme, competition and related processes tend to reduce microbial diversity (Updegraff, 1991). In anoxic environments sulphate-reducing bacteria and methanogenic bacteria compete for hydrogen. As the former are thermodynamically advantaged, they outcompete methanogens in every environment providing sufficient sulphate.

Most populations that are dispersed within a given physical space may not interact strongly or directly. Groups of species which have mutually strong interactions frequently live in a matrix consisting of the physical, chemical, and biological aspects of the environment (Fenchel, 1987). This is the case for polymer-producing microorganisms commonly found in soils and in biofilms that spread over submerged surfaces.

Relationships Between Organisms and Their Environment

The small size and simple structure of bacteria imply a greater degree of contact with their environment than is the case for larger organisms. Bacteria have a relatively high surface area:volume ratio, with more direct communication between events in the environment at the cell surface and in the intracellular matrix.

Environmental conditions influence species composition of populations and some of these properties may be vital to the survival of individual species. Moreover, the shortness of bacterial generation times and, therefore the intense selection pressures for efficient occupation of the whole range of potential ecological niches have resulted in the development of diverse metabolic capabilities, with highly refined physiological and genetic control mechanisms (Hopwood and Chater, 1989).

Physiological Diversity of Microbial Communities: Random Distribution or Functional Necessity?

The 'competitive exclusion principle' is a general principle of ecology (Fenchel, 1987) that expresses the idea that, given a universe with only one resource, only one species population persists.

At the present time, the competitive exclusion principle has been combined with a theory of ecological niches to allow prediction of differential utilization of common resources necessary for stable coexistence of diversified populations (Fenchel, 1987). The niche is considered to have three dimensions: resources, habitat, and time. The differential exploitation of any of these dimensions may explain regional coexistence of different species. This theory appears to be more suitable for microbial ecology of natural environments, and provides a functional basis for microbial diversity. That is, in nature, energy and carbon sources for heterotrophs occur over a wide spectrum and infinite diversity of organic molecules. This allows for several coexisting species, each of which exploits a certain range of organic compounds and occurs because mechanisms for trapping energy are most efficient only for a restricted range of nutrient.

Is the Microbial Community Managed in a Hierarchical Model?

Major communities include producers, macroconsumers and microconsumers. Not all organisms in the community are equally important in determining the nature and function of the entire community. A relatively few species (or species groups) generally exert a major controlling influence by virtue of their numbers, size, metabolism or other activities. Within these groups, species (or species groups) which largely control energy flow and strongly affect the environment of all other species are known as the ecological dominant. Generally, dominants are those species in their trophic groups which have the largest productivity. The degree to which dominance is concentrated in one, several, or many species, can be expressed by an appropriate index of dominance that sums each species' importance in relation to the community as a whole. Several indices are available for describing the state of ecological diversity, the most widely used being the Shannon index, but these indices do not provide direct information about physiological functioning of a community (Atlas, 1984).

It is important to remember that diversity indexes and dominant species, as far as bacteria are concerned, refer only to the portion of the community able to be cultured (Bianchi and Bianchi, 1982). As this cultivable part is often limited to only a few per cent of the total community representatives, we must keep in mind the fact that the described dominant organism may be only poorly representative of a natural community.

The more numerous, but rare, species determine specific diversity, an equally important aspect of community structure (Odum, 1971). In a poorly diversified community, removal of the dominant species results in important changes, not only in the biotic community but also in the physical environment. At the opposite end, in a well-diversified community, any organism dying out, even if the dominant one, can be replaced by a 'hidden' species functionally able to fill in for the lost species to complete the neglected pathway. In this way, biodiversity is an important factor in ecosystem maintenance.

The ratio between the number of species and 'importance values' (numbers, biomass, productivity, etc.) of individuals is called the species diversity index. Species diversity indices tend to be low in physically controlled ecosystems, i.e., those subjected to strong physicochemical limiting factors, and high in biologically controlled ecosystems. In general, diversity increases with decrease in the ratio of antithermal maintenance to biomass (the R/B, or 'Schrödinger ratio', or ecological turnover). It is directly correlated with stability, but it is not clear to what extent this relationship is cause-and-effect.

Why is a Species Predominant in a Given Habitat?

Competition is one of the major interactions which occurs between different populations in nature and for which some of the basic kinetic principles of competitive growth have been elucidated. Competition is a fundamental mechanism acting in natural selection. It occurs when two (or more) populations growing in the same habitat or niche are limited, either in terms of specific growth rate or final population size, as the result of a common dependence on an external factor for growth (Lynch and Poole, 1979).

Natural habitats contain many kinds of organisms which interact in complex ways resulting in positive (increased growth rate) or negative (decreased growth rate) response. These interactions depend on bacterial density. When bacterial density is very low, there is little or no interaction between populations. At a higher density, positive interactions may predominate, and the growth rate may increase. When the number of organisms reaches maximum, negative interactions (nutrient exhaustion, accumulation of toxic metabolites) begin to predominate over the positive interactions. Cumulative effects of these interactions provoke the bell-shaped growth curve of a population versus time in an ecosystem.

Randomly dispersed species, adapted species and selected species

As a general rule, microorganisms are well equipped to produce a very large distribution, with ubiquity the general trend. This phenomenon is based on three major properties (Schlegel and Jannasch, 1981): (i) their small size providing easy dispersal by air and water; (ii) metabolic versatility and flexibility, making them able to obtain energy and synthesize cell constituents from diverse sources; and (iii) their ability to tolerate unfavourable conditions, using specific resistance forms, or, more commonly, using very simple cellular adaptation.

As a consequence of these properties, bacteria can be found almost everywhere, including extreme environments with respect to conditions for growth and cell division.

When studying bacteria present in an ecosystem, it is important to distinguish between: (i) organisms usually occurring in other ecosystems and introduced only accidentally via air, dust, incoming water flow, or migrating animals, etc.; and (ii) organisms typically adapted to the particular habitat and not occurring in any other, except in the form of survival stages. *Escherichia coli* is an example of accidentally introduced bacteria in polluted waters, but no physiologically adapted to growth in the natural aquatic habitat. Cyanobacteria present in deep ocean water or in mine water is also an example of randomly dispersed organisms unable to grow in this environment. The white sulphur bacterium is an example of a typically adapted bacterium since its requirement for dissolved oxygen and hydrogen sulphide at the same time requires high mobility combined with chemotactic orientation in an aquatic oxic/anoxic interface (Fenchel, 1987).

In open systems, bacterial diversity can result from allochthonous origin, i.e., diversity is managed by external input of bacterial species (opportunistic bacteria) dispersed by air or water fluxes entering the system. These inputs can lead to rapid and drastic changes in microbial community structure. If the allochthonous bacteria appear functionally well adapted to prevailing conditions with a nutritional advantage (i.e., better growth rate), it could overgrow indigenous populations and enter as a permanent member of the bacterial community, unlike randomly dispersed species that rarely become permanent members of the community. If an allochthonous organism does not find nutritional advantage in its new location, it would be in a negative position to compete with autochthonous species. Its growth rate would be slower, and as predators usually do not differentiate between allochthonous and indigenous strains, it will disappear (top-down control of the community). The most ubiquitous bacterial contaminants known, *Bacillus* and *Streptococcus*, are common in environments directly in contact with air, such as soil surface or air–water interface, but these bacteria do not persist as permanent members of the bacterial community deep in the sediment or in the water column (Bianchi et al., 1987).

In closed systems, like deep subsurface aquifers or sediments deeply buried under the earth's surface, usually without external inputs, bacterial diversity depends on the ability of the initial community to manage two antagonist processes: physiological adaptation to physicochemical conditions resulting from confinement, that can provoke the development of new varieties or races and thereby act to increase diversity; and selection for survival of the most competitive bacterial species, which can provoke a decrease in functional diversity. In such a case, microbial diversity is due only to the intrinsic ability of the system for specific adaptation.

As a general ecological law, in closed ecosystems, specific diversity tends to decrease with time, which is why, in geological deposits several hundred thousands or millions of years old, microbial communities, when present, tend to be monospecific (Bensoussan and Bianchi, 1983; Bianchi et al., 1987).

Bacterial strategy for nutritional adaptation: K- and r-strategists

Excluding those environments offering extreme physiochemical conditions, a response to change always finds its origin at the metabolic level. The overall result is generally reflected in a change in specific growth rate, which can be studied at the population level (Konings and Veldkamp, 1980).

Winogradsky (1949) described two main groups with respect to preferred substrate concentration, the autochthonous and the zymogenous bacteria. The former can use low energy components of the natural organic reserve occurring at low concentrations in soils and water. The latter become dominant when a readily degradable organic material is available. Koch (1979) used, in a similar way, the terms K- and r-strategists: K-strategists show steady growth at very low concentrations of nutrient and have a very low threshold, with respect to utilization of the nutrient, and r-strategists exhibit a higher threshold and grow relatively fast at higher nutrient concentrations.

At present, it is not known what are the biochemical differences leading to these two kinds of bacteria. From a practical point of view, it can be assumed that the zymogenous bacteria are those able to grow on conventional peptone-based culture media (that is, commonly 1/100 or less of direct counts for the sea-water sample plated), and the autochthonous bacteria those able to grow only on more oligotrophic culture media such as natural water without enrichment (that is, 1/10 or more of the direct count for the given sample). The latter appear to be the most closely adapted to ecosystem conditions, but, unfortunately being unable to recover them in culture using conventional culture methods, we cannot, at present, appreciate their taxonomic or their functional diversity and, therefore, their role in ecosystem maintenance.

Bacterial strategy for optimal utilization of microniche volume

Heterogeneous distribution of substrates in natural systems means that microbial activity is distributed similarly, leading to:

- patchy distribution of monospecific microcolonies: these organisms find nutritional value in a very close association with cells of the same species (ectoenzyme-producing species for example) and so struggle for a private microniche generally by producing antagonist substances;
- melting-pot habitats: these organisms are more sociable and accept, or need, to share their habitat to pool resources with other species. The best example we can propose is the bacterial consortium working on and inside particles floating in an aquatic environment, associating at the micrometre scale with strongly aerobic organisms, such as nitrifying bacteria with methanogens, the anaerobes being the most sensitive to oxygen (Sieburth, 1987, 1993; Bianchi *et al.*, 1992).

The habitats that are very densely colonized by microorganisms, such as the rhizosphere, biofilm attached to a submerged surface, rumen, or faecal pellets, will contain many different species, some of them with completely opposite physiological requirements. These will often grow as monospecific microcolonies, but also as random distribution of individual cells or small groups of cells.

Inversely, in most aquatic environments the free-living microorganisms are too sparsely distributed for any direct interaction to occur. Usually, in natural aquatic habitats, even those with relatively high bacterial density, that is, around 10^6 cells ml^{-1}, bacterial cells occupy only a very limited part, *ca.* 10^{-6} of the total volume in which they occur (Azam and Ammerman, 1984). In such habitats bacteria do not have to struggle for space. In these aquatic environments, it is only under certain conditions that allow increased cell concentrations that the organisms can exert any influence on one other. There is long and unresolved debate to determine what is the most common way of life for bacteria in natural water communities, namely as free-living bacteria or as cells fixed on every kind of submerged surfaces, like floating particles or other living organisms.

Another main characteristic influencing bacterial behaviour in natural ecosystems is the presence of numerous and complex physicochemical gradients. These gradients include organic and inorganic nutrients, toxics, temperature, etc. Most of the microorganisms we know currently that live in natural environments (not including those found in extreme conditions) are not strictly adapted to a narrow range of nutrient, salt concentration, or temperature. Therefore, these gradients can favour optimal colonization of the whole volume of the habitat. Some microorganisms are very sensitive to gradient concentrations and motile forms can behave positively or negatively toward these gradients. Some bacteria are particularly well adapted to occupying a very specific niche due to cumulative effects of two or more gradients. A very good example is that of some spirilla which are at the same time microaerophiles, oligocarbonophiles and nitrogen-fixing bacteria.

Bacterial Diversity: A Response of Small Size Organisms to Diversified and Changing Environmental Conditions

A common statement is that bacterial volume is too small to contain very diversified metabolic pathways in a single cell. Certainly this is true and could be an argument for the interest (and the necessity) of cometabolism in microbial consortia, the associating of several bacterial species in order to degrade complex materials. As an example, pure cultures of naturally occurring bacteria can degrade only a few of the many hydrocarbon molecules constituting a crude oil. In nature, crude oil degradation requires concomitant or successive activity of a microbial consortium. Chakrabarty et al. (1978) have clearly demonstrated that it is possible to produce a genetically manipulated strain containing all the enzymes necessary to degrade all components of a crude oil. These 'super-bacteria' work perfectly under artificial laboratory conditions, i.e., without natural competitors, but their efficiency under natural conditions, interacting with natural inhabitants of an ecosystem, is poor.

This provides evidence, first, that the small size of bacteria is not the only reason for the necessity of having a diversified microbial population. Second, the possession of a nutritional advantage does not necessarily give a decisive advantage in mixed population competition. We could also conclude that it is easier to introduce new genetic information into a bacterial cell, than to introduce a new population in an established community.

Bacterial Diversity in Respect to Physical or Chemical Stress

Kuenen et al. (1977) showed that the competitive advantage spirilla have in very dilute media is due to the enhanced surface area to cell volume ratio, compared to rod forms able to grow in the same medium, allowing the spirilla to compete more successfully for available nutrients. The role of prosthecae, polymorphism and membranous lamellae in survival in very dilute environments (e.g. groundwater) is that the bacteria can increase, by these means, their surface area to volume ratio. For example, *Ancalomicrobium* under very dilute conditions expresses prosthecae fully, whereas under nutrient-rich conditions these are fully suppressed. The role is, therefore, to greatly increase the absorptive area of the cell, in relation to its volume, thereby ensuring survival and growth under near-starvation conditions.

Biodiversity as a Weapon Against Invaders

In theory, communities can reach a steady state that prevents invasion by a new component. In fact open environments are permanently submitted to fluxes, and for this reason steady state is seldom obtained in nature. For many years, selected forms of microorganisms have been released without any effect on the natural population. Indeed, when strains of *Rhizobium* have been introduced into cultivated soils with natural populations of the same species, the introduction fails to become established (Stacey, 1985). Another example is the failure of the introduction of a mixed culture of selected hydrocarbon-degrading bacteria in the treatment of oil spills to enhance degradation (Tagger et al., 1983).

Evolution of Microbial Communities in Extreme Environments

As a rule, the most common ecosystems, without any special physicochemical or nutritional characteristic, such as soils or sea water, usually carry low numbers of culturable microorganisms, but a high diversity of species. At the opposite extreme, ecosystems submitted to extreme physical or chemical conditions can exert a strong pressure for selection and therefore bacterial diversity is low. Good examples can be given by very acidic mine water, salt brine, and hot springs. When the energy source is abundant, these extreme environments may support a relatively high cell density, but relatively few species.

In general, microorganisms that proliferate and grow under the most extreme conditions are obligately adapted to their particular environment (Horikoshi and Grant, 1991), and fail to grow at lower intensities of the same environmental factor. Such organisms have acquired the ability to grow in one extreme environment at the expense of ability to grow in others. Less rigid or extreme environmental conditions are tolerated by a greater number of organisms. Some of these are obligately bound to these conditions and others grow facultatively. There are several examples of low numbers of species and high specialization in habitats of extreme conditions, such as low and high temperature, high salt concentrations, low moisture, low pH, low nutrient concentrations, and others. In these extreme conditions, ecosystem maintenance obligately relies upon the very highly diversified functional abilities of the microbial world.

Role of Microbial Heterotrophs in Ecosystem Maintenance

The very complex processes of degradation of organic matter by heterotrophs control a number of important functions in every ecosystems:

1. recycling of nutrients through mineralization of dead organic matter;
2. production of food for a sequence of organisms in the detritus food chain (grazing of bacteria by the protozoa resulting in the microbial loop);
3. production of regulatory 'ectocrine' substances;
4. modification of the global ecosystem, to produce the complex system known as 'soil' (Odum, 1971), or to modify the Earth's climate (global change, Rambler *et al.*, 1989).

Homeostasis: the Immune Response of Natural Ecosystems

Ecosystems are capable of self-maintenance and self-regulation as are their component populations and organisms. Homeostasis is the term generally applied to the tendency of biological systems to resist change and to remain in a state of equilibrium (Odum, 1971).

Equilibrium between organisms and environment may also be maintained by factors which resist change in the system as a whole. The interplay of material cycles and energy flows in large ecosystems generates a self-correcting homeostasis with no outside control required. The possible role of 'ectocrine' substances in coordinating units of the ecosystem has been mentioned. Homeostasis mechanisms have limits beyond which unrestricted positive feedback leads to death (Odum, 1971).

In low-diversity, physically stressed ecosystems, or in those subject to irregular or unpredictable extrinsic perturbations, populations tend to be regulated by physical components such as temperature, water currents, chemical limiting factors, pollution, and so forth. In high-diversity ecosystems, or in those not physically stressed, populations tend to be biologically controlled. In all ecosystems, there is a strong tendency for all populations to evolve through natural selection towards self-regulation, since overpopulation is not in the best interests of any population (Odum, 1971).

The short-term and long-term evolution of ecosystems is shaped by: (i) allogenic forces such as geological and climatic changes, and (ii) autogenic processes resulting from activities of the living components of the ecosystem.

Answer to Allogenic Perturbations

Communities can adapt themselves to perturbations of the physical environment. If the community is diversified (mature systems), one of the species can be adapted to new environmental conditions and thereby take advantage, tending to be dominant. Such factor compensation is particularly effective at the community level of organization, but also occurs within the species. Species with wide geographical ranges almost always develop locally adapted populations called ecotypes that have optima and limits of tolerance adjusted to local conditions. Compensation along gradients of temperature, light, or other factors may involve genetic races or merely physiological acclimatization.

The physical conditions may not only be limiting factors in the detrimental sense but also regulatory factors in the beneficial sense, that adapted organisms respond to these factors in such a way that the community of organisms achieves maximum homeostasis possible under the given conditions. Effects of actual biochemical relationships between organisms added to the reciprocal natural selections between species (coevolution) permit the whole community to respond to most of the allogenic perturbation (Odum, 1971).

Answer to Autogenic Perturbations

In complex ecosystems containing a diversity of organisms the metabolic activity of any individual organism is likely to have an effect on its neighbours. The organism fulfilling a certain function in a particular ecosystem is determined by diverse physical and chemical factors and by all of the organisms constituting the community. As an example, in some rumen bacteria, the metabolic processes are either impeded or modified if a product of their fermentation, hydrogen, is not removed by methanogenic bacteria. In this case interspecies hydrogen transfer may be involved as demonstrated for the so-called *Methanobacterium omelianskii*, which is a consortium of two bacteria. One can grow only if the hydrogen it produces as a waste product is efficiently removed from its surroundings by the second, which in turn can grow only when it can use hydrogen free of charge.

Interactions of many types are continually taking place between components of the microbiota. Examples can be found in the hydrolysis of polymers with subsequent use of low-molecular-weight compounds, grazing of a bacterial community by protozoans and so forth. These interactions are largely self-regulating so that an equilibrium is established between the various beneficial, mutualistic, or antagonistic activities.

Pioneering Aseptic Systems and Colonizing Mature Systems

The most violent bacterial blooms occur when a species is introduced into a new area where there are both unexploited resources and a lack of negative interactions. Such bacterial blooms can be observed when the introduced bacteria occur in aseptic (or only poorly colonized) deep subsurface aquifers or oil fields. New ecosystems tend to be less able to resist outside perturbations, compared with established systems in which the components have had a chance to make mutual adjustments to each other (Odum, 1971).

As a general rule, diversity is low in newly established communities and tends to increase in established systems. When the R/B ratio is low, more of the community energy can go into diversity. Higher diversity means longer food chains and more cases of symbiosis (mutualism, parasitism, commensalism, etc.), and greater possibilities for negative feedback control, which reduces oscillations and hence increases stability.

Conclusion

The ecosystem is the basic functional unit in ecology, since it includes both the organisms and their abiotic environment, each influencing the properties of the other. As long as the major components are present and operate together to achieve functional stability, the entity may be considered an ecosystem. Bacteria can be considered as the key components of most natural ecosystems, principally due to their metabolic versatility and their physiological adaptability. Whereas the behaviour of a microbial cell is determined by the characteristics of its genome, the behaviour of a bacterial community is determined by the physiological characteristics of its specific diversity.

References

Atlas, R.M. (1984) Use of microbial diversity measurements to assess environmental stress. In: Klug, M.J. and Reddy, C.A. (eds) *Current Perspectives in Microbial Ecology*. American Society for Microbiology, Washington, DC, pp. 540–545.

Azam, F. and Ammerman, J.W. (1984) Cycling of organic matter by bacterioplankton in pelagic marine ecosystems: microenvironmental considerations. In: Fasham, M.J.R. (ed.) *Flows of Energy and Materials in Marine Ecosystems. Theory and Practice*. Plenum Press, New York, pp. 345–360.

Bensoussan, M. and Bianchi, A. (1983) Distribution et activité catabolique potentielle des communautés bactériennes des eaux et des sédiments profonds prélevés sur diverses marges continentales. In: Pelet, R. (ed.) *Géochimie*

Organique des Sédiments Marins Profonds, d'ORGON à MISEDOR. Editions du CNRS, Paris, pp. 39–72.

Bianchi, A. and Bianchi, M. (1991) Ecologie microbienne et environnement global. *Bulletin de la Société Française de Microbiologie* 6(4), 7–10.

Bianchi, A., Hinojosa, M., Garcin, J., Delebassée, M., Normand, M., Ralijoana, C., Sohier, L., Vianna Doria, E. and Villata, M. (1987) Etude bactériologique des sédiments quaternaires et pliocène supérieur du delta de la Mahakam (Kalimantan, Indonésie). In: Pelet, R. (ed.) *Géochimie Organique des Sédiments Plioquaternaires du Delta de la Mahakam (Indonésie). Le Sondage MISEDOR.* Ed. Technip, Paris.

Bianchi, M. and Bianchi, A. (1982) Statistical sampling of bacterial strains and its use in bacterial diversity measurement. *Microbial Ecology* 8, 61–69.

Bianchi, M., Marty, D., Teyssié, J.L. and Fowler, S.W. (1992) Strictly aerobic and anaerobic bacteria associated with sinking particulate matter and zooplankton fecal pellets. *Marine Ecology Progress Series* 88, 55–60.

Chakrabarty, A.M., Friello, D.A. and Bopp, L.H. (1978) Transposition of plasmid DNA segments specifying hydrocarbon degradation and their expression in various microorganisms. *Proceedings of the National Academy of Sciences of the USA* 75, 3109–3112.

Fenchel, T. (1987) Ecology – potentials and limitations. In: Kinne, O. (ed.) *Excellence in Ecology*, Vol. 1. Ecological Institute, Oldendorf/Luhe, 186 pp.

Hopwood, D.A. and Chater, K.F. (1989) *Genetics of Bacterial Diversity.* Academic Press, London, 449 pp.

Horikoshi, K. and Grant, W.D. (eds) (1991) *Superbugs. Microorganisms in Extreme Environments.* Springer-Verlag, Berlin, 299 pp.

Hurst, C.J. (1991) *Modeling the Environmental Fate of Microorganisms.* American Society for Microbiology, Washington, DC, 292 pp.

Koch, A.L. (1979) Microbial growth in low concentrations of nutrients. In: Shilo, M. (ed.) *Strategies of Microbial Life in Extreme Environments.* Verlag Chemie, Weinheim, pp. 261–279.

Konings, W.N. and Veldkamp, H. (1980) Phenotypic response to environmental changes. In: Ellwood, D.C., Hedger, J.N., Latham, M.J., Lynch, J.M. and Slater, J.H. (eds) *Contemporary Microbial Ecology.* Academic Press, London, pp. 161–191.

Kuenen, J.G., Boonstra, J., Schroder, H.G.J. and Veldkamp, H. (1977) Competition for inorganic substrates among chemoorganotrophic and chemolithotrophic bacteria. *Microbial Ecology* 3, 119–130.

Lynch, J.M. and Poole, N.J. (1979) *Microbial Ecology. A Conceptual Approach.* Blackwell Scientific Publications, Oxford, 266 pp.

Odum, E.P. (1971) *Fundamentals of Ecology*, 3rd edn. W.B. Saunders, Philadelphia, 574 pp.

Rambler, M.B., Margulis, L. and Fester, R. (eds) (1989) *Global Ecology. Towards a Science of the Biosphere.* Academic Press, Boston, 204 pp.

Schlegel, H.G. and Jannasch, H.W. (1981) Prokaryotes and their habitats. In: Starr, M.P., Stolp, H., Trüper, H.G., Balows, A. and Schlegel, H.S. (eds) *The Prokaryotes. A Handbook on Habitats, Isolation, and Identification of Bacteria,* Vol. 1. Springer-Verlag, Berlin, pp. 4–43.

Sieburth, J.McN. (1987) Contrary habitats for redox-specific processes: methanogenesis in oxic waters and oxidation in anoxic waters. In: Sleigh, M.A. (ed.) *Microbes in the Sea*. Ellis Horwood, Chichester, pp. 11–38.

Sieburth, J.McN. (1993) C_1 bacteria in the water column of Chesapeake Bay, USA. I. Distribution of subpopulations of O_2-tolerant, obligately anaerobic, methylotrophic methanogens that occur in microniches reduced by bacterial consorts. *Marine Ecology Progress Series* 95, 67–80.

Stacey, G. (1985) The *Rhizobium* experience. In: Halvorson, H.O., Pramer, D. and Roguls, M. (eds) *Engineered Organisms in the Environment: Scientific Issues*. American Society for Microbiology, Washington, DC, pp. 109–121.

Tagger, S., Bianchi, A., Julliard, M., Le Petit, J. and Roux, B. (1983) Effect of microbial seeding of crude oil in seawater in a model system. *Marine Biology* 78, 13–20.

Updegraff, D.M. (1991) Background and practical applications of microbial ecology. In: Hurst, C.J. (ed.) *Modeling the Environmental Fate of Microorganisms*. American Society for Microbiology, Washington, DC, pp. 1–20.

Winogradsky, S. (1949) *Microbiologie du Sol. Problèmes et Méthodes. Cinquante Ans de Recherches. Oeuvres Complètes*. Masson, Paris, 861 pp.

Ecological Role of Microphytic Soil Crusts in Arid Ecosystems

S.D. WARREN

US Army Construction Engineering Research Laboratories, Natural Resources Division, P.O. Box 9005, Champaign, Illinois 61826-9005, USA

Introduction

Microphytic soil crusts, synonymously referred to as cryptogamic, cryptobiotic, cyanobactieral, organogenic, microfloral and biological soil crusts, are common in many arid ecosystems. As the various names imply, the crusts are formed by tiny, sometimes microscopic, non-vascular 'plants' living on or in the surface few millimetres of the soil. Among the microphytes contributing to soil crusts are algae, lichens, fungi, mosses, liverworts, and bacteria. Microphytic crusts dominated by lichens, mosses, and liverworts are usually readily visible. Crusts dominated by algae, fungi, and bacteria can be difficult to detect, but are often identifiable by distinct microtopographic characteristics or darkening of the soil surface (Fig. 11.1).

The presence of microphytic soil crusts is inversely proportional to the presence of vascular plant cover. This general trend is true for successional gradients as well as climatic gradients. Microphytes, particularly algae, are often among the first colonizers of severely disturbed lands (Booth, 1941; Rayburn et al., 1982; Pluis and de Winder, 1989). More importantly for the context of this chapter, microphytic crusts are often the dominant vegetative component in all seral stages of arid ecosystems where vascular plants are naturally sparse. Extensive and diverse microphytic communities have been reported from nearly all desert regions of the world (Harper and Marble, 1988; West, 1990). At least 16 species of algae occur in the soils of Death Valley, California, where the surface soil temperature can reach 88°C (Durrell, 1962). Algal crusts are often the only living plants in areas of the

© 1995 CAB INTERNATIONAL. *Microbial Diversity and Ecosystem Function*
(eds D. Allsopp, R.R. Colwell and D.L. Hawksworth)

Fig. 11.1. Microphytic crusts are often identifiable by distinct microtopographic characteristics and darkening of the soil surface. For scale, the coin is a US penny, approximately 2 cm in diameter.

Atacama Desert where average annual precipitation is as low as 1 mm yr^{-1} (Rauh, 1985). Despite the rigours of such harsh environments, biomass estimates as high as 1.4 t ha^{-1} have been reported for desert microphytic communities (Novichkova-Ivanova, 1972).

Microphytic soil crusts contribute substantially to the biodiversity of desert ecosystems in terms of numbers of species, abundance of individuals and contributions to total biomass. However, the magnitude of their contribution becomes more significant when one considers their roles in ecosystem function and stability. Microphytic soil crusts are critical elements in soil stability and nutrient cycling in arid regions. They are also important to the establishment and growth of vascular plants, and they have been implicated in the survival of some desert animals. Each of these ecological roles will be discussed below. The effects of anthropogenic disturbances on microphytic crusts will also be considered, as well as the potential to restore the crusts following disturbance.

Soil Stability

Microphytic soil crusts contribute to soil stability in at least three ways. First, the presence of vegetative structures on the soil surface tends to dissipate the kinetic energy of wind, raindrops and overland flow of water, thus reducing the susceptibility of the soil to erosion. Second, even where larger vegetative structures may be absent, many microphytic crusts tend to create an uneven microtopography (Fig. 11.1). This roughened surface decreases the velocity of both wind and water. Finally, microphytes contribute to mechanical and chemical aggregation of soil particles. Moss rhizoids, fungal hyphae and filamentous algae often form a dense fibrous mesh that holds soil particles in place (Fletcher and Martin, 1948; Bond and Harris, 1964; Schulten, 1985; Belnap and Gardner, 1993). Some microphytic species, particularly blue–green algae, exude polysaccharide compounds that form glue-like bonds between the filament sheaths and the soil particles (Schulten, 1985; Belnap and Gardner, 1993).

Some researchers have argued that microphytic crusts contribute to soil sealing or hydrophobicity (Rozanov, 1951; Walker, 1979), thus reducing infiltration while increasing erosion (Bolyshev, 1962). Most quantitative evidence, however, indicates that microphytic crusts have a positive or neutral influence on rainfall infiltration and a negative correlation with net erosion (Booth, 1941; Faust, 1970; Loope and Gifford, 1972; MacKenzie and Pearson, 1979; Harper and St Clair, 1985; Kinnell et al., 1990). Even where the rate of infiltration has been shown to decline, the rough microtopography associated with the crust may enhance surface ponding, thus facilitating greater net infiltration and deeper water penetration (Brotherson and Rushforth, 1983).

The effects of microphytic crusts on wind erosion have been less studied, but there appears to be a near consensus that soils colonized by microphytic crusts are less susceptible to wind erosion (MacKenzie and Pearson, 1979; van den Ancker et al., 1985; Pluis and de Winder, 1989).

Nutrient Cycling

The presence of microphytic soil crusts is positively correlated with the abundance of a variety of micro- and macronutrients in the soil. Numerous studies have documented nitrogen fixation by microphytic crusts, particularly those dominated by blue–green algae or lichens possessing blue–green algal symbionts. However, the significance of their contribution has been questioned (Snyder and Wullstein, 1973; West, 1990). Rates of nitrogen fixation and total quantities of fixed nitrogen vary widely depending on light, temperature, moisture availability and the extent of microphytic cover. In desert ecosystems, optimal conditions may exist for only a few days of the year. Estimates of nitrogen input by microphytic crusts have ranged as high as 25–100 kg ha^{-1} yr^{-1} (West and Skujiņš, 1977; Rychert et al., 1978). However, controlled experimentation under representative conditions indicates that more realistic values probably seldom exceed 10 kg

Table 11.1. Estimates of average annual nitrogen input by microphytic soil crusts in desert ecosystems.

Desert	Estimated annual nitrogen fixation (kg ha^{-1} yr^{-1})	Source
Sonoran, USA	8.5	Mayland et al. (1966)
Sonoran, USA[a]	2.6	MacGregor and Johnson (1971)
Great Basin, USA[b]	10.8	Rychert and Skujiņš (1974)
Great Basin, USA	2.2	Jeffries et al. (1992)
Kalahari, Botswana	1.9	Skarpe and Henriksson (1987)

[a] Based on same assumptions used by Mayland et al. (1966).
[b] Assumes 100% microphytic cover and normal field conditions.

ha^{-1} yr^{-1} (Table 11.1). This represents a significant source of nitrogen in desert ecosystems. Vascular plants in the Great Basin Desert, USA, take up approximately 10–12 kg nitrogen ha^{-1} yr^{-1} (West and Skujiņš, 1977). Precipitation may provide 1–6 kg ha^{-1} yr^{-1} (West, 1978; Schlesinger, 1991), whereas heterotrophic nitrogen fixers account for 2 kg ha^{-1} yr^{-1} or less (Steyn and Delwiche, 1970; Rychert et al., 1978). The balance can be attributed to microphytic crusts.

Much of the nitrogen fixed by microphytic crusts is retained in the surface few centimetres of the soil (Fletcher and Martin, 1948; Skujiņš, 1984) where it is available to vascular plants. The nitrogen content of soils with microphytic crusts may be up to seven times that of similar soils lacking microphytes (Shields, 1957). Other essential elements have also been shown to accumulate in the surface layer of soil occupied by microphytic crusts (Kleiner and Harper, 1972; Loope and Gifford, 1972; Reynaud, 1987; Harper and Pendleton, 1993). Although this has often been attributed to the accumulation of fine soil particles by the roughened microtopography of the crusts (Kleiner and Harper, 1972), other mechanisms may also play a significant role. Microphytic organisms accelerate the weathering of rocks, thus speeding the genesis of soil and adding important minerals (Metting, 1991). It has been demonstrated that the polysaccharide exudates of blue–green algae include chelating agents that concentrate essential nutrients (Lange, 1974, 1976). In addition, negatively charged clay particles may be bound to, or incorporated into, the polysaccharide exudates of some blue–green algae, thus attracting and bonding with positively charged essential elements (Belnap and Gardner, 1993).

Vascular Plants

Soil moisture is very important to the germination and establishment of vascular plants. From the discussion of soil stability, it is apparent that microphytic crusts frequently improve the infiltration of water into the soil. Researchers have shown that microphytic crusts also impede evaporation and retain moisture in the soil (Booth, 1941; Cornet, 1981). This may be accomplished as the crusts form a surface seal or mulch layer (Fritsch, 1922; Booth, 1941), by an increase in the organic matter content of the soil (Metting, 1981), or by the absorption of moisture into the polysaccharide sheaths of filamentous blue–green algae (Belnap and Gardner, 1993).

Dark-coloured crusts absorb more solar radiation than adjacent uncrusted soil, raising the temperature near the surface by as much as 5°C (Harper and Marble, 1988). Elevated soil temperatures, when coupled with adequate soil moisture, may accelerate the germination and initial growth of vascular plants (Harper and Pendleton, 1993).

Evidence of increased seed germination and seedling establishment in the presence of microphytic soil crusts has been seen in a number of studies (Nebeker and St Clair, 1980; St Clair et al., 1984; Harper and St Clair, 1985; Harper and Marble, 1988; Belnap, 1994). In the few studies that have reported poorer germination on crusted soil than on bare soil, it has been noted that subsequent growth and long-term survival were greater where the crust was present (McIlvanie, 1942; Schlatterer and Tisdale, 1969; Lesica and Shelly, 1992). There is evidence that some degree of mechanical disturbance on very heavily crusted soils may be beneficial in terms of increasing seed to soil contact (Sylla, 1987).

The contribution of microphytic soil crusts to nutrient-rich substrates on which vascular plants can grow has already been discussed. Using nitrogen isotopes, it has been conclusively demonstrated that nitrogen fixed by microphytic crusts is assimilated by vascular plants (Mayland et al., 1966; Stewart, 1967). Due to the increased supply of nitrogen and other essential elements, several species of plants grown on soil with microphytic crusts have produced significantly more biomass and accumulated significantly greater quantities of most essential elements than seedlings grown in the same soil without a crust (Harper and Pendleton, 1993).

Although many desert plants are capable of taking up nutrients directly from the soil, the process is often enhanced by the presence of vesicular arbuscular mycorrhizas or rhizobia that infect the root system of the vascular plants. Very little information is available regarding potential interactions of microphytic crusts with microorganisms of the rhizosphere. However, Harper and Pendleton (1993) present evidence that root colonization by rhizosymbionts is significantly enhanced where microphytic crusts are present.

Desert Fauna

Greater accumulations of vascular plant biomass and essential elements have significant implications for desert animals that depend on the plants as a food source. Both the quantity and quality of biomass are important. As discussed in the previous section, microphytic crusts have been related to increased biomass and increased elemental accumulation in vascular plants. It is also important to note that the concentration of some essential elements has been found to be higher in the tissue of plants grown in the presence of microphytic crusts (Marble, 1990; Harper and Pendleton, 1993; Belnap, 1994). A potential link has been suggested between microphytic soil crusts and the decline of the desert tortoise, an endangered species of the Mojave Desert, USA (Harper and Pendleton, 1993). The decline of the tortoise has been attributed to dietary deficiencies leading to osteoporosis (Jarchow, 1984). Higher concentrations of calcium, phosphorus, and magnesium, three elements important to bone formation, have been detected in plants grown in soils dominated by microphytic crusts (Harper and Pendleton, 1993).

Desert animals may also depend on microphytes as direct sources of food. Wombats, guanacos, pronghorn antelope and sheep have been observed consuming lichens in various arid regions of the world (West, 1990). Desert-dwelling snails (Shachak and Steinberger, 1980), woodlice (Steinberger, 1989), tenebrionid beetles (Rogers et al., 1988) and ants (Loria and Herrnstadt, 1980) are also known to use microphytes as a food source. It has even been conjectured that the 'manna' referred to in the Bible was a lichen blown into windrows (Donkin, 1981).

Decline and Recovery of Microphytic Crusts

A decline in the abundance and diversity of microphytic crust cover on desert soils has been reported as a result of livestock trampling (Rogers and Lange, 1971; Kleiner and Harper, 1972; Anderson et al., 1982; Brotherson et al., 1983; Johansen and St Clair, 1986; Jeffries and Klopatek, 1987), fire (Johansen et al., 1982, 1984; Callison et al., 1985; Greene et al., 1990; Johansen et al., 1993), human trampling (Cole, 1990), tillage (Terry and Burns, 1987), off-road recreational vehicle traffic (Harper and Marble, 1988) and military training manoeuvres (Marston, 1986). Natural events such as intense rainstorms (Johansen, 1984) or volcanic ash fallout (Harris et al., 1987) may also damage or kill microphytic crusts. Any decline in cover is accompanied by a decline in the benefits afforded by a more stable crust.

Anderson et al. (1982) estimated that 14–18 years were adequate for recovery of a microphytic crust following disturbance by livestock in the Great Basin Desert, USA. However, there were no signs of recovery almost

20 years after burning of a shrub community in the Mojave Desert, USA (Callison *et al.*, 1985). Belnap (1993) estimated that full recovery at a Great Basin Desert site, including visual as well as functional characteristics, could take as long as 30–40 years for blue–green algae, 45–85 years for lichens and 250 years for mosses. The time for recovery is greatly influenced by the nature and periodicity of the disturbance, the size of the disturbed area, the proximity of microphytic propagules, and edaphic and climatic characteristics of the site.

Artificial Recovery

Due to the prolonged time required for natural recovery, the activities of humans in arid ecosystems have been largely exploitative or aimed at ameliorating the environment. Very little attention has been given to the development of land reclamation technologies that restore natural components of the ecosystem. Nevertheless, the potential to use microphytic species in desert reclamation has been demonstrated. On a site disturbed by fire, St Clair *et al.* (1986) treated small plots with a soil crust slurry made by stripping areas of microphytic crusts and mixing with water. Belnap (1993) used stripped crusts as a dry inoculant for small plots where the original microphytic crusts had been removed. In both studies, inoculation significantly hastened recovery of the microphytic crusts.

Although the destruction of microphytic crusts in one area to provide inoculants for other areas is counterproductive for large-scale desert reclamation, the results of these inoculation studies are encouraging. They represent a first step toward reestablishing ecosystem stability, function, and biodiversity in damaged desert ecosystems. Further research should consider the development of cost-effective techniques to mass-culture, store, transport and apply microphytic inoculants on disturbed sites.

References

Anderson, D.C., Harper, K.T. and Rushforth, S.R. (1982) Recovery of cryptogamic soil crusts from grazing on Utah winter ranges. *Journal of Range Management* 35, 355–359.

Belnap, J. (1993) Recovery rates of cryptobiotic crusts: inoculant use and assessment methods. *Great Basin Naturalist* 53, 89–95.

Belnap, J. (1994) Potential role of cryptobiotic soil crusts in semiarid rangelands. In: Monsen, S.B. and Kitchen, S.G. (eds) *Proceedings of the Symposium on Ecology, Management, and Restoration of Intermountain Annual Rangelands.* US Forest Service General Technical Report INT-GTR-313, pp. 179–185.

Belnap, J. and Gardner, J.S. (1993) Soil microstructure in soils of the Colorado

Plateau: the role of the cyanobacterium *Microcoleus vaginatus*. *Great Basin Naturalist* 53, 40–47.

Bolyshev, N.N. (1962) Role of algae in soil formation. *Soviet Soil Science* 1964, 630–635.

Bond, R.D. and Harris, J.R. (1964) The influence of microflora on physical properties of soils. I. Effects associated with filamentous algae and fungi. *Australian Journal of Soil Research* 2, 111–122.

Booth, W.E. (1941) Algae as pioneers in plant succession and their importance in erosion control. *Ecology* 22, 38–46.

Brotherson, J.D. and Rushforth, S.R. (1983) Influence of cryptogamic crusts on moisture relationships of soils in Navajo National Monument, Arizona. *Great Basin Naturalist* 43, 73–78.

Brotherson, J.D., Rushforth, S.R. and Johansen, J.R. (1983) Effects of long-term grazing on cryptogam crust cover in Navajo National Monument, Ariz. *Journal of Range Management* 36, 579–581.

Callison, J., Brotherson, J.D. and Bowns, J.E. (1985) The effects of fire on blackbrush [*Coleogyne ramosissima*] community of southwestern Utah. *Journal of Range Management* 38, 535–538.

Cole, D.N. (1990) Trampling disturbance and recovery of cryptomanic soil crusts in Grand Canyon National Park. *Great Basin Naturalist* 50, 321–325.

Cornet, A. (1981) *Le bilan hydrique et son role dans la production de la strate herbacée de quelques Phytocénoses Sahéliennes au Sénégal*. Languedoc University of Science and Technology, Languedoc, France.

Donkin, R.A. (1981) The manna lichen *Lecanora esculenta*. *Anthropos* 76, 562–572.

Durrell, L.W. (1962) Algae of Death Valley. *Transactions of the American Microscopical Society* 81, 267–273.

Faust, W.F. (1970) The effect of algal-mold crusts on the hydrologic processes of infiltration, runoff, and soil erosion under simulated rainfall conditions. MS Thesis, University of Arizona, Tucson.

Fletcher, J.E. and Martin, W.P. (1948) Some effects of algae and molds in the raincrust of desert soils. *Ecology* 29, 95–100.

Fritsch, F.E. (1922) The terrestrial algae. *Journal of Ecology* 10, 220–236.

Greene, R.S.B., Chartres, C.J. and Hodgkinson, K.C. (1990) The effects of fire on the soil in a degraded semi-arid woodland. I. Cryptogam cover and physical and micromorphological properties. *Australian Journal of Soil Research* 28, 755–777.

Harper, K.T. and Marble, J.R. (1988) A role for nonvascular plants in management of arid and semiarid rangelands. In: Tueller, P.T. (ed.) *Vegetation Science Applications for Rangeland Analysis and Management*. Kluwer Academic Publishers, Dordrecht, pp. 135–169.

Harper, K.T. and Pendleton, R.L. (1993) Cyanobacteria and cyanolichens: can they enhance availability of essential minerals for higher plants? *Great Basin Naturalist* 53, 59–72.

Harper, K.T. and St Clair, L.L. (1985) *Cryptogamic Soil Crusts on Arid and Semiarid Rangelands in Utah: Effects on Seedling Establishment and Soil Stability*. Report to the Bureau of Land Management, Salt Lake City, Utah.

Harris, E., Mack, R.N. and Ku, M.S.B. (1987) Death of steppe cryptogams under the ash from Mount St Helens. *American Journal of Botany* 74, 1249–1253.

Jarchow, J. (1984) Veterinary management of the desert tortoise, *Gopherus agassizi*, at the Arizona–Sonoran Desert Museum: a rational approach to diet. In: Trotter, M.W. (ed.) *Proceedings, 1984 Symposium of the Desert Tortoise Council.* Long Beach, California, pp. 83–94.

Jeffries, D.L. and Klopatek, J.M. (1987) Effects of grazing on the vegetation of the blackbrush association. *Journal of Range Management* 40, 390–392.

Jeffries, D.L., Klopatek, J.M., Link, S.O. and Bolton, H. Jr (1992) Acetylene reduction by cryptogamic crusts from a blackbrush community as related to resaturation and dehydration. *Soil Biology and Biochemistry* 24, 1011–1105.

Johansen, J.R. (1984) Response of soil algae to a hundred-year storm in the Great Basin Desert, U.S.A. *Phykos* 23, 51–54.

Johansen, J.R. and St Clair, L.L. (1986) Cryptogamic soil crusts: recovery from grazing near Camp Floyd State Park, Utah, USA. *Great Basin Naturalist* 46, 632–640.

Johansen, J.R., Adchara, J. and Rushforth, S.R. (1982) Effects of burning on the algal communities of a high desert soil near Wallsburg, Utah. *Journal of Range Management* 35, 598–600.

Johansen, J.R., St Clair, L.L., Webb, B.L. and Nebeker, G.T. (1984) Recovery patterns of cryptogamic soil crusts in desert rangelands following fire disturbance. *Bryologist* 87, 238–243.

Johansen, J.R., Ashley, J. and Rayburn, W.R. (1993) Effects of rangefire on soil algal crusts in semiarid shrub-steppe of the lower Columbia Basin and their subsequent recovery. *Great Basin Naturalist* 53, 73–88.

Kinnell, P.I.A., Chartres, C.J. and Watson, C.L. (1990) The effects of fire on the soil in degraded semi-arid woodland. II. Susceptibility of the soil to erosion by shallow rain-impacted flow. *Australian Journal of Soil Research* 28, 779–794.

Kleiner, E.F. and Harper, K.T. (1972) Environment and community organization in grasslands of Canyonlands National Park. *Ecology* 53, 299–309.

Lange, W. (1974) Chelating agents and blue–green algae. *Canadian Journal of Microbiology* 20, 1311–1321.

Lange, W. (1976) Speculations on a possible essential function of the gelatinous sheath of blue–green algae. *Canadian Journal of Microbiology* 22, 1181–1185.

Lesica, P. and Shelly, J.S. (1992) Effects of cryptogamic soil crust on the population dynamics of *Arabis fecunda* (Brassicaceae). *American Midland Naturalist* 128, 53–60.

Loope, W.L. and Gifford, G.F. (1972) Influence of a soil microfloral crust on select properties of soils under pinyon-juniper in southeastern Utah. *Journal of Soil and Water Conservation* 27, 164–167.

Loria, M. and Herrnstadt, J. (1980) Moss capsules as food of the harvest ant, *Messor*. *Bryologist* 83, 524–525.

MacGregor, A.N. and Johnson, D.E. (1971) Capacity of desert algal crusts to fix atmospheric nitrogen. *Soil Science Society of America Proceedings* 35, 843–844.

MacKenzie, H.J. and Pearson, H.W. (1979) Preliminary studies on the potential use of algae in the stabilization of sand wastes and wind blow situations. *British Phycology Journal* 14, 126.

Marble, J.R. (1990) Rangeland microphytic crust management: distribution, grazing impacts, and mineral nutrient relations. PhD dissertation, Brigham Young University, Provo, Utah.

Marston, R.A. (1986) Maneuver-caused wind erosion impacts, south central New

Mexico. In: Nickling, W.G. (ed.) *Aeolian Geomorphology*. Allen & Unwin, Boston, pp. 273–290.
Mayland, H.F., McIntosh, T.H. and Fuller, W.H. (1966) Fixation of isotopic nitrogen on a semiarid soil by algal crust organisms. *Soil Science Society of America Proceedings* 30, 56–60.
McIlvanie, S.K. (1942) Grass seedling establishment, and productivity – overgrazed vs. protected range soils. *Ecology* 23, 228–231.
Metting, B. (1981) The systematics and ecology of soil algae. *Botanical Review* 47, 195–312.
Metting, B. (1991) Biological surface features of semiarid lands and deserts. In: Skujiņš, J. (ed.) *Semiarid Lands and Deserts*. Marcel Dekker, New York, pp. 257–293.
Nebeker, G.T. and St Clair, L.L. (1980) Enhancement of seed germination and seedling development by cryptogamic soil crusts. *Botanical Society of America Miscellaneous Series Publication* 158.
Novichkova-Ivanova, L.N. (1972) Soil algae of Middle Asian deserts. In: Rodin, L.E. (ed.) *Ecophysiological Foundation of Ecosystem Productivity in the Arid Zone*. Academy of Science, Leningrad, pp. 180–182.
Pluis, J.L.A. and de Winder, B. (1989) Spatial patterns in algae colonization of dune blowouts. *Catena* 16, 499–506.
Rauh, W. (1985) The Peruvian–Chilean deserts. In: Evenari, M., Noy-Meir, I. and Goodall, D.W. (eds) *Hot Deserts and Arid Shrublands*, Vol. 12A, *Ecosystems of the World*. Elsevier, Amsterdam, pp. 239–268.
Rayburn, W.R., Mack, R.N. and Metting, B. (1982) Conspicuous algal colonization of the ash from Mount St. Helens. *Journal of Phycology* 18, 537–543.
Reynaud, P.A. (1987) Ecology of nitrogen-fixing cyanobacteria in dry tropical habitats of West Africa: a multivariate analysis. *Plant and Soil* 98, 203–220.
Rogers, L.E., Fitzner, R.E., Cadwell, L.L. and Vaughn, B.E. (1988) Terrestrial animal habitats and population responses. In: Rickard, W.H., Rogers, L.E., Vaughn, B.E. and Leibetrau, S.F. (eds) *Shrub-Steppe: Balance and Change in a Semi-arid Terrestrial Ecosystem*. Elsevier, Oxford, pp. 181–256.
Rogers, R.W. and Lange, R.T. (1971) Lichen populations on arid soil crusts around sheep watering places in South Australia. *Oikos* 22, 93–100.
Rozanov, A.N. (1951) *Serozems of Central Asia*. Israel Program of Scientific Translations, Jerusalem.
Rychert, R.C. and Skujiņš, J. (1974) Nitrogen fixation by blue–green algae–lichen crusts in the Great Basin Desert. *Soil Science Society of America Proceedings* 38, 768–771.
Rychert, R.C., Skujiņš, J., Sorenson, D. and Porcella, D. (1978) Nitrogen fixation by lichens and free-living microorganisms in deserts. In: West, N.E. and Skujiņš, J. (eds) *Nitrogen in Desert Ecosystems*. Dowden, Hutchinson and Ross, Stroudsburg, PA, pp. 20–30.
St Clair, L.L., Webb, B.L., Johansen, J.R. and Nebeker, G.T. (1984) Cryptogamic soil crusts: enhancement of seedling establishment in disturbed and undisturbed areas. *Reclamation and Revegetation Research* 3, 129–136.
St Clair, L.L., Johansen, J.R. and Webb, B.L. (1986) Rapid stabilization of fire-disturbed sites using soil crust slurry: inoculation studies. *Reclamation and Revegetation Research* 4, 261–269.

Schlatterer, E.F. and Tisdale, E.W. (1969) Effects of litter of *Artemisia, Chrysothamnus* and *Tortula* on germination and growth of three perennial grasses. *Ecology* 50, 869–873.
Schlesinger, W.H. (1991) *Biogeochemistry: An Analysis of Global Change*. Academic Press, New York.
Schulten, J.A. (1985) Soil aggregation by cryptogams of a sand prairie. *American Journal of Botany* 72, 1657–1661.
Shachak, M. and Steinberger, Y. (1980) An algae-desert snail food chain: energy flow and soil turnover. *Oecologia (Berlin)* 46, 402–411.
Shields, L.M. (1957) Algal and lichen floras in relation to nitrogen content of certain volcanic and arid range soils. *Ecology* 38, 661–663.
Skarpe, C. and Henriksson, E. (1987) Nitrogen fixation by cyanobacterial crusts and by associative-symbiotic bacteria in western Kalahari, Botswana. *Arid Soil Research and Rehabilitation* 1, 55–60.
Skujiņš, J. (1984) Microbial ecology of desert soils. In: Marshall, C.C. (ed.) *Advances in Microbial Ecology*. Plenum Press, New York, pp. 49–91.
Snyder, J.M. and Wullstein, L.H. (1973) The role of desert cryptogams in nitrogen fixation. *American Midland Naturalist* 90, 257–265.
Steinberger, Y. (1989) Energy and protein budgets in the desert isopod *Hemilepistus reaumuri*. *Acta Oecologia* 10, 117–134.
Stewart, W.D.P. (1967) Transfer of biologically fixed nitrogen in a sand dune slack region. *Nature* 214, 603–604.
Steyn, P.L. and Delwiche, C.C. (1970) Nitrogen fixation by nonsymbiotic microorganisms in some California soils. *Environmental Science and Technology* 4, 1122–1128.
Sylla, D. (1987) Effect of microphytic crust on emergence of range grasses. MS thesis, University of Arizona, Tucson.
Terry, R.E. and Burns, S.J. (1987) Nitrogen fixation in cryptogamic soil crusts as affected by disturbance. In: Everett, R.E. (ed.) *Proceedings of the Pinyon-Juniper Conference*, US Forest Service General Technical Report INT-215, pp. 369–372.
van den Ancker, J.A.M., Jungerius, P.D. and Mur, L.R. (1985) The role of algae in the stabilization of coastal dune blowouts. *Earth Surface Processes and Landforms* 10, 189–192.
Walker, B.H. (1979) Game ranching in Africa. In: Walker, B.H. (ed.) *Management of Semi-arid Ecosystems*. Elsevier, Amsterdam, pp. 55–81.
West, N.E. (1978) Physical inputs of nitrogen to desert ecosystems. In: West, N.E. and Skujiņš, J. (eds) *Nitrogen in Desert Ecosystems*. Dowden, Hutchinson and Ross, Stroudsburg, PA, pp. 165–170.
West, N.E. (1990) Structure and function of microphytic soil crusts in wildland ecosystems of arid to semi-arid regions. *Advances in Ecological Research* 20, 179–223.
West, N.E. and Skujiņš, J. (1977) The nitrogen cycle in North American cold-winter semi-desert ecosystems. *Oecologia Plantarum* 12, 45–53.

The Diversity of Microorganisms Associated with Marine Invertebrates and Their Roles in the Maintenance of Ecosystems

12

D.L. SANTAVY

Gulf Breeze Environmental Research Laboratory, US Environmental Protection Agency, Sabine Island, Gulf Breeze, Florida 32561–5299, USA

Introduction

Marine environments are characterized by very different physical, chemical, and biological parameters compared to terrestrial and freshwater environments. Many unique microorganisms have already been discovered living within the depths of the seas, possessing properties unlike any previously studied. Coral reef environments are distinctly characterized by great species diversity and abundance which is paralleled by tropical rainforests from terrestrial systems. In an environment where space, nutrients, and light are at a premium, competition to survive is fierce. Survival and effective competition in nature often require specialized strategies, and may impart unique physiological, chemical, and ecological capabilities to the organism.

On coral reefs, a great number of organisms form assemblages such as symbioses, partition resources by spatial separation, or develop other mechanisms to increase the efficiency for utilizing scarce resources from oligotrophic waters. This is especially true among the sessile or immobile invertebrate populations. Symbiotic microorganisms associated with invertebrate populations are often essential for maintaining the fitness and survival of both associates. Mutualistic symbionts can provide required energy for survival of the host, protection from pathogenic or opportunistic organisms, protection from ultraviolet radiation, or chemical defence acting as effective deterrents against other organisms. Consequently, the marine environment, especially coral reef environments, offers many unique niches for opportunities to explore for and discover novel microorganisms and the significant roles they assume. The search for increasing our knowledge of microbial biodiversity should include an examination of mutualistic symbioses, pathogenic associations, and other categories of assemblages. A few intriguing relationships among marine invertebrates and bacterial associates will be presented to illustrate the vast biodiversity of marine microorganisms and their importance in the maintenance of large ecosystems.

Microorganisms Associated with Marine Sponges

Marine sponges harbour microbial symbionts including bacteria, cyanobacteria, and unicellular algae. The most prolific sponge populations, with respect to biomass and diversity, are found in the Caribbean basin. Many of these associations remain undescribed, a few have been investigated in ultrastructure studies primarily discussing sponge cytology. Often very high concentrations of microorganisms comprise the overall biomass of the sponge and symbiote complex, occasionally with the biomass of the microorganisms exceeding the biomass of the sponge. Comparison of microorganisms from adjacent microenvironments can elucidate important properties and verify the specificity of associated bacteria in symbiosis. This is particularly important for bacteria found in marine sponges, because sponges filter planktonic bacteria from ambient sea water for food. Two examples will be presented to gain an appreciation of the importance of these associations in considering the biodiversity of microorganisms within the marine environment.

Microbial Community Associated with Ceratoporella nicholsoni

A community of morphologically diverse bacteria is found associated with the Caribbean sclerosponge, *Ceratoporella nicholsoni*, with large numbers of sponge-associated bacteria comprising up to 57% of the cellular composition of the mesohyl, the region between the thin external epithelium and internal flagellated epithelium (Willenz and Hartman, 1989). The bacteria are located primarily extracellularly, in the mesohyl regions and rarely are found associated with the dermal layers attached to the surface and lining of the aquiferous system. The size and morphology of the bacteria are highly variable, with the majority of bacteria being rod- or coccoid-shaped (Fig. 12.1). Most bacteria possess cell walls typical of Gram-negative bacteria, with an additional diffuse, loosely bound layer encapsulating most of the cells (Santavy *et al.*, 1990).

Examination of the diversity of the bacterial community with transmission electron microscopy shows distinct morphological forms indicative of

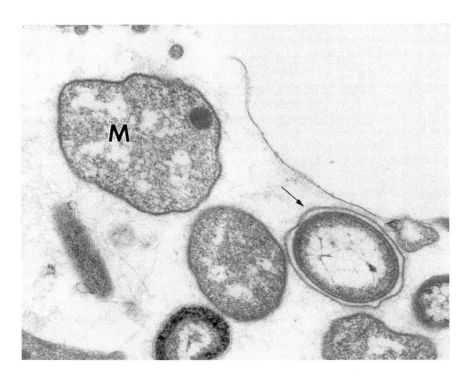

Fig. 12.1. Transmission electron micrograph (TEM) of microorganisms found in the mesohyl of the sclerosponge *Ceratoporella nicholsoni*. Typical Gram-negative bacterial cells contain an additional diffuse, loosely bound layer which encapsulates most of the bacterial cells (arrow). Mollicute-like or mycoplasma bacterial cell (M), pleomorphic form distinctly lacks any distinguishable cell wall and contains an electron-dense vacuole. (Magnification × 11,000.)

certain groups of bacteria. The most abundant cell type contains a typical Gram-negative cell wall with an additional undulating membrane with varying degrees of attachment to the cell wall. It is unclear whether this membrane is derived from the symbiont or host. Various forms of Gram-negative bacteria contain multiple invaginations of the cytoplasmic membrane of additional electron-dense layers about the cell wall. Another form is mollicute-like or resembles a mycoplasma, its pleomorphic form lacks any distinguishable cell wall and contains an electron-dense vacuole (Fig. 12.1). An unusual elliptical prokaryote found at the base of the calices, contains dense parallel lamellae in the cytoplasmic region.

Physiological, metabolic and morphological attributes of 80 bacterial strains isolated and purified from the sponge *C. nicholsoni*, and 48 bacterial strains were obtained from the sea water in the same vicinity. All collections were made from submarine caves from the Bahamas during August 1985.

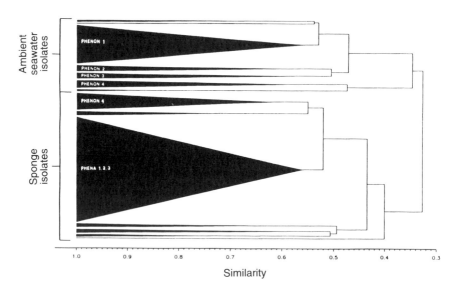

Fig. 12.2. Dendrogram showing clustering patterns for numerical taxonomic analysis of ambient sea water and sponge isolates. The dendrogram was constructed using similarity values employing the Jaccard Similarity Coefficient (S_J) and unweighted average linkage. Clusters are described as groups of strains possessing average S_J values of 55% or greater.

Phenotypic profiles from the bacterial isolates cultured from the sponge and ambient sea water were compared, to establish the specificity of the bacteria for their sponge habitat. Isolates from both habitats were analysed by numerical taxonomy, using 85 phenotypic traits for each isolate permitting comparison of all attributes (Santavy and Colwell, 1990). Morphometric methods were used to determine that 3–11% of the total bacteria inhabiting the sponge were culturable by the methods employed (Santavy et al., 1990).

Bacteria isolated from the sponge only were found associated with the sponge and there was no evidence that they were simply bacteria filtered from the sea water by *Ceratoporella nicholsoni*. None of the sponge-associated isolates were detected among isolates from the surrounding sea water. These conclusions are supported by phenotypic (Santavy and Colwell, 1990) and serological data, where species-specific antibodies were prepared to each of four species of sponge symbionts and were used to establish serological relationships among isolates from the sponge and sea water (Santavy, 1988). In general, the sea-water bacteria displayed a greater phenotypic heterogeneity among the strains, whereas 70% of the sponge isolates comprised a large cluster (Fig. 12.2). All bacteria isolated from the same habitat clustered most closely with other bacteria isolated from the same habitat, in phenotypic analyses. The two clusters of bacteria isolated from the two different habitats linked at an average similarity value of 32.7% (Santavy and Colwell, 1990).

Physiological profiles of the isolates generally reflected the nutrient profile of the habitat and were highly correlated with metabolic capabilities of bacteria isolated from each environment. Sponges can provide a relatively nutrient-rich environment for bacteria by concentrating particulate organic matter filtered from the water column and by supplying nitrogenous excretory products accumulated by the organism. During periods of low pumping rates, anaerobic pockets may develop inside the sponges, which allow fermentative bacteria to thrive. Sponge-associated bacteria demonstrated greater metabolic capabilities than the sea-water isolates and were able to catabolize a large number of substrates, both oxidatively and fermentatively. Fermentative abilities and amino acid decarboxylation by the sponge isolates were the most prominent metabolic attributes distinguishing the sponge symbionts from sea-water bacteria. Bacteria from sponges were most readily able to catabolize arabinose, fucose, galactose, and sucrose anaerobically, but not glucose. Most of the sponge isolates demonstrated the ability to degrade large organic molecules, such as DNA and chitin. Pigmentation was not observed to occur among the sponge-associated bacteria. Taxonomic affiliations among the sponge isolates included two undescribed species of the genus *Vibrio*, one undescribed species of *Aeromonas* and a distant relative of the coryneform bacteria (Santavy *et al.*, 1990).

In comparison, sea-water isolates were potentially more restricted in their dissimilatory abilities, being capable only of oxidative metabolism, with respect to the substrates examined. The water column overlying coral reefs is classified as oligotrophic, with available nutrients quickly assimilated, especially nitrogenous compounds. The sea-water isolates appeared to be metabolically restricted by the reduced concentration of nutrients in the water column. Bacteria isolated from sea water were generally characterized as Gram-negative rods, but included coccoid forms; and strains were oxidative in respiration, limited in their ability to metabolize high-molecular-weight organic compounds, pigmented, and able to reduce nitrate. Typically shared traits of the bacteria isolated from both environments included Gram-negative rods exhibiting little polymorphism, with an inability to grow in a medium without sodium chloride, and tolerances to 6% sodium chloride concentrations (Santavy and Colwell, 1990).

Blue-pigmented Filamentous Bacterium in Marine Sponges

Another example of an unusual symbiont is the bacterium associated with the Hawaiian sponge *Terpios granulosa*, the Californian sponge, *Hymenamphiaster cyanocryta*, and several other sponge species. These sponges contain

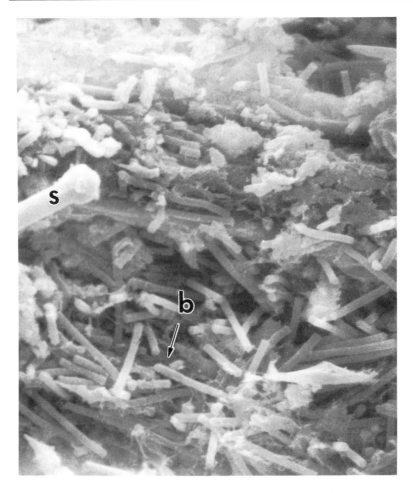

Fig. 12.3. Scanning electron micrograph of *Terpios granulosa* with filamentous bacterial symbiont (b) densely packed in sponge mesohyl. Membrane sheets of collagen form layers, with sponge spicule head exposed (s). (Magnification × 2500.)

a monoculture of a symbiotic bacterium, which occupies 60–80% of the extracellular region of the sponge (Fig. 12.3). The bacterium contains a characteristic cobalt blue pigmentation which is not related to any photosynthetic pigments, but is reported to be a carotenoprotein biochrome (Castro, 1979). The pigment is contained within birefractive inclusions located throughout individual cells. The multicellular trichomes superficially resemble filamentous cyanobacteria but do not contain photosynthetic pigments or any discernible thylakoid structure. Hormogonia type of reproduction and septal centripetal annual ingrowth from the cytoplasmic membrane are characteristics in common with cyanobacteria. The sponge and bacterium appear to engage in an obligate symbiotic relationship, as efforts to remove the symbiont from the sponge, ultimately result in sponge death (Santavy, 1985).

Ultrastructural studies of the bacterium reveal a morphology most closely resembling *Methanosaeta soehngenii* (Whitman et al., 1991). Individual trichomes may reach up to 30 µm in length. The cell wall is similar to a Gram-negative cell wall, but is more extensive. The cytoplasmic membrane is beneath an electron-dense thickened, periplasmic space, separated

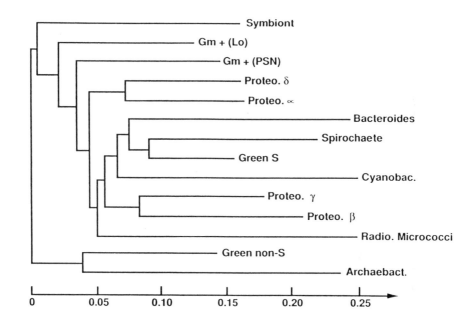

Fig. 12.4. Phylogenetic tree based upon 16S rRNA sequence comparison for the blue-pigmented symbiont with each of the eubacterial phyla and an *Archaebacteria*. An alignment was constructed using the eubacterial representative sequence mask. The sequences used represented the eubacterial phyla as follows: symbiont from *Hymenamphiaster cyanocrypta*; Gram-positive low G&C species, *Bacillus subtillis*; Gram-positive photosynthetic species, *Heliobacterium chlorum*; *Proteobacteria* delta subdivision, *Desulfovibrio desulfuricans*; *Proteobacteria* alpha subdivision, *Agrobacterium tumefaciens*; bacteroides, flavobacteria and relatives, *Bacteroides fragilis*; spirochaetes and relatives, *Leptospira illini*; green sulphur bacteria, *Chlorobium vibrioforme*; cyanobacterium and chloroplast, *Anacystis nidulans*; *Proteobacteria* gamma subdivision, *E. coli*; *Proteobacteria* beta subdivision, *Pseudomonas testosteroni*; radioresistant micrococcus and relatives, *Deinococcus radiodurans*; green non-sulphur bacteria and relatives, *Thermomicrobium roseum*; Archaebacteria, *Sulfolobus solfataricus*.

from the outermost membrane. The terminal cell is characterized by a flattened conical tip, with a thickened periplasmic space. The cell wall is this region contains two electron-dense membranes separated by an electron-opaque region. This region varies in the amount of separation between the outermost membrane and the cytoplasmic membrane (Santavy, 1986).

Although attempts to culture the bacterium have been unsuccessful, molecular methods have been used to determine its phylogenetic relationships. Symbiont cells were dissociated from the sponge and separated on a density gradient. DNA was extracted from the symbiont, purified, and used to amplify the 16S rRNA gene using the polymerase chain reaction (PCR) (Saiki et al., 1988) employing eubacteria and eukaryotic specific primers (Giovannoni et al., 1988). The amplified product was cloned into a plasmid vector transformed into *E. coli* (Bluescript, Strategene) and a partial 16S rRNA gene sequence was determined using double-stranded DNA dideoxynucleotide sequencing technique employing Sequenase (US Biochem. Corp.). Approximately 60% of the sequence was determined, aligned and compared to the 16S rRNA database held by the University of Illinois Ribosomal Database Project using maximum parsimony analysis (Olsen et al., 1991). The bacterium did not closely resemble any of the major groups within the *Eubacteria* (Fig. 12.4), but appears to be distantly related and deeply rooted in the eubacterial phylogenetic tree.

Microorganisms Associated with Scleractinian Corals

Coral reef ecosystems are major contributors to subtropical and tropical marine coastal and oceanic processes. Hard corals or scleractinia sustain the architectural structure for reefs and provide habitats for many organisms at all trophic levels. Scleractinian or hard corals constitute one of the classic examples of symbiosis, the association between the coral host and symbiotic dinoflagellate algae or zooxanthellae. An obligate mutualism is derived from the intracellular algae for the nutritional requirements of the host in exchange for providing the algae a protected habitat. Bacteria have also been found associated with Scleractinia ranging from pathogenic invasions to ovoid bodies containing bacteria in localized areas of coral tissue.

Black Band Disease of Brain Corals

The relationship described as black band disease affects many of the brain corals in the Caribbean and Central Atlantic Ocean regions. The disease symptoms for the coral are characterized by rapid loss of tissue and skeletal bleaching. The most dominant microorganism has been identified as the motile filamentous cynobacterium, *Phormidium corallyticum* (Rützler and Santavy, 1983). The disease only infects specific scleractinian species, its occurrence is seasonal and correlates with higher sea-water temperatures (Rützler et al., 1983). Once the disease is established on the coral colony it advances very rapidly, up to 6.2 mm maximum linear increase over a 24 h cycle (Rützler et al., 1983). It can kill an entire coral colony, measuring over a metre in diameter, within weeks, if the disease persists. This devastating disease is responsible for coral mortality over vast regions of coral reef environments in some locations in the Florida Keys and USA.

Black band disease is characterized by a dense thick black band which separates the healthy uninfected tissue from the tissue-depleted white coral skeleton. The black band is composed of a dense cyanobacterial mat, a complex microbial consortium containing many heterotrophic bacteria, *P. corallyticum*, other cyanobacteria such as *Spirulina*, the sulphur-oxidizing bacterium *Beggiatoa*, ciliates, flagellates and other microeukaryotes. *Phormidium corallyticum* has very distinct cell morphology, with a tapering terminal apical cell at one end of the trichome and a blunt end at the other terminus (Rützler and Santavy, 1983). The individual cells are almost isodiametric, with a unique thylakoid structure of straight radiating lamellae when viewed in cross-section, and lack heterocysts (Rützler and Santavy, 1983). To date, this cyanobacterium has only been found affiliated with the black band condition of corals.

Bacteria in the Tentacular Tissue of Porites astreoides

Undescribed ovoid bodies have been found in the tentacle tissue of the scleractinian coral, *Porites astreoides* and occasionally with other tissues including the oral disc, stomadeom, and mesenterial filaments. No pathobiosis was observed, either grossly or microscopically, in the *P. astreoides* colonies containing these structures. The ovoid bodies are up to 50 μm in diameter, contain densely packed bacteria-like cells which are not membrane-bound and cause compression of adjacent cells. The occurrence of these structures is highly variable within each polyp, ranging in number from 0 to 13 (Santavy and Peters, 1991).

The bacteria-like cells in the ovoid bodies are separable, indicating they are not tightly packed. The bacteria are variable in size and shape, with one type larger, pleomorphic, and often greater than 4 μm in diameter. The other type is smaller, about 1.5 μm in length, and rod-shaped. Contradictory results were obtained using two histological stains specific for Gram reaction. The Tworts method of staining cells indicated the cells were Gram-positive, whereas the Brown–Hopps method of staining cells indicated the cells were Gram-negative or variable. An extensive investigation of different *Porites* species throughout the Caribbean basin was carried out for the presence of these ovoid bodies. The results revealed that their occurrence was consistent along phylogenetic lines appearing in the same host species, independent of geographic location. Fluorescently labelled kingdom-specific probes prepared to the small subunit of the 16S rRNA hybridized to the bodies indicating a phylogenetic affiliation to the *Eubacteria* (Santavy and Peters, 1991).

Efforts were made to culture the bacteria contained within the ovoid bodies. Bacteria were obtained from macerated tissue and inoculated into enrichment media for marine bacteria, ammonium-oxidizing bacteria, and oxygen-tolerant methanogenic bacteria. No bacteria grew in media specific for ammonium-oxidizing or oxygen-tolerant methanogenic bacteria. The cultured bacteria were characterized using the BIOLOG Gram-Negative Microplate Identification System (Hayward, CA). The BIOLOG system incorporates 95 unique carbon sources in a microplate format and assays for the ability of bacteria to utilize them. Approximately 40% of the isolates obtained could not utilize any of the carbon sources in the assay, therefore only 19 isolates were characterized using the BIOLOG system. Of the isolates analysed with the BIOLOG system 32% were classified as different *Moraxella* species. Identification of isolates from *P. astreoides* using the BIOLOG system yielded different relationships among the isolates than the clustering patterns formed using traditional phenotypic analysis (Jacobs, 1990). Most of the isolates were Gram-negative, oxidase-positive, pigmented, facultative anaerobes and capable of fermentative metabolism. Approximately 50% of the isolates were capable of siderophore production. It was noteworthy that the ability to utilize specific carbon sources which segregated the groups were glycoses found in coral mucus (fucose, galactose, and *N*-acetylglucosamines) (Santavy and Peters, 1991).

Role of Microorganisms in the Maintenance of Ecosystem Biodiversity

Microorganisms are fundamentally responsible for nutrient recycling and, ultimately, the health of coral reef ecosystems. Evaluating short-term microbial community responses to obvious perturbed hosts may provide a first alert to signal the potential of long-term debilitating effects on the coral ecosystem. Since coral reefs may be most sensitive to climatic and anthropogenically induced stress, they may serve as early indicators of change which may eventually affect more resilient coastal ecosystems.

Microbial communities associated with the mucus of scleractinian corals were employed in an attempt to develop a biological diagnostic technique to quantify threshold responses of reef ecosystems to environmental stresses. The potential was assessed for developing a microbial assay which can be used to predict the coral's response to stress for defining *in situ* sublethal or chronic impacts that may lead to long-term ecological damage in reef ecosystems. This research is based on the hypothesis that increased stress to coral hosts, such as infection by disease or other obvious tissue abnormalities, will result in a change of the microbial community structure and productivity associated with coral mucus. Thus short-term microbial community responses to perturbation on coral hosts can be used to predict the survival and fitness of the host invertebrate population, thereby assessing their use as a bioindicator for the coral reef ecosystem.

It has been reported that corals exposed to environmental perturbations often have increased mucus production and higher numbers of bacteria in the mucus (Ducklow and Mitchell, 1979). Detecting shifts in both microbial community structure and activity may provide an indication of stress not visually obvious. To assess this, the microbial community structure and productivity in different coral mucus samples were determined and compared to overlying water samples, and compromised corals, focusing on those infected with black band disease. Microbial responses were correlated to the fitness of the coral, to ascribe each measured microbial response's predictive value to determine the survival and fitness of the host invertebrate population. Initial efforts targeted extremes in stressed and healthy colonies to determine what resolution this approach may have across an obvious spectrum of coral physiological conditions.

The microbial communities associated with the mucus of different physiological states of the hard coral, *Colpophyllia natans*, were studied off the Florida Keys in July 1992. Mucus samples were obtained from individual coral colonies ranging from good physiological condition to obvious infections by black band disease to a mottling condition with swollen tissues. Mucus samples were taken at tissue sites on the host which had not yet experienced any damage from disease. Sea-water samples were taken simultaneously at each location.

Coral mucus was collected using a suction device employing a pneumatic pump driven by compressed air from a scuba tank. The outlet hose was connected to a sterile, collapsed 4 l IV serum bag secured to the diver's chest for collection of the sample and described in more detail elsewhere (Santavy, 1995). All samples were collected using scuba. Sterile sea-water samples were collected using pneumatic pumps from the boat. Estimates for total number of microorganisms were obtained by direct counts of bacteria, zooxanthellae and eukaryotic nuclei (Porter and Feig, 1980; Caron, 1983). Heterotrophic bacterial productivities were measured by incorporation of [^{3}H]thymidine into DNA (Chin-Leo and Kirchman, 1988), [^{3}H]uridine into RNA (Karl, 1982), and [^{14}C]leucine into proteins (Chin-Leo and Kirchman, 1988). Primary production was estimated by the incorporation of $Na^{14}CO_3$ (Parson *et al.*, 1985). Total alkaline phosphatase activity was determined by fluorimetric methods (Ammerman and Azam, 1991). Total biomass was determined by particulate DNA measurements (Paul *et al.*, 1985).

Differences in microbial community structure were determined by phospholipid fatty acid profiles (GC-FAME). Phospholipid fatty acid profiles were obtained from both concentrated coral mucus and water samples. Cells were extracted using a modified Bligh and Dyer extraction procedure (Guckert et al., 1985). The phospholipid, ester-linked fatty acids were prepared for gas chromatography (GC) by a mild alkaline transesterification, and separated, quantified, and identified using gas chromatography (Parker et al., 1984) and compared against Bacterial Acid Methyl Esters CP Mix standards (Matreya, Inc.). Phospholipid components of cell membranes from microbial communities were evaluated using principal component analysis (Jacobs, 1990).

The strategy used in this research was to measure an exhaustive suite of microbial parameters indicative of biomass and productivity over a relatively short time frame and on limited hosts. The goal was to provide data to be used in selecting those measurements most useful in detecting changes in the microbial community that would later reflect a reduced fitness of the coral hosts. These measurements would be applied in future studies to complete a more comprehensive sampling regime.

All the microbial parameters and productivity measurements analysed were at least one order of magnitude greater in the coral mucus compared to those of the overlying sea water. The microbial indicators for productivities indicative of DNA, RNA, and protein synthesis rates were significantly higher when adjusted for both rates per volume and rates per bacterial cell (Fig. 12.5). The microbial parameter measurements for number of bacteria, the number of eukaryotic nuclei, the carbon fixation rate, µg of chlorophyll a and µg of DNA per volume were significantly higher, and alkaline phosphatase activity was significantly higher for samples obtained from the coral mucus compared to sea water (Santavy et al., 1993).

Microbial productivity measurements made among all hosts with different fitness states were significantly higher for all hosts infected with black band disease as compared to healthy and other compromised states studied (Fig. 12.6). All microbial productivity measurements adjusted per bacterial cell were significantly higher for the black band diseased corals as compared to the healthy. The uridine incorporation rate per cell was also significantly less for the mottled condition than for both the black band infected corals and healthy corals. The microbial parameter measurements which correlated well with different fitness states include total eukaryotic nuclei and alkaline phosphatase turnover rate. The microbial parameter measurements which did not vary between different fitness states include the number of bacteria and zooxanthellae per unit surface area. The following parameters were not significantly different between the healthy and black band infected corals: µg DNA per unit area, non-zooxanthellae nuclei, and µg chlorophyll a per unit area. Primary production was significantly less in the black band infected colonies and mottled colonies as compared to the healthy colonies. Some microbial parameters varied significantly between different locations (Santavy et al., 1993).

The microbial community structures determined by preliminary data

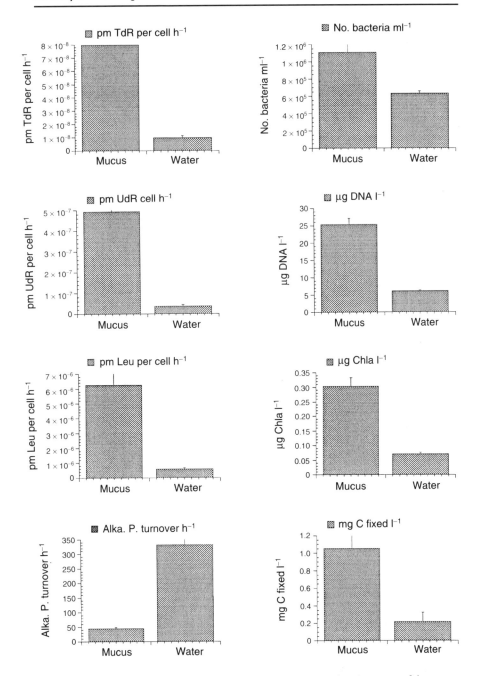

Fig. 12.5. Analysis of variance methods (ANOVA) were used to determine if these microbial productivity measurements and microbial parameters were significantly different ($P<0.05$) between coral mucus and sea water. Abbreviations: ALKAP, alkaline phosphate; CHLA, chlorophyll a; mgC, mg carbon; TdR, [^{3}H]thymidine; UdR, [^{3}H]uridine.

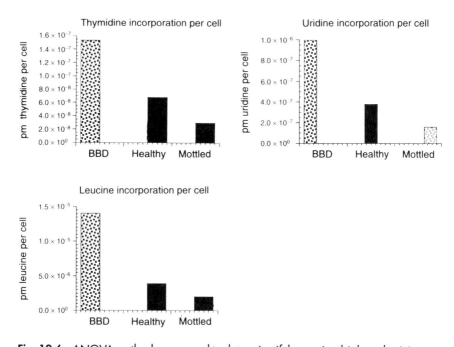

Fig. 12.6. ANOVA methods were used to determine if these microbial productivity measurements were significantly different ($P<0.05$) among different fitness states. Bar graphs of the same pattern denote fitness states which are not significantly different from each other based on results from Duncan's multiple range tests (alpha = 0.05).

generated by GC-FAME analysis were significantly different in the flora residing in the mucus of healthy and diseased corals (Fig. 12.7). The data were analysed using chromatograms generating phospholipid profiles of each colony or sea-water sample and comparing them with a commercially available mix of standard bacterial fatty acid methyl esters characteristic of bacterial cell membranes. Principal component analysis of the samples obtained from healthy, diseased, and mottled colonies as well as sea-water habitats was used in the analysis. Two dominant clusters were formed (Fig. 12.7), with the first cluster comprising 55% of samples from the healthy coral hosts and 36% of the mottled hosts. Only 9% of the samples were infected with black band disease. The second cluster comprised 54% of samples from black band infected colonies and 30% of samples obtained from healthy hosts. The percentage of overlap between the clusters showed more 'healthy corals' possessing profiles similar to the black band infected corals than the converse. A possible explanation is perhaps that those corals which appear to be visually healthy may already be in the early stages of a

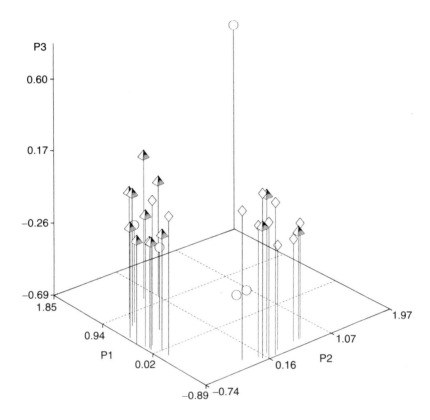

Fig. 12.7. Principal components analysis of GC-FAME data indicates that the microbial community structure in the mucus of healthy and diseased coral hosts were different. Symbols present the phospholipid analysis for each community sampled. Samples obtained from the water column are represented by circles, those obtained from healthy corals are represented by diamonds, and those obtained from diseased or compromised corals are represented by pyramids.

declining physiological condition with no detectable signs of stress. Signature phospholipids implicated changes in specific groups of microorganisms (Dobbs and Guckert, 1988; Findlay et al., 1990), with an increase in the fatty acids characteristic of anaerobic bacteria and microeukaryotes associated in the mucus collected from healthy tissue on a colony infected with black band disease. The microbial community structure of the flora from the water column was highly variable especially with regard to different locations indicating significant spatial heterogeneity across the reef track (Santavy, 1995).

In conclusion, all microbial parameters and productivity measurements were significantly greater in the coral mucus than in the overlying water. The microbial productivity measurements, whether determined on a cellular or per unit area basis, were all significantly higher in mucus collected from corals infected with black band disease as compared to healthy or 'mottled' corals. The microbial biomass measurements were more varied among the three fitness states than were productivity measurements. Significant relationships varied depending on the parameter measured and the relative health of the coral host and the type of injurious condition. Preliminary data suggest that microbial activity parameters may be most sensitive in detecting potential physiological (e.g. disease) stress in coral hosts. Microbial biomass parameters were less sensitive in detecting differences in coral host fitness states but may be more sensitive in detecting other forms of stress (e.g. environmental). The microbial community structure found associated with the mucus of healthy corals was different from those found in black band diseased corals as well as those found in the water column. A more complete sampling regime of many more hosts and water samples could generate results which may eventually be used to formulate a predictive model for coral health based on the microbial population inhabiting the coral mucus. In principle, the model may allow determination of the coral's response to stresses for defining *in situ* sublethal impacts that may lead to long-term ecological damage.

Conclusions

All proposed initiatives aimed at investigating and expanding our knowledge of microbial diversity must include examination of the marine environment. Marine environments are characterized by very different physical, chemical, and biological parameters compared to terrestrial and freshwater environments. Consequently, the marine environment, especially coral reef environments with its vast biodiversity, offers many unique niches for opportunities to explore and discover novel microorganisms and the significant roles they assume. A number of microbial floras have already been described and shown to yield genetically novel microorganisms, although most are non-culturable at this time. The search for increasing our knowledge of microbial biodiversity should not exclude microorganisms associated with higher organisms, including descriptions and examination of functional roles of mutualistic symbioses, pathogenic associations, and other categories of assemblages.

Microorganisms are fundamentally important to the cycling of nutrients among different trophic levels, remineralization of recalcitrant substrates, and survival of higher organisms by imparting unique survival qualities. All of these organisms are necessary to maintain the ecological balance of marine environments; the disruption or loss of certain components of the community may permanently alter entire ecosystems. One of the most germane issues to address is the immense scale, intensity and irreversibility of habitat modification and subsequent loss of species diversity. Synergistic effects of global climate change and burgeoning anthropoganic stresses induced by increased development in coastal regions are difficult to discern and often compounded. The need to assess entire ecosystem responses to perturbations is critical since long-term consequences may affect overall reproduction, recruitment, community structure and composition of the entire species assemblages and perhaps, ultimately the survival and fate of entire complex and diverse living communities. Potential causes and effects of species decline and the concomitant loss of biodiversity and habitat should be identified.

Acknowledgements

Partial support for research was provided by the US EPA Biotechnology Risk Assessment Program, NOAA National Undersea Research Program, NCI grant 2R01CA17256-10, a Mellon Fellowship at Scripps Institution of Oceanography. Shipboard time was funded by NSF grant BSR-84-01397. Collaborators for portions of the research include Dr Philippe Willenz, Royal Belgium National Institute of Natural Sciences; Dr Rita Colwell, Maryland Biotechnology Institute; Dr Esther Peters, Tetra Tech, Inc.; Dr Richard A. Snyder and Dr Wade Jeffrey, University of West Florida, CEDB; Dr Norman Pace and Dr Anna Louise Reysenbach, Indiana University; and Dr Tom Schmidt, Michigan State University. Technical assistance was provided by Mike Anakis (University of Maryland), Jeddai Campbell (Technical Resources Inc.), Leslie Cole (UWF) and Penny Malouin (EPA). Transmission electron micrographs of symbiotic bacteria from *Ceratoporella nicholsoni* were provided by Dr Willenz. Statistical assistance was provided by George Ryan (CSC) and Dr Dan Jacobs (University of Maryland). This information reflects the findings of the author's research and does not represent any policy views of the US EPA.

References

Ammerman, J.W. and Azam, F. (1991) Bacterial nucleosidase activity in estuarine and coastal marine water: Characterization of enzyme activities. *Limnology and Oceanography* 36, 1427–1436.

Caron, D.A. (1983) Technique for enumeration of heterotrophic and phototrophic nanoplankton, using epifluorescence microscopy, and comparison to other procedures. *Applied and Environmental Microbiology* 46, 491–498.

Castro, P. (1979) Studies on the symbiosis between a filamentous microorganism and *Hymenamphiaster cyanocrypta*, a sponge from California. In: Levi, C. and

Boury-Esnault, N. (eds) *Biologie des Spongiaires.* Coll. Int. CNRS 291, Paris, pp. 365–371.

Chin-Leo, G. and Kirchman, D.L. (1988) Estimating bacterial production in marine waters from the simultaneous incorporation of thymidine and leucine. *Applied and Environmental Microbiology* 54, 1934–1939.

Dobbs, F.C. and Guckert, J.B. (1988) Microbial food resources of the macrofaunal-deposit feeder *Ptychodera bahamensis* (Hemichordata: Enteropneusta). *Marine Ecology Progress Series* 45, 127–136.

Ducklow, H.W. and Mitchell, R. (1979) Bacterial populations and adaptations in the mucus layers on living corals. *Limnology and Oceanography* 24, 715–725.

Findlay, R.H. Trexler, M.B., Guckert, J.B. and White, D.C. (1990) Laboratory study of disturbance in marine sediments: response of a microbial community. *Marine Ecology Progress Series* 62, 121–133.

Giovannoni, S.J. DeLong, E.F., Olsen, G.J. and Pace, N.R. (1988) Phylogenetic group-specific oligonucleotide probes for identification of single microbial cells. *Journal of Bacteriology* 170, 720–726.

Guckert, J.B., Antworth, C.P., Nichols, P.D. and White, D.C. (1985) Phospholipid ester-linked fatty acid profiles as reproducible assays for changes in prokaryotic community structure of estuarine sediments. *FEMS Microbiology and Ecology* 31, 147–158.

Jacobs, D. (1990) SAS/GRAPH software and numerical taxonomy. In: *Proceedings of the 15th Annual Users Group Conference.* SAS Institute, Inc., Cary, NC, pp. 1413–1418.

Karl, D.W. (1982) Selected nucleic acid precursors in studies of aquatic microbial ecology. *Applied and Environmental Microbiology* 44, 891–902.

Olsen, G.J., Larsen, N. and Woese, C.R. (1991) The ribosomal RNA database project. *Nucleic Acids Research* 19 (Supplement), 2017–2021.

Parker, J.H. Nickels, J.S., Martz, R.F., Gehron, M.J., Richards, N.L. and White, D.C. (1984) Effect of well-drilling fluids on the physiological status and microbial infection of the reef building coral *Montastrea annularis. Archives of Environmental Contamination and Toxicology* 13, 113–118.

Parson, P.R., Maita, Y. and Lalli, C.M. (1985) Photosynthesis as measured by the uptake of radioactive carbon. In: *A Manual of Chemical and Biological Methods for Seawater Analysis.* Pergamon Press, New York, pp. 115–120.

Paul, J.H., Jeffrey, W.H. and DeFlaun M.F. (1985) Particulate DNA in subtropical oceanic and estuarine planktonic environments. *Marine Biology* 90, 95–100.

Porter, K.G. and Feig, Y.S. (1980) The use of DAPI for identifying and counting aquatic microflora. *Limnology and Oceanography* 25, 943–948.

Rützler, K. and Santavy, D.L. (1983) The black band disease of Atlantic reef corals. I. Description of the cyanophyte pathogen. *P.S.Z.N.I. Marine Ecology* 4, 301–319.

Rützler, K., Santavy, D.L. and Antonius, A. (1983) The black band disease of Atlantic reef corals. III. Distribution, ecology, and development. *P.S.Z.N.I. Marine Ecology* 4, 329–358.

Saiki, R.K., Gelfand, D.H., Stoffel, S., Scharf, S.J., Higuchi, R., Horn, G.T., Mullis, K.B. and Erlich, H.S. (1988) Primer-directed enzymatic amplification of DNA with a thermostable DNA polymerase. *Science* 239, 487–491.

Santavy, D.L. (1985) A blue-pigmented bacterium symbiotic with *Terpios granulosa,* a coral reef sponge. In: Harmelin-Vivien, M. and Salvat, D. (eds) *Proceedings of the Fifth International Coral Reef Congress, Tahiti* vol. 2, pp. 135–140.

Santavy, D.L. (1986) A blue-pigmented bacterium symbiotic with *Terpios granulosa*, a coral reef sponge. In: Jokiel, P.L., Richmond, R.H. and Rogers, R.A. (eds) *Coral Reef Population Biology.* UNIHI-SEAGRANT-CR-86-01, University of Hawaii. HIMB Tech. Report No. 37, pp. 380–393.

Santavy, D.L. (1988) Marine bacteria–invertebrate symbiosis, the Caribbean sclerosponge *Ceratoporella nicholsoni* as a paradigm. PhD dissertation, University of Maryland, College Park.

Santavy, D.L. (1995) An environmental assessment of microbial community responses to a comprised host. In: *Proceedings of the 5th symposium for EPA's Risk Assessment Research Program for Release of Biotechnology Products.* 14–17 June 1993, Duluth, Minnesota. EPA Centre for Biotechnology Series (in press).

Santavy, D.L. and Colwell, R.R. (1990) Comparison of the bacterial community associated with the Caribbean sclerosponge *Ceratoporella nicholsoni* and bacteria from ambient seawater. *Marine Ecology Progress Series* 67, 73–82.

Santavy, D.L. and Peters, E.S. (1991) Bacteria found in the tentacle tissue of the scleractinian coral *Porites astreoides*. In: *Abstracts of the Annual Meeting of the Americal Society of Microbiology, 91st Meeting, Dallas, Texas.* Washington, DC, Abstract Q-122.

Santavy, D.L., Willenz, P. and Colwell, R.R. (1990) Phenotypic study of bacteria associated with the Caribbean sclerosponge, *Ceratoporella nicholsoni. Applied and Environmental Microbiology* 56, 1750–1762.

Santavy, D.L., Jeffrey, W.H. and Snyder, R.A. (1993) Changes in the microbial community structure and productivity affiliated the mucus of healthy and stressed corals. In: *Abstracts on the Annual Meeting of the American Society of Microbiology. 93rd General Meeting, Atlanta, GA.* Washington, DC, Abstract Q-287.

Whitman, W.B., Bowen, T.L. and Boone, D.R. (1991) The methanogenic bacteria. In: Balows, A., Truper, H.G., Dworkin, M., Harder, W. and Schliefer, K.H. (eds) *The Prokaryotes*, 2nd edn. Springer-Verlag, New York, pp. 719–767.

Willenz, P. and Hartman, W.D. (1989) Micromorphology and ultrastructure of Caribbean sclerosponges. I. *Ceratoporella nicholsoni* and *Stromatospongia norae* (Ceratoporellidae – Porifera). *Marine Biology* 103, 387–402.

Fungi, a Vital Component of Ecosystem Function in Woodland

13

A.D.M. RAYNER

School of Biological Sciences, University of Bath, Claverton Down, Bath BA2 7AY, UK.

Introduction: Superstructure and Infrastructure

Think of a city. Impressive though its superstructure of buildings and the toings and froings of its human inhabitants might be, it could not function as a coherent system without its underlying infrastructure of communicating pipelines and cables. Though there may be outward signs of this infrastructure – lamp standards, fire hydrants, manhole covers, telephone kiosks, etc. – it takes prior knowledge to recognize them as such.

Similar outward signs, in the form of fungal fruit bodies, occur in woodlands. To the uninformed, the relative infrequency of these signs, combined with their seasonality, implies unimportance and transience. They are passed by, attracting little more than temporary curiosity. So it is that the infrastructure of the ecosystem, the mainstay of its vitality and heterogeneity, is overlooked. For the lonely mycologist who knows this infrastructure, the inability of others to see the fungi for the trees is frustrating, to say the least. To complain of neglect seems like special pleading. On the other hand, that same neglect impedes the research which is needed to substantiate fungal roles in the ecosystem, and fundamental issues are left unaddressed; neglect begets neglect in a cloud of uncertainty. Yet by resolving these same issues if may be possible greatly to enrich understanding of the mechanisms and processes by which biodiversity itself is generated and maintained.

For underlying the fruit bodies, and a vast variety of spore-producing structures of more microscopic dimensions, are living networks that maintain cycles of growth, death and decay in the ecosystem, interconnecting the lives of plants and animals in innumerable and often surprising ways (Rayner, 1993a).

Mycelial Interconnectedness

These underlying networks are mycelia, teams of branching, protoplasm-filled hyphal tubes with deformable tips and rigid side walls that convert organic and mineral nutrients and water into biomass on scales measurable in units varying from micrometres to kilometres. Knowledge of the real spatiotemporal distribution patterns and functioning of these networks in natural ecosystems is vital to comprehension of the importance of fungi in the generation and maintenance of biodiversity. However, due to a combination of neglect, misconception and inappropriate methodology, such knowledge is fragmentary. Moreover, understanding of the pattern-generating processes within mycelia themselves has been hampered by the paradigm that these systems can be regarded ideally as homogeneously assimilative, modular assemblies of duplicating hyphal growth units (e.g. Prosser, 1993).

This paradigm implies easily calculable, and hence predictable, dynamics under such narrowly prescribed, uniform conditions as may occur in artificial culturing systems, where heterogeneity is commonly regarded as imperfection. However, by disregarding the influence of feedback processes and intrahyphal communication, it may result in serious misrepresentation of the more complex dynamics of mycelial networks in natural populations. Here heterogeneity is the rule in the environment and adaptive in the organism. If the organism were to be homogeneously assimilative, and so be equally susceptible to losing resources as it is capable of gaining them, it would be unable to traverse resource-poor domains without incurring immense energy costs. It would also be incapable of expanding outwardly in more than one spatial dimension without becoming more sparse or slowing its extension rate asymptotically. On the other hand, a heterogeneously assimilative system can sustain indefinite outward expansion and maintain an energy-efficient balance between exploitation where and when there is plenty, and exploration, conservation and redistribution when and where there is shortage.

It has been suggested that mycelia achieve this balance by acting as 'self-plumbing' systems. This implies that they are able, via a metabolic feedback response to changes in internal energy charge, to open their hyphal boundaries to active transport when assimilable resources are available, and to seal these boundaries with hydrophobic materials when resources are unavailable (Rayner, 1994; Rayner *et al.*, 1994a, b). Moreover, they can regulate the internal distribution of resources by means of closable septal partitions and degenerative responses that can serve to isolate redundant or damaged parts of the system.

According to this model, mycelial systems are not additive and modular. Rather they are non-linear and fluid-dynamical: expansive processes, due to uptake of resources and associated deformation and permeability of hyphal boundaries, are counteracted by constraining processes involving rigidification and sealing ('insulation'). Mycelia can therefore generate variable degrees of hydraulic thrust and a consequent diversity of patterns simply by changing their boundary parameters, both via adaptive alterations in genetic specification and by feedback-driven epigenetic responses to varying circumstances.

The pattern-generating processes begin when a spore germinates, usually first swelling isotropically and then breaking symmetry as one or more germ hyphae emerge. As uptake occurs into the hyphae, so they expand at their tips, allowing an initial exponential increment in their absorptive surface. If the rate of uptake comes to exceed the rate of throughput to existing tips, as will apply in energy-rich environments, branching will be promoted, so generating an exploitative system. However, if the hyphae become insulated, while remaining deformable at their tips, branching will tend to be suppressed and exploration promoted, with the hyphae acting as distributive conduits. In higher fungi, some of the branches may fuse, transforming the initially dendritic system at least partially into a true network, increasing overall throughout capacity, inhibiting further branch proliferation and conserving resources.

As networks enlarge, there may come a point where their demand for internal maintenance becomes so powerful as to resist further expansion at the margins. In such cases it is necessary, if the system is not to stagnate, for autodegenerative processes, analogous to inner city decay, to ensue so that resources can be redistributed from the interior to growing margins. This could explain how and why fairy rings are formed, with the processes of exploration, exploitation, conservation and redistribution occurring in distinctive zones from front to rear of expanding annuli (Dowson *et al.*, 1989a).

Sooner or later, the hyphae within mycelial networks may also combine into aggregates. This can result in the formation of cable-like structures, mycelial cords and rhizomorphs, allowing the mycelium to change its operational scale – rather like establishing international telephone links once national networks are in place. Alternatively, or additionally, local aggregations of hyphae may expand into visible fruit bodies that by drawing from the network are able to produce and liberate genetic survival and dispersal units in the form of spores.

An effective way of examining in the laboratory how mycelial systems build functional structures at different scales is to grow them in heterogeneous matrices of discrete but intercommunicating microenvironmental sites. An example, which clearly illustrates the sensitive interdependence of dissociative and associative processes governed by feedback, is shown in Fig. 13.1.

Given such dynamic properties, it should come as no surprise that fungal mycelia permeate the woodland scene. Some, the lichens, are readily

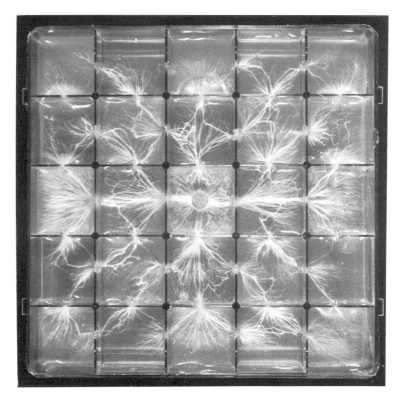

Fig. 13.1. Formation of a mycelial network by the basidiomycete, *Coprinus picaceus*, within a matrix of alternating high nutrient (2% malt agar) and low nutrient (water agar) sites, following growth from a central high nutrient site. Communication channels have been cut through the centre of each partition. Notice the production of more diffuse mycelium on high nutrient sites, distributive mycelial cords on low nutrient sites, and alignment of cords with communication pathways. (Courtesy of Louise Owen and Erica Bower.)

visible even on exposed surfaces, having made a sandwich of photosynthetic cells. Others can only grow where they are protected from desiccation and there is a suitable supply of organic nutrients to fulfil their energy needs, and so are less visible on exposed surfaces but abound elsewhere. Almost 70,000 m of hyphae have been estimated to reside in a gram (dry weight) of the organic horizon of a Swedish coniferous forest soil (Bååth and Söderström, 1979). There may be several tonnes of mycelium in a hectare of forest. Although the mycelia of some fungal species may be restricted to local sites where their specialized requirements for particular microenvironmental conditions are fulfilled, e.g. by the presence of a suitable host plant, others are more wide-ranging. A single individual mycelium of

Armillaria bulbosa has been found to occupy an area of about 15 ha and is estimated to be 1500 years old and to weigh 100 tonnes (Smith *et al.*, 1992) and even larger individuals are suspected.

All this weight of mycelium contrasts with a maximum production of some tens of kilograms of fruit bodies per hectare per year (Richardson, 1970). These outward signs of infrastructure are deceptive indeed. Far from being on the margins of forest life, fungi are central to it, creating and exploiting a huge variety of niches, some very narrowly and others more broadly bounded, throughout the ecosystem. An exploration of these diverse niches and their relation to biodiversity and ecosystem functioning now follows.

Permeating Foliage

Both the surfaces and interior of living leaves can be colonized by fungi, with varied consequences for the host plants. Like any other parts of a plant, leaves can appropriately be thought of as mini-ecosystems supporting potentially unstable microbial communities. Superficial inhabitants, the 'phylloplane fungi', depend on a thin film of nutrients that either originate from the leaf itself or are deposited there from the atmosphere. On recently emerged leaves these fungi, which generally exhibit little host selectivity, are predominantly yeasts; mycelial forms increase later on. Some of these organisms can degrade surface waxes and cuticular materials, hence permeabilizing and reducing the durability of the leaves, but at the same time may compete with actively pathogenic fungi. The latter penetrate leaf tissue where they obtain nutrients either biotrophically or necrotrophically, respectively without killing or having first killed plant cells. Biotrophic pathogens create heterogeneity by keeping infected parts alive while hastening senescence of adjacent regions, so forming 'green islands'; necrotrophic fungi cause leaf spot diseases in angiosperms and needle diebacks in conifers which, in sufficiently severe cases, may result in defoliation. Whereas biotrophic pathogens invariably exhibit a degree of host selectivity, necrotrophs can either be specialized, hence exhibiting selectivity, or unspecialized, having a wide host range and tending to colonize young, senescent or damaged organs.

Another group of fungi, the foliar endophytes, live inside leaves without causing obvious symptoms, at least not until the leaf ages (Carroll, 1988). These fungi, which show varying degrees of host selectivity, are probably widespread in plants of all kinds, but until recently have been neglected as 'inconsequential'. However, it is now recognized that they can produce toxins which deter feeding by herbivores and may also compete against pathogens. One species, *Rhabdocline parkeri*, which inhabits Douglas fir, is known to attack insect galls, killing the larvae within. As with all endophytes (of which more will be said below) such potentially beneficial effects may be obviated and more pathogenic activity induced if the balance of their relationship with the plant is disturbed, for example through imposition of some external stress such as drought. Indeed the fineness of this balance is indicated by the fact that some endophytic species are closely related to pathogenic species, e.g. *Rhabdocline parkeri* to *R. pseudotsugae* in Douglas fir and *Lophodermium pinastri* to *L. seditiosum* in pine. The potential for conversion of innocuous or even beneficial endophytes into pathogens is therefore a particularly concerning aspect of stress to plants induced by such factors as environmental change.

Corticolous Communities

Bark insulates a tree's internal plumbing against loss to or ingress from the external environment. Corticolous fungal communities therefore have a critical influence on tree health and like foliar communities may harbour actual or potential pathogens as well as superficial inhabitants and endophytes. Superficial inhabitants can form immensely rich communities in their own right and include: lichens; fungi that grow on lichens, algae and bryophytes (or in them as endophytes); fungi that grow on bark secretions; and members of the *Septobasidiaceae* that associate mutualistically with scale insects. The interactive influence of these communities on woodland life is evident in the fact that patterns of colonization of sycamore (*Acer pseudoplatanus*) bark by ivy (*Hedera helix*) and by endophytic fungi are affected by lichen cover (Rayner, 1993a; L. Whitehouse and A.D.M. Rayner, unpublished).

Biotrophic inhabitants of bark include certain rust fungi (*Cronartium* and *Peridermium* species) and the ascomycete, *Ascodichaena rugosa*, which causes the horizontally elongated patches often seen on *Fagus sylvatica* (Speer, 1981). Necrotrophic inhabitants cause dead patches or strips of bark known as cankers (e.g. Sinclair *et al.*, 1987). However, many cankers are probably also caused by endophytic fungi that change their pattern of activity when trees are stressed. For example, diamond-bark canker of sycamore (Fig. 13.2), caused by *Dichomera saubinetii*, was first recorded in Britain following the drought summers of 1975 and 1976 (Bevercombe and Rayner, 1978, 1980).

Root Connections

As is the case with foliage and bark, fungi associated with roots form relationships with plant tissues which can be ranked in order of increasing intimacy. In the soil around the roots, exudates and sloughed off cells

Fig. 13.2. Diamond cankers formed by *Dichomera saubinetii* on sycamore (*Acer pseudoplatanus*) following the drought summer of 1976. (From Bevercombe and Rayner, 1980.)

provide nutrients supporting free-living rhizosphere fungi. At the surface of the roots are rhizoplane fungi. Within the roots, endophytic, necrotrophic and biotrophic fungi may occur. The necrotrophs cause root disease, and commonly only become invasive after producing a superficial, non-assimilative mycelial phase, sometimes – perhaps inaptly – referred to as 'ectotrophic', which obviates any pre-existing or induced physical or chemical barriers protecting the internal tissues (cf. Garrett, 1970). On the other hand, biotrophic infection by mycelial fungi results in the intimate and often mutually beneficial associations known as mycorrhizas (Harley and Smith, 1983; Allen, 1991).

Mycorrhizas are of incalculable importance to ecosystem function, particularly on soils that are not over-endowed with supplies of nitrogen and phosphorus. Two types of mycorrhizal association are of particular importance in woodlands. In boreal and temperate forests containing members of the *Fagaceae, Salicaceae, Betulaceae, Pinaceae* or, in southern latitudes, *Myrtaceae*, sheathing or ectomycorrhizas predominate. Vesicular–arbuscular (VA) mycorrhizas, although present in boreal and temperate forests, are especially prevalent in tropical forests (Read, 1984). With both kinds of association, the presence of the fungus is thought greatly to enhance uptake of water and mineral nutrients, whereas the fungus obtains photosynthetically derived organic compounds. Mycorrhizal fungi may additionally give protection against root pathogens and sequester toxic metal ions.

The importance of mycorrhizal infection often becomes evident in attempts to establish trees in exotic locations, which are unsuccessful if the appropriate fungi are not transferred at the same time. The production of suitable mycorrhizal inoculum for such purposes is attracting increasing interest in applied mycological research (Jeffries and Dodd, 1991).

Sheathing mycorrhizas are formed by a variety of ascomycetes and, more especially, basidiomycetes. Their mycelium forms a tight-knit envelope or sheath around absorptive short roots, causing them to change colour, thicken and branch. Connected to this sheath are hyphae that ramify among the cortical cells of the root, forming a communications interface with these cells, known as the Hartig net (e.g. Duddridge, 1985).

VA mycorrhizas are formed by a family of *Zygomycetes*, the *Endogonaceae*. The fungal hyphae form only a loose covering over the surface of infected roots, and within the interior penetrate both between and within cortical cells, where they may form inflated vesicles having a possible role as storage organs and finely branched arbuscules which serve as interfacial organs.

Hyphae of both VA and sheathing mycorrhizal fungi can extend out for some distance into soil from infected roots. In sheathing mycorrhizas, these hyphae can associate into mycelial cord systems spanning many centimetres (Read, 1984, 1992).

The importance of these migratory hyphal systems is potentially profound. As well as greatly extending the range over which mineral nutrients may be absorbed from soil, they provide a means of establishing connections between roots of separate plants, both of the same and, depending on the host selectivity of the fungus (which varies), different species. The resultant opening of communication channels may enable adult plants to 'nurse' seedlings (Read, 1984), reduce competition between plant species and enhance overall community uptake of soil nutrients (Perry *et al.*, 1989). On the other hand it provides a route whereby some plants may draw resources from the network at the expense of others. The yellow bird's nest, *Monotropa hypopitys*, clearly does just this (Björkmann, 1960; Duddridge and Read, 1982). Also, some plants may benefit more by associating with certain mycorrhizal fungi than with others and the cost of supporting a fungal partner can be very high, e.g. as much as 25% of a tree's assimilate (Newman, 1978). Such possibilities provide an adaptive explanation for the occurrence of incompatibility systems which cause particular plants only to accept connections made by specific fungi and to reject others (Molina and Trappe, 1982). Otherwise the rejection of a potential ally would seem counterproductive (Harley and Smith, 1983).

Not only plants, but some fungi can also piratize mycorrhizal networks, and hence in effect parasitize host plants attached to the networks. An example is *Cordyceps capitata*, which parasitizes ectomycorrhizal truffles. A slightly different situation occurs with certain orchids, which by forming mycorrhizas with certain parasitic and decomposer fungi effectively parasitize them, at least up until the stage at which photosynthesis becomes established (if it does) (Arditti, 1979; Alexander and Hadley, 1985). Clearly, once fungal communication channels are in place, the scope for complex interplays between ecosystem components multiplies.

Avenues to Decay

Estimates of biomass in woodland and forest ecosystems commonly fall within the range 200–500 t ha^{-1}, of which more than 90% consists of woody material (Swift *et al.*, 1979). Probably about 40% of annual net primary production in forests is woody (Swift, 1977). By any standards, wood is therefore a major component of the biosphere and the recycling of its store of energetic and mineral resources is among the planet's most important ecosystem processes. Not only do fungi play a predominant role in such recycling, but the abundance and relative durability of wood ensure that it provides sustenance for fungal communities that are rich both in the complexity of their organization and the diversity of their species.

The importance of fungi as agents of bioconversion of wood may be traced to the fundamental nature of this tissue as an array of lignocellulose-bounded conduits which serve to connect the photosynthesizing canopy with subterranean sources of mineral nutrients and water. What better means of invading such a plumbing system than a plumbing system of a smaller scale, a mycelium, able efficiently to infiltrate avenues of least resistance and powered by nutrients extractable from the very walls of those avenues? Moreover, those enzymes, peroxidases and phenoloxidases, which, via the generation of free radicals, enable mycelial systems to degrade the lignocellulosic boundaries of woody conduits may also be those which permit them to insulate and deinsulate their own boundaries (Rayner *et al.*,

1994b). In degrading woody cell walls, fungi cause a variety of types of decay. Although these decays range across a spectrum with respect to the sites and rates of degradation and woody cell wall components predominately affected, they are commonly, and perhaps too rigidly, allocated to one or other of three distinct categories: white rot, brown rot and soft rot (e.g. Blanchette *et al.*, 1985; Rayner and Boddy, 1988).

White rot involves degradation of lignin and bleaching of the wood fibre. It is sometimes further subdivided according to whether lignin and cellulosic components are degraded at equivalent rates or lignin is degraded preferentially. Simultaneous white rot results in a macroscopically homogeneous pattern of decay, whereas selective delignification results in a heterogeneous alternation of sound and decayed tissue known as white pocket rot. Individual species (and individual genotypes) of decay fungi may cause one or other or both types of degradation.

Brown rot involves the preferential degradation of cellulosic components, leaving lignin relatively unaffected. Degradation occurs in the inner part (S2 layer) of secondarily thickened walls, away from the immediate vicinity of hyphae, implicating the diffusion of a small molecule, hydrogen peroxide, rather than local enzyme action, in the initiation of degradation (Koenigs, 1974).

Soft rot, at least in its classical sense, also predominantly involves degradation of cellulosic components, but is localized to the immediate vicinity of hyphae growing, by means of a remarkable oscillatory process, within the woody cell walls, so forming spiralling chains of spindle-shaped cavities (Hale and Eaton, 1985).

Whereas white rot and brown rot have traditionally been regarded as the province of basidiomycetes, and soft rot predominantly attributed to ascomycetes and mitosporic fungi, there are reasons for viewing this generalization as oversimplified and artefactual (Rayner and Boddy, 1988).

Before any of these types of decay processes can be initiated, however, wood has to become accessible to fungal colonists. The complexity and diversity of its fungal communities arise largely because wood is by no means uniformly accessible and the variety of ways in which it can be made accessible result in distinctive patterns and consequences of colonization.

Colonizing the Wood of Standing Trees

In natural woodlands, the majority of wood that enters the decomposition cycle probably does so while it is still a part of standing trees (Swift, 1977; Rayner and Boddy, 1988). Here, conditions in functional sapwood are inimical to fungal colonization probably due primarily to the high moisture contents which cause oxygen-diffusion paths to be greatly extended, so inhibiting fungal respiration and extracellular enzyme action (Rayner and

Boddy, 1988; Rayner, 1993b). Nonetheless, some fungi are able to multiply in functional tissues, particularly if they can produce a yeast phase which is distributed by the sap stream, as is testified by the occurrence of vascular wilt diseases, most notoriously Dutch elm disease caused by *Ophiostoma novo-ulmi* (Brasier, 1991).

For other fungi, including those that cause decay, dysfunction of the xylem as a water-carrying tissue is a prerequisite for actively exploitative mycelial growth. Such dysfunction can be brought about in a variety of ways, so defining distinctive modes of fungal establishment that have been characterized as a set of five 'colonization strategies'; heartrot, unspecialized opportunism, specialized opportunism, active pathogenesis and desiccation tolerance (Rayner and Boddy, 1988; Rayner, 1993b). It needs to be appreciated that only in the case of active pathogenesis can the dysfunction and resultant colonization patterns be regarded unequivocally as detrimental to the fitness of individual trees. Under other circumstances they may even have beneficial effects, quite apart from being the instrument of fungal biodiversity.

Heartrot, the decay of the inner cylinder of non-conductive heartwood which arises either progressively or abruptly through the cavitation (air embolism) of sapwood conduits, is a characteristic condition in mature ('post-mature', from the forester's perspective) trees. Despite the dysfunction, conditions in heartwood have a strongly selective influence on prospective fungal colonists. This is due to the accumulation of fungitoxic phenolic and terpenoid compounds and the occurrence of gaseous regimes liable to have relatively high contents of gases such as ethylene and carbon dioxide and low content of oxygen. Opportunities for gaining access to heartwood may also be restricted by the presence of intact layers of sapwood and bark. Given the resulting relative lack of competitors, individual mycelia of heartrot fungi may in time become very extensive, and give rise to correspondingly large fruit bodies on the outside of the tree, such as those of *Inonotus dryadeus* illustrated in Fig. 13.3. Individual species of heartrot fungi also often exhibit strong preferences for particular host trees, as *I. dryadeus* does for oak (*Quercus* spp.).

The activity of heartrot fungi results in hollowing of the trunk, while leaving an intact zone of functional sapwood around the outside. In itself, this may be more detrimental for the timber merchant than the tree, as decay of the non-functional tissue allows recycling of the resources it contains (Janzen, 1976) and creates a light cylinder resistant to storm damage. It is not uncommon to find roots proliferating within a hollowed interior, and decays caused by certain heartrot fungi, e.g. *Fromitopsis pinicola* provide favourable habitats for development of ectomycorrhizas and nitrogen-fixing bacteria (Sinclair *et al.*, 1987). Practices such as pollarding, by exposing the inner wood to colonization, may actually encourage heartrot and so prolong the life of the tree! Many of Britain's characteristically most ancient trees are indeed hollow pollards (Rackham, 1986).

Fig. 13.3. Fruit bodies of the heartrot fungus, *Inonotus dryadeus*, at the base of an oak tree (*Quercus robur*). (From Rayner and Boddy, 1988.)

Death or removal of bark, due to disease or injury, exposes underlying sapwood to aeration and drying, and so to invasion by a variety of unspecialized opportunistic fungi that tend to lack host selectivity. On the other hand, environmental stresses and internal competition resulting in systemic withdrawal of water supplies to the whole or parts of trunks and branches are thought to activate exploitative mycelial development of specialized opportunists. The latter are considered to establish endophytically (Boddy and Griffith, 1989), in fully functional tissues, and often exhibit high degrees of host selectivity as well as producing extensive individual mycelial genotypes, e.g. the basidiomycete, *Piptoporus betulinus*, in birch and the ascomycete, *Daldinia concentrica*, in ash. Depending on circumstances, such fungi may have a beneficial role in the 'natural pruning' (Peace, 1962) of redundant branches or contribute to rapid declines in trees

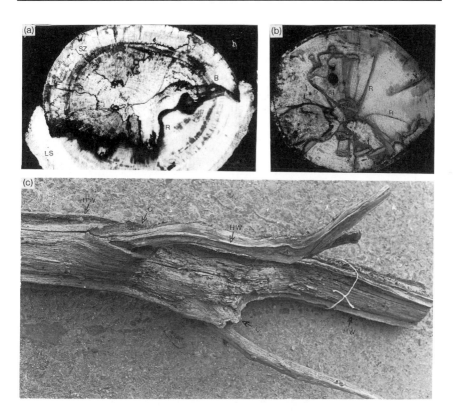

Fig. 13.4. Patterns resulting from the dynamic interplay between dysfunction, sealing off, fungal and animal invasion in attached tree branches. (a) Transverse section from the base of a partly living beech (*Fagus sylvatica*) branch in which there have been two rounds of damage to bark, the second resulting in loss of bark cover. A reaction zone (R) demarcates decayed wood from living sapwood (LS). An old barrier zone (B) formed following initial damage is traversed in one place by a fungal zone line (SZ) which passes straight across the barrier, and in another place by a line (DZ) which is deflected by the barrier. (From Rayner and Boddy, 1988.) (b) Transverse section through a beech branch showing complex patterns of reaction zone formation (R) and animal invasion (AI) (From Rayner and Boddy, 1988.) (c) Part of a long-dead branch of oak (*Quercus robur*) from which all the bark and sapwood has been lost. The residue consists of original heartwood and associated ridges ('heartwood wings', HW) and collars (C) of induced heartwood formed at boundaries between living and dysfunctional sapwood respectively within and at junctions between branches of different order. (From Boddy and Rayner, 1981.)

predisposed by environmental stresses.

For both kinds of opportunism, the extent of colonization depends on the degree of spread of dysfunction, which depends in turn on the rate and extent to which the tree is able to seal off functional tissues, primarily through deposition of water-repellent layers. These layers are referred to as reaction zones if formed in wood extant at the time of dysfunction, and as barrier zones when they isolate tissues formed subsequently. The spatio-temporal location of these zones can be used to tell of the dynamic and sometimes complex historical relationships between dysfunction, isolation and patterns of decay and discoloration (Fig. 13.4). The zones are often described, somewhat prejudicially, as a defensive means by which trees 'compartmentalize' decay fungi (Shigo, 1984), especially in view of the fact that they happen to contain toxic phenolics and terpenoids. However, it may be better to regard them as having a reparative or sealant function, involving activation of generally protective rather than specifically defensive metabolic pathways (Rayner and Boddy, 1988; Rayner, 1993b), in much the same way as applies to fungi (see above). Whichever explanation is most apposite, the mechanisms that allow trees to isolate dysfunctional xylem tissues are critical to their survival and complex relationships with fungi.

Rather than depending on other factors, some fungi gain access to woody tissues as a direct consequence of their own ability to bring about dysfunction, as active pathogens. Some, such as *Armillaria* spp. and *Heterobasidion annosum*, are 'ectotrophic' root pathogens, gaining access to the wood cylinder by first killing bark and cambium. An equivalent mechanism in tree

Fig. 13.5. Mycelial bridges (arrowed) formed by *Hymenochaete corrugata* across contacts between infected and uninfected hazel (*Corylus avellana*) stems (left) and between a hazel stem and *Clematis vitalba* vine (right). (From Ainsworth and Rayner, 1990.)

canopies is exhibited by certain members of the *Hymenochaetaceae* (Ainsworth and Rayner, 1990; Fig. 13.5). Other fungi establish an inoculum base in heartwood, at wounds and in stress-predisposed tissues by other strategies, prior to broaching functional sapwood.

Finally, since major dysfunction leading to loss of bark cover will cause the moisture content of wood in affected trunks and branches to fluctuate with atmospheric humidity, fungi colonizing under these circumstances may have to withstand periodic desiccation. Little is known about this group, but they may have considerable ecological importance.

Colonizing Cut or Fallen Wood

Wood that becomes detached from standing trees may already contain established decay communities, as will be normal in unmanaged forest, in which case subsequent patterns of colonization will depend strongly on the ability of invasive fungi to replace residents. Alternatively, wood felled by storm or forester will lack resident communities, and so potentially be available to a wide range of colonists. In either case, the emphasis switches from the dynamic relationships between fungi and trees to the equally dynamic relationships between different fungi, both of the same and of different species (Rayner and Todd, 1979; Rayner *et al.*, 1987a). Decaying wood can therefore often be likened to a complex battleground where disputed territorial rights between different individuals are mapped by demarcation zones whose persistence or impermanence respectively reflect the relative stability and instability of the community (Fig. 13.6). The scope

Fig. 13.6. Surface (left) and underlying cross-section through a beech (*Fagus sylvatica*) log showing sporophores of *Coriolus versicolor* and mosaic pattern arising from inter- and intraspecific mycelial interactions in the wood. (Courtesy of David Coates.)

for diversity of identity, relationship and scale in these communities is epitomized by a group of microfungi which specialize in inhabiting the 'no-man's land' between warring basidiomycete neighbours (Rayner, 1976) and the occurrence of 'temporary parasites'. The latter include the basidiomycetes *Pseudotrametes gibbosa*, *Lenzites betulina* and *Phanerochaete magnoliae* which selectively take over territory pioneered by *Bjerkandera* spp., *Coriolus* spp. and *Datronia mollis*, respectively, paralleling the behaviour of certain ant species and strangler figs (Rayner et al., 1987b; Ainsworth and Rayner, 1991).

Among the most combative colonists of detached wood are certain mycelial cord-forming basidiomycetes, able to grow out into soil and so interconnect spatially discontinuous resource units (Thompson, 1984; Dowson et al., 1986). By their very nature, these fungi possess the ability to replace resident fungi established from air-borne spores in the canopy or soon after detachment, and they can indeed be introduced directly into woodland soil by means of wood block inocula (Dowson et al., 1988a). They form both migratory systems, which locate resources by active foraging in space, and persistent networks from which colonization can be effected whenever suitable units fall within range (Dowson et al., 1988b, 1989b). Their foraging patterns closely parallel those produced by, for example, army ants (Rayner and Franks, 1987), and cover a spectrum from short-range to long-range, depending on the probability of locating resources within the vicinity of an existing base. Where resources are frequent but small, a short-range strategy is apt whereby an initially dense, radiating search pattern is superseded by redistribution of effort away from unsuccessful paths and along and out from successful paths, as in *Hypholoma fasciculare*. Where resources are large but sparse, long-range foraging involves outgrowth of relatively few, independent search lines, as in *Phanerochaete velutina*. Generally, these distinctive foraging strategies beautifully illustrate the principles of maintaining an energy-efficient dynamic balance between exploration, exploitation, conservation and redistribution alluded to earlier.

Litter Clearance

Besides wood, a range of smaller debris – leaves, flowers, fruits, etc. – rains from the canopy to the woodland floor, either seasonally or continuously, to form a layer of litter. These remains would accumulate indefinitely if it were not for the activity of decomposer fungi that render them down to humus. The contribution of fungi to this process depends partly on the relationship between fungi and soil type. Trees such as ash and elm with palatable foliage tend to give rise to 'mull' soils rich in mineral nutrients and animal activity. The leaves have a short residence time at the soil surface before being incorporated, especially by earthworms, into humus, reducing the relative importance of mycelial fungi in decomposition. Trees with less palatable foliage produce soils with a more persistent litter layer in which mycelial fungi, with their penetrative organization and battery of lignin- and cellulose-degrading enzyme systems are essential to the decomposition process. As with wood-inhabiting forms, these litter fungi can be classified ecologically according to whether they are resource unit-restricted, with mycelial individuals being confined within the boundaries of discrete litter components, or non-unit-restricted if capable of producing migratory mycelium (Cooke and Rayner, 1984). Unit-restricted fungi are unable to commit resources to large fruiting structures and include both relatively ubiquitous and more specialized species, the latter including several small species of the agaric genera *Marasmius* and *Mycena* (Rayner *et al.*, 1985). Non-unit-restricted fungi form dense patches of mycelium, as with *Marasmius wynnei*, or fairy rings, as with *Clitocybe nebularis*. The latter can be thought of as having an extremely short-range foraging strategy among small, densely distributed resource units, and correspondingly come to 'avoid' bare patches of greater width than the mycelial annulus (Dowson *et al.*, 1989a).

Relationships with Animals

So far, fungal roles in woodland have been largely discussed in terms of their relationships with the living parts and dead remains of primary producers, i.e. plants. However, it is also important to appreciate that they have similarly varied relationships with animals, both through direct interactions and via their influence on supply of plant resources to herbivores (e.g. Anderson *et al.*, 1984). It is not possible to detail these relationships here, and so only a brief summary of some of the more significant direct interactions follows.

Fungi are used as food, either relatively casually or more selectively and in extreme cases involving actual cultivation, as by wood wasps, ambrosia beetles, macrotermites and attine ants. Fungi can grow nectrotrophically, biotrophically and saprotrophically within or upon the surface of animals, resulting in parasitic, neutralistic or mutualistic symbioses. Fungal decomposition of hard materials such as wood enables animals ranging from microfauna to nesting birds and mammals to invade. Such invasion may have dramatic effects on the composition and activity of the fungal communities themselves through comminution of the material and other factors. Animal faeces, urine and remains provide nutrient sources for fungi. In Japanese forests, the basidiomycete, *Hebeloma vinosophyllum*, produces its fruit bodies on the ground above buried mammalian carcasses and has even been suggested to provide a clue as to the whereabouts of murder victims (Sagara, 1976).

Conclusion

It should by now be clear that the functioning and biodiversity of woodland ecosystems in indeed strongly bound up with the ecological preferences, interconnectedness and diversity of their fungal inhabitants. Whereas mycorrhizal fungi feed the roots of the trees and other plants, the parasites and decomposers recycle the woodland's products. Moreover, woodlands are not unique in their dependence on fungal infrastructures, though it is here that these infrastructures reach their zenith of complexity. Other ecosystems, both terrestrial and aquatic, have similar dependencies (Carroll and Wicklow, 1992). The distributive systems of fungi are the mainstream of ecosystem function, not incidental to it.

References

Ainsworth, A.M. and Rayner, A.D.M. (1990) Aerial mycelial transfer by *Hymenochaete corrugata* between stems of hazel and other trees. *Mycological Research* 94, 263–266.

Ainsworth, A.M. and Rayner, A.D.M. (1991) Ontogenetic stages from coenocyte to basidiome and their relation to phenoloxidase activity and colonization processes in *Phanerochaete magnoliae*. *Mycological Research* 95, 1414–1422.

Alexander, C. and Hadley, G. (1985) Carbon movement between host and mycorrhizal endophyte during the development of the orchid, *Goodyera repens*. *New Phytologist* 101, 657–665.

Allen, M.F. (1991) *The Ecology of Mycorrhizae*. Cambridge University Press, Cambridge.

Anderson, J.M., Rayner, A.D.M. and Walton, D.W.H. (eds) (1984) *Invertebrate–Microbial Interactions*. Cambridge University Press, Cambridge.

Arditti, J. (1979) Aspects of the physiology of orchids. *Advances in Botanical Research* 7, 421–655.

Bääth, E. and Söderström, B. (1979) Fungal biomass and fungal immobilization of plant nutrients in Swedish coniferous forest soils. *Revue d'Ecologie et de Biologie du Sol* 16, 477–489.

Bevercombe, G.P. and Rayner, A.D.M. (1978) *Dichomera saubinetii* and bark-diamond canker formation in sycamore. *Transactions of the British Mycological Society* 71, 505–507.

Bevercombe, G.P. and Rayner, A.D.M. (1980) Diamond-bark diseases of sycamore in Britain. *New Phytologist* 86, 379–392.

Björkman, E. (1960) *Monotropa hypopitys* L. an epiparasite on tree roots. *Physiologia Plantarum* 13, 308–329.

Blanchette, R.A., Otjen, L., Effland, M.J. and Eslyn, W.E. (1985) Changes in structural and chemical components of wood delignified by fungi. *Wood Science and Technology* 19, 35–46.

Boddy, L. and Griffith, G.S. (1989) Role of endophytes and latent invasion in the development of decay communities in sapwood of angiospermous trees. *Sydowia* 41, 41–73.

Boddy, L. and Rayner, A.D.M. (1981) Fungal communities and formation of heartwood wings in attached oak branches undergoing decay. *Annals of Botany* 47, 271–274.

Brasier, C.M. (1991) *Ophiostoma novo-ulmi* sp. nov., causative agent of current Dutch Elm disease pandemics. *Mycopathologia* 115, 151–161.

Carroll, G.C. (1988) Fungal endophytes in stems and leaves: from latent pathogen to mutualistic symbiont. *Ecology* 69, 2–9.

Carroll, G.C. and Wicklow, D.T. (eds) (1992) *The Fungal Community*, 2nd edn. Marcel Dekker, New York.

Cooke, R.C. and Rayner, A.D.M. (1984) *Ecology of Saprotrophic Fungi*. Longman, London.

Dowson, C.G., Rayner, A.D.M. and Boddy, L. (1986) Outgrowth patterns of mycelial cord-forming basidiomycetes from and between woody resource units in soil. *Journal of General Microbiology* 132, 203–211.

Dowson, C.G., Rayner, A.D.M. and Boddy, L. (1988a) Inoculation of mycelial cord-forming basidiomycetes into woodland soil and litter. II. Resource capture and persistence. *New Phytologist* 109, 343–349.

Dowson, C.G., Rayner, A.D.M. and Boddy, L. (1988b) Foraging patterns of *Phallus impudicus*, *Phanerochaete laevis* and *Steccherinum fimbriatum* between discontinuous resource units in soil. *FEMS Microbiology Ecology* 53, 291–298.

Dowson, C.G., Rayner, A.D.M. and Boddy, L. (1989a) Spatial dynamics and interactions of the woodland fairy ring fungus, *Clitocybe nebularis*. *New Phytologist* 111, 699–705.

Dowson, C.G., Springham, P., Rayner, A.D.M. and Boddy, L. (1989b) Resource relationships of foraging mycelial systems of *Phanerochaete velutina* and *Hypholoma fasciculare* in soil. *New Phytologist* 111, 501–509.

Duddridge, J.A. (1985) A comparative ultrastructural analysis of the host–fungus interface in mycorrhizal and parasitic associations. In: Moore, D., Casselton, L.A., Wood, D.A. and Frankland J.C. (eds) *Developmental Biology of Higher Fungi*. Cambridge University Press, Cambridge, pp. 141–173.

Duddridge, J.A. and Read, D.J. (1982) An ultrastructural analysis of the development of mycorrhizas in *Monotropa hypopitys* L. *New Phytologist* 92, 203–214.

Garrett, S.D. (1970) *Pathogenic Root-Infecting Fungi*. Cambridge University Press, Cambridge.

Hale, M.D. and Eaton, R.A. (1985) Oscillatory growth of hyphae in wood cell walls. *Transactions of the British Mycological Society* 84, 277–288.

Harley, J.L. and Smith, S.E. (1983) *Mycorrhizal Symbiosis*. Academic Press, London.

Janzen, D.H. (1976) Why tropical trees have hollow cores. *Biotropica* 8, 110.

Jeffries, P. and Dodd, J.C. (1991) The use of mycorrhizal inoculants in forestry and agriculture. In: Arora, D.K., Rai, B., Mukerji, K.G. and Knudsen, G.R. (eds) *Handbook of Applied Mycology*, Vol. 1. Marcel Dekker, New York, pp. 155–185.

Koenigs, J.W. (1974) Hydrogen peroxide and iron: a proposed system for decomposition of wood by brown rot basidiomycetes. *Wood Fibre* 6, 66–79.

Molina, R. and Trappe, J.M. (1982) Patterns of ectomycorrhizal host specificity and

potential among Pacific Northwest conifers and fungi. *Forest Science* 28, 423–458.
Newman, E.I. (1978) Root microorganisms: their significance in the ecosystem. *Biological Reviews of the Cambridge Philosophical Society* 53, 511–554.
Peace, T.R. (1962) *Pathology of Trees and Shrubs.* Clarendon Press, Oxford.
Perry, D.A., Margolis, H., Choquette, C., Molina, R. and Trappe, J.M. (1989) Ectomycorrhizal mediation of competition between coniferous tree species. *New Phytologist* 112, 501–511.
Prosser, J.I. (1993) Growth kinetics of mycelial colonies and aggregates of ascomycetes. *Mycological Research* 97, 513–528.
Rackham, O. (1986) *The History of the British Countryside.* J.M. Dent, London.
Rayner, A.D.M. (1976) Dematiaceous hyphomycetes and narrow dark zones in decaying wood. *Transactions of the British Mycological Society* 67, 546–549.
Rayner, A.D.M. (1993a) The fundamental importance of fungi in woodlands. *British Wildlife* 4, 205–215.
Rayner, A.D.M. (1993b) New avenues for understanding processes of tree decay. *Arboricultural Journal* 17, 171–189.
Rayner, A.D.M. (1994) Pattern-generating processes in fungal communities. In: Ritz, K., Dighton, J. and Giller, K.E. (eds) *Beyond the Biomass: Compositional and Functional Analysis of Soil Microbial Communities.* John Wiley & Sons, Chichester, pp. 247–255.
Rayner, A.D.M. and Boddy, L. (1988) *Fungal Decomposition of Wood.* John Wiley & Sons, Chichester.
Rayner, A.D.M. and Franks, N.R. (1987) Ecological and evolutionary parallels between ants and fungi. *Trends in Ecology and Evolution* 2, 127–133.
Rayner, A.D.M. and Todd, N.K. (1979) Population and community structure and dynamics of fungi in decaying wood. *Advances in Botanical Research* 7, 333–420.
Rayner, A.D.M., Watling, R. and Frankland, J.C. (1985) Resource relations – an overview. In: Moore, D., Casselton, L.A., Wood, D.A. and Frankland, J.C. (eds) *Developmental Biology of Higher Fungi.* Cambridge University Press, Cambridge, pp. 1–40.
Rayner, A.D.M., Boddy, L. and Dowson, C.G. (1987a) Genetic interactions and developmental versatility during establishment of decomposer basidiomycetes in wood and tree litter. In: Fletcher, M., Gray, T.R.G. and Jones, J.G. (eds) *Ecology of Microbial Communities.* Cambridge University Press, Cambridge, pp. 83–123.
Rayner, A.D.M., Boddy, L. and Dowson, C.G. (1987b) Temporary parasitism of *Coriolus* spp. by *Lenzites betulina*: a strategy for domain capture in wood decay fungi. *FEMS Microbiology Ecology* 45, 53–58.
Rayner, A.D.M., Griffith, G.S. and Ainsworth, A.M. (1994a) Mycelial interconnectedness. In: Gow, N.A.R. and Gadd, G.M. (eds) *The Growing Fungus.* Chapman & Hall, London, pp. 21–40.
Rayner, A.D.M., Griffith, G.S. and Wildman, H.G. (1994b) Differential insulation and the generation of mycelial patterns. In: Ingram, D.S. (ed.) *Shape and Form in Plants and Fungi.* Academic Press, London, pp. 291–310.
Read, D.J. (1984) The structure and function of the vegetative mycelium of mycorrhizal roots. In: Jennings, D.H. and Rayner, A.D.M. (eds) *The Ecology*

and Physiology of the Fungal Mycelium. Cambridge University Press, Cambridge, pp. 215–240.

Read, D.J. (1992) The mycorrhizal community with special reference to nutrient mobilization. In: Carroll G.C., Wicklow, D.T. (eds) *The Fungal Community*, 2nd edn. Marcel Dekker, New York, pp. 631–652.

Richardson, M.J. (1970) Studies on *Russula emetica* and other agarics in a Scots pine plantation. *Transactions of the British Mycological Society* 55, 217–229.

Sagara, N. (1976) Presence of a buried mammalian carcass indicated by fungal fruiting bodies. *Nature* 262, 816.

Shigo, A.L. (1984) Compartmentalization: a conceptual framework for understanding how trees grow and defend themselves. *Annual Review of Phytopathology* 22, 189–214.

Sinclair, W.A., Lyon, H.H. and Johnson, W.T. (1987) *Diseases of Trees and Shrubs*. Comstock Publishing Associates, Cornell University Press, New York.

Smith, M.L., Bruhn, J.N. and Anderson, J.B. (1992) The fungus *Armillaria bulbosa* is among the largest and oldest living organisms. *Nature* 356, 428–431.

Speer, E.O. (1981) Recherches sur la biologie des champignons corticole des ligneux. PhD Thesis, l'Université Louis Pasteur de Strasbourg.

Swift, M.J. (1977) The ecology of wood decomposition. *Science Progress, Oxford* 64, 179–203.

Swift, M.J., Heal, O.W. and Anderson, J.M. (1979) *Decomposition in Terrestrial Ecosystems*. Blackwell Scientific, Oxford.

Thompson, W. (1984) Distribution, development and functioning of mycelial cord systems of decomposer basidiomycetes of the deciduous woodland floor. In: Jennings, D.H. and Rayner, A.D.M. (eds) *The Ecology and Physiology of the Fungal Mycelium*. Cambridge University Press, Cambridge, pp. 185–214.

MICROORGANISMS IN EXTREME ENVIRONMENTS   V

Molecular Biology of Alkaliphiles 14

T. HAMAMOTO AND K. HORIKOSHI

The Institute of Physical and Chemical Research (RIKEN), Wako, Saitama 351-01, Japan, and DEEPSTAR Program, Japan Marine Science and Technology Center, c/o The Institute of Physical and Chemical Research (RIKEN), Wako, Saitama 351-01, Japan.

Alkaliphiles are microorganisms which can grow under alkaline conditions (pH >10). Since the first alkaliphilic microorganisms were isolated in 1969 and an intensive study on them was started, these alkaliphiles have shown a number of interesting physiological characteristics and, in addition, have given rise to significant biotechnological applications (Horikoshi and Akiba, 1982; Horikoshi, 1991; Hamamoto and Horikoshi, 1992). This chapter describes the molecular biology of the mechanism of alkaliphily.

Physiology of Alkaliphilic Microorganisms

Internal pH

Most alkaliphiles have their optimal growth pH at around 10, which is the most significant difference from well-investigated neutrophilic microorganisms (Horikoshi and Akiba, 1982). Differences between internal pH and environmental pH affect the transportation of ions and substances through the cell membrane, one of the important parameters in maintaining vital biological activities. Since it is difficult to measure the internal pH of microorganisms directly, several indirect methods have been applied. One of the methods is to measure the distribution in the cells of compounds which are not actively transported by cells and which have an ionized form which is pH dependent. A weakly alkaline compound, methylamine, can be used for this purpose under alkaline pH. By calculating the distribution of methylamine, it has been shown that the alkaliphiles have a lower internal

Table 14.1. Internal pH and external pH of alkaliphilic microorganisms.

	Optimal growth pH	Internal pH*	External pH
Bacillus alcalophilus	10.5	7.4	7.0
		8.0	8.0
		7.6	9.0
		8.6	10.0
		9.2	11.0
Bacillus sp. no. 38-2	9.5	7.8–8.2	7–10

*Internal pH of alkaliphiles was measured by the distribution of methylamine.

Table 14.2. Optimal pH of the intracellular enzymes produced by alkaliphiles.

Enzyme	Optimal pH	Stable pH range	Producer
α-Galactosidase	7.5	7.5–8.0	Micrococcus sp. no. 31-2
β-Galactosidase	6.5	5.5–9.0	Bacillus sp. no. C-125

pH than that in the external environment (Guffanti et al., 1978; Kitada et al., 1982) (Table 14.1). The internal pH was maintained at around 8, in spite of a high external pH (8–11). In addition to this method, the internal pH can also be estimated from the optimal pH of the intracellular enzymes. For example, α-galactosidase from an alkaliphile, *Micrococcus* sp. no. 31-2, has an optimal catalytic pH at 7.5, suggesting that the internal pH is around neutral (Akiba and Horikoshi, 1976) (Table 14.2). Furthermore, the cell-free protein synthesis machinery, consisting of ribosomes and tRNA from alkaliphiles, incorporated amino acids into protein at pH 8.2–8.5 optimally (Ikura and Horikoshi, 1978), which was only 0.5 pH units higher than that of neutrophilic *Bacillus subtilis*. Therefore, one of the key features in alkaliphily is associated with the cell surface, which discriminates between the intracellular neutral environment and the extracellular alkaline environment.

Na^+ Ion and Membrane Transport

As mentioned above, alkaliphilic microorganisms grow vigorously at pH 9–11, and require Na^+ for growth. The presence of sodium ions in the surrounding environment has been shown to be essential for effective solute transport through the membranes of alkaliphilic *Bacillus* spp. (Kitada and Horikoshi, 1980a). According to the chemiosmotic theory, the proton motive force in the cells is generated by excreted H^+ derived from ATP metabolism by ATPase, or respiration. Then, the H^+ is reincorporated into the cells with cotransport of various substrates. In the case of Na^+-dependent transport systems, the H^+ is exchanged with Na^+ by Na^+/H^+ antiport systems, thus generating a sodium motive force which drives substrates accompanied by Na^+ into the cells (Krulwich, 1986). The incorporation of AIB (alpha-aminoisobutyrate, used as a non-metabolizable amino acid analogue) increased twice as the external pH shifted from 7 to 9, and the presence of sodium ion significantly enhanced incorporation; 0.2 M NaCl produced an optimum which corresponded to 20 times the rate of that observed in the absence of NaCl (Kitada and Horikoshi, 1980b). Other cations, including K^+, Li^+, NH_4^+, Cs^+, Rb^+, did not show any effect nor did their counter anions (Horikoshi, 1991).

In Vitro *Membrane Transport Systems*

It is generally accepted that Na^+ enhances the affinity between Na^+ carrier proteins and substrates in active transport systems. These carrier proteins are also required in the passive excretion of substrates from the cells. In alkaliphiles, the incorporation of AIB and serine was shown to be enhanced by the presence of an Na^+ concentration gradient through the membrane, as well as by the membrane electron potential in experiments with *in vitro* membrane particle systems (Kitada and Horikoshi, 1980a). Therefore, it is suggested that sodium motive force and sodium ion-dependent membrane transport systems function in both the incorporation of nutrients and in internal pH adjustment, whereas neutrophiles on the other hand utilize a proton motive force, in the same way. Since alkaline environments contain high concentrations of Na^+ ions in contrast to neutral environments, alkaliphiles may have Na^+/H^+ antiporter systems which share the same subunits with nutrient transport systems.

Flagella of Alkaliphiles

Some alkaliphiles, such as *B. alkalophilus*, *B. firmus* RAB, and *Bacillus* sp. no. 8-1, are motile by means of flagella. Motility caused by flagella is considered to be driven by a sodium motive force instead of a proton motive force as shown by neutrophiles (Hirota *et al.*, 1981). These alkaliphiles are most motile at pH 9–10.5, whereas no motility is observed at pH 7 and, in addition, they require Na^+ for motility. Motility even in the presence of Na^+ is inhibited after abolishing their membrane potential by adding ionophoric substances (Hirota *et al.*, 1981).

Outer Surface of Alkaliphiles

Cell Wall Composition

In spite of the external alkaline pH, the intracellular pH of alkaliphilic microorganisms is neutral (pH 7.5–8). Therefore, the pH difference must be due to cell surface components, namely, the cytoplasmic membrane and the cell wall. Since the protoplasts of alkaliphilic *Bacillus* strains lose their stability in alkaline environments (unpublished data), it is suggested that the cell wall may play a certain role in protecting the cell from alkaline environments. Components of the cell walls of several alkaliphilic *Bacillus* spp. have been investigated in comparison with those of the neutrophilic *Bacillus subtilis* (Horikoshi, 1991). In addition to the peptidoglycan which was essentially similar in composition to that of *B. subtilis*, alkaliphilic *Bacillus* spp. contain acidic compounds. Hydrolytic analysis of the acidic compounds yielded galacturonic acid, gluconic acid, glutamic acid, aspartic acid and phosphoric acid (Aono, 1987). A clear relationship was observed between cell wall composition and growth–pH characteristics. The alkaliphiles can be divided into three groups, as described in Table 14.3 (Hamamoto and Horikoshi, 1992).

Composition of the Peptidoglycans

The peptidoglycans of alkaliphilic *Bacillus* appeared to be similar to that of *Bacillus subtilis*. However, their composition was characterized by excess of hexosamines and amino acids in the cell walls. Glucosamine, muramic acid, D- and L-alanine, D-glutamic acid, *meso*-diaminopimeric acid, and acetic

Table 14.3. Cell wall composition of alkaliphilic *Bacillus* strains.

Group	Components of cell walls	Growth pH	Ion requirement
1	High in glucuronic acid, teichuronic acid, and hexosamine	No growth at pH 7	Na^+ (essential)
2	High in glutamic acid, aspartic acid, galacturonic acid and glucuronic acid	Capable of growth at pH 7	Na^+ (essential)
3	Presence of phosphoric acid, similar to *B. subtilis*	Capable of growth at pH 7 and 10 in the presence of Na^+ and K^+	

acid were found in the hydrolysate (Aono, 1987). Although some variation in amide content was observed among the peptidoglycans from alkaliphilic *Bacillus* strains, the variation in pattern was similar to that known for the neutrophilic *Bacillus* species.

Acidic Polymers in the Cell Walls

As described in Table 14.3 (Aono, 1989, 1990; Hamamoto and Horikoshi, 1992), the alkaliphilic *Bacillus* strains in group 1 contain teichuronic acid in the cell walls and cannot grow at neutral pH. On the other hand, the cell walls of group 2 strains contain large amounts of acidic amino acids and uronic acids. In contrast, the walls of group 3 organisms contain teichoic acid. Even though most of the strains in group 2 can grow at neutral pH, the amount of acidic amino acids and uronic acids found in cell walls from bacteria grown at neutral pH was much smaller than that from an alkaline pH. Therefore, it is likely that the acidic components in the outermost layer of the group 2 bacteria have a function in supporting growth at alkaline pH. The negative charges on the acidic non-peptidoglycan components may give the cell surface its ability to absorb sodium and hydronium ions and repulse hydroxide ions, and as a consequence, enable the cells to grow in alkaline environments.

Genetic Analysis of Alkaliphily

Characteristics of Alkali-sensitive Mutants

Two alkali-sensitive mutants of alkaliphilic *Bacillus* C-125 have been obtained by chemical mutagenesis (Kudo *et al.*, 1990). Since strain C-125 grows well in Horikoshi minimal medium (supplemented with 0.2% glutamate and 0.5% glycerol as nutrients in place of glucose/starch polypeptone and yeast extract) over the pH range of 7.5–10, it is convenient to use this strain to obtain mutants defective in alkaline growth. Moreover, the strain closely resembles *B. subtilis*, and genes from the strain are known to be expressed well in *Escherichia coli*. Therefore, the strain was considered suitable for genetic analysis in *E. coli*. The two mutants showed different properties from each other. One of the mutants, no. 18224, cannot grow above pH 9. The internal pH of the mutants, calculated by methylamine incorporation across the membrane, was 8.7 in the presence of Na_2CO_3. This value is close to that of the parent strain C-125 (pH 8.6). However, this mutant cannot maintain a low internal pH in the presence of K_2CO_3. It was also found that the parent alkaliphilic strain can maintain a low internal pH in the presence of Na_2CO_3 but not in the presence of K_2CO_3, whereas another mutant strain, 38154, was unable to sustain low internal pH in the presence of either Na_2CO_3 or K_2CO_3. The internal pH was 10.4, which was the same as that for all strains in the presence of K_2CO_3. To summarize, it is suggested that mutant 38154 was defective in the regulation of internal pH, whereas mutant 18224 appeared to show normal regulation of internal pH values and, at the same time, Na^+ ion plays some role in the pH homeostasis mechanism, either directly or indirectly. In the case of mutant 18224, inability to grow at alkaline pH is not because of its inability to generate a pH gradient but as a result of a defect in some other function.

Molecular Cloning of DNA Fragments Conferring Alkaliphily

We attempted to isolate DNA fragments that would transform alkaline-sensitive mutants into alkaliphilic strains (Kudo *et al.*, 1990). DNA of the parental strain was digested with *BcI*I, ligated into the *BcI*I site of plasmid

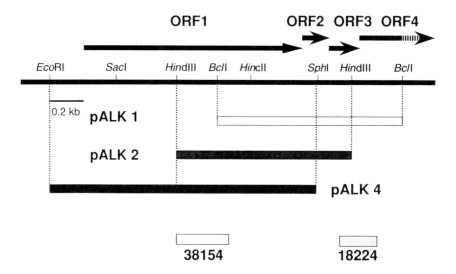

Fig. 14.1. Map of DNA fragments in chromosomal DNA which confer alkaliphily on alkali-sensitive mutants of alkaliphilic *Bacillus* strain C-125 and the locations of the smallest DNA fragments responsible for the recovery of alkaliphily in the mutants. The DNA fragment (3.7 kb) cloned from the chromosomal DNA of alkaliphilic *Bacillus* strain C-125 is shown with restriction enzyme sites above the corresponding cleavage sites. Regions of DNA corresponding to the cloned DNA fragments in pALK1, pALK2 and pALK4 are indicated with blank, dotted and black bars, respectively. Open reading frames in the DNA fragment are indicated by arrows including ORF4 which lacks the 3'-region. Subcloned DNA fragments which complement alkali-sensitive mutations are shown as bars with the number of the corresponding mutant.

pHW1, and used to transform the alkali-sensitive mutant, 18224. Transformants having alkaliphily harboured a recombinant plasmid (pALK1) containing a 2.0 kb DNA insert in the *Bcl*I site. Another 2 kb obtained by *Hin*dIII digestion could transform mutant 38154 to an alkaliphilic strain (carrying plasmid pALK2). Furthermore, transformation of mutant 38154 by pALK2 restored low internal pH levels to that of the parent strain. Southern hybridization and restriction mapping indicated about 1.6 kb of common sequence between pALK1 and pALK2 (Fig. 14.1).

Nucleotide Sequence of the Open Reading Frames Responsible for the Alkaliphily

Nucleotide sequence analysis of pALK1 and pALK2 revealed that the DNA fragment did not contain the 5'-region of one of the open reading frames in the fragments (unpublished data). Therefore, another recombinant plasmid, pALK4, which harboured the full open reading frame in a 3.0 kbp DNA fragment of the parent bacterial strain, was obtained. The whole

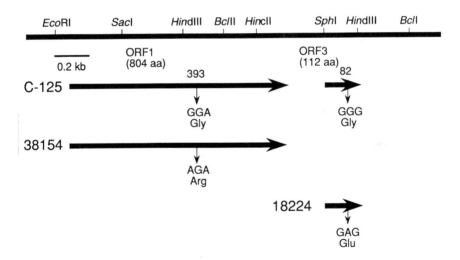

Fig. 14.2. Mutation responsible for the recovery of alkaliphily in the mutants. The nucleotide substitutions revealed by nucleotide sequencing of the chromosomal DNA of the parent strain C-125 and the corresponding DNA regions of the alkali-sensitive mutants amplified by polymerase chain reaction were shown. The single nucleotide substitution is shown with the resulting amino acid change in putative ORFs (arrows) in the parental strain and the mutants as indicated. The numbers above the amino acids indicate the numbers of amino acids from the N-termini of the products of the corresponding ORFs.

cloned DNA fragment contained three complete (ORF1-3) and an open reading frame (ORF4) which lacked the 3'-region (Fig. 14.1). The smallest DNA fragments for the recovery of two alkali-sensitive mutants were obtained by subcloning (Fig. 14.1). Nucleotide sequencing of the corresponding DNA fragments of the mutants was carried out by amplifying the corresponding DNA fragments by polymerase chain reactions. Comparison of the sequences revealed single nucleotide substitution mutations which were suspected to be the cause of the loss of alkaliphily in the mutants (Fig. 14.2). The whole ORF is not necessary to recover the alkali-sensitive mutations probably because homologous recombinations occur in the host mutant cells after introduction of the parental DNA.

The putative polypeptide deduced from preliminary determined nucleotide sequence of ORF1 is calculated to have a molecular weight of *ca.* 80 kDa and is extremely hydrophobic. Further analysis of the product may reveal the nature of alkaliphily.

References

Akiba, T. and Horikoshi, K. (1976) Properties of α-glucosidases of alkalophilic bacteria. *Agricultural Biological Chemistry* 40, 1851–1855.
Aono, R. (1987) Characterization of structural component of cell walls of alkalophilic strain of *Bacillus* sp. C-125. *Biochemical Journal* 245, 467–472.
Aono, R. (1989) Characterization of cell wall components of the alkalophilic *Bacillus* strain C-125: Identification of a polymer composed of polyglutamate and polyglucuronate. *Journal of General Microbiology* 135, 265–271.
Aono, R. (1990) The poly-α- and -β-1,4-glucuronic acid moiety of teichuronopeptide from the cell wall of the alkalophilic *Bacillus* strain C-125. *Biochemistry* 270, 363–367.
Guffanti, A.A., Susman, P., Blanco, R. and Krulwich, T.A. (1978) The proton-motive force and α-aminoisobutyric acid transport in an obligately alkalophilic bacterium. *Journal of Biological Chemistry* 253, 708–715.
Hamamoto, T. and Horikoshi, K. (1992) Alkaliphiles. In: Lederberg, J. (ed.) *Encyclopedia of Microbiology*. Academic Press, New York, pp. 81–87.
Hirota, N., Kitada, M. and Imae, Y. (1981) Flagellar motors of alkalophilic *Bacillus* are powered by an electrochemical potential gradient of Na^+. *FEBS Letters* 132, 278–280.
Horikoshi, K. (1991) *Microorganisms in Alkaline Environments*. VCH Verlag AG, Basel.
Horikoshi, K. and Akiba, T. (1982) *Alkalophilic Microorganisms. A New Microbial World*. Springer-Verlag, Berlin.
Ikura, Y. and Horikoshi, K. (1978) Cell free protein synthesizing system of alkalophilic *Bacillus* No. A-59. *Agricultural Biological Chemistry* 42, 753–756.
Kitada, M. and Horikoshi, K. (1980a) Sodium ion-stimulated amino acid uptake in membrane vesicles of alkalophilic *Bacillus* no. 8-1. *Journal of Biochemistry* 87, 1757–1764.

Kitada, M. and Horikoshi, K. (1980b) Further properties of sodium ion-stimulated α-(1-^{14}C)aminoisobutyric acid uptake in alkalophilic *Bacillus* species. *Journal of Biochemistry* 87, 1279–1284.

Kitada, M., Guffanti, A A. and Horikoshi, K. (1982) Bioenergetic properties and viability of alkalophilic *Bacillus firmus* RAB as a function of pH and Na$^+$ contents of the incubation medium. *Journal of Bacteriology* 152, 1096–1104.

Krulwich, T.A. (1986) Bioenergetics of alkalophilic bacteria. *Journal of Membrane Biology* 89, 113–125.

Kudo, T., Hino, M., Kitada, M. and Horikoshi, K. (1990) DNA sequences required for the alkalophily of *Bacillus* sp. strain C-125 are located close together on its chromosomal DNA. *Journal of Bacteriology* 172, 7282–7283.

15 Thermophilic Fungi in Desert Soils: A Neglected Extreme Environment

J. MOUCHACCA

Laboratoire de Cryptogamie, Muséum National d'Histoire Naturelle, 12, rue Buffon, 75005 Paris, France.

Introduction

Temperature is one of the extremely important environmental variables that play a key and decisive role in the survival, growth, distribution and diversity of microorganisms on the surface of the earth. Brock (1969, 1970) stressed that thermophilic microorganisms are inhabitants of extreme environments where the temperature rises above the prevailing normal temperature in the area. The increase in temperature may be caused by geothermal heat, microbial thermogenesis of organic-rich material and solar heating of substrates such as soils, litter and rock.

In most of these extreme habitats if other conditions such as pH, nutrients, etc. are appropriate, some photosynthetic and non-photosynthetic organisms can develop. Photosynthetic prokaryotes (blue–green algae and some bacteria) are usually unable to grow at temperatures as high as non-photosynthetic chemolithotrophic and heterotrophic bacteria (Tansey and Brock, 1978). The upper temperature limit for photosynthetic microorganisms is thus about 70–73°C. The upper temperature limit for eukaryotic microorganisms is even lower, approximately 60–62°C, at which only a few species of fungi can grow.

Temperature and Fungi: Some Definitions

The response of fungi to temperature varies between the two extremes of obligatorily thermophilic through thermotolerant to psychrophilic species.

Table 15.1. Minimum and maximum temperatures for growth of some thermophilic and thermotolerant fungi.

	Temp. (°C)
Thermophilic	
Thermomyces lanuginosus[a]	30–60
Thermoascus crustaceus[b]	30–58
Acremonium alabamense[b]	30–55
Acremonium strictum[b]	30–55
Talaromyces thermophilus[a]	27–59
Chaetomium thermophile[a]	27–58
Malbranchea cinnamomea[a]	27–56
Thielavia terrestris[b]	27–55
Talaromyces emersonii[b]	27–55
Melanocarpus albomyces[a]	26–57
(*Melanocarpus albomyces*)[b]	18–52
Rhizomucor miehei[a]	25–57
Scytalidium thermophilum	23–58
(= *Humicola grisea* var. *thermoidea*)[a]	24–56
(= *Humicola insolens*)[a]	23–55
Thermoascus aurantiacus[a]	22–55
Thermomyces stellatus	22–50
Rhizopus micro. var. *microsporus*[b]	22–55
Rhizomucor pusillis[a]	20–55
Thermotolerants	
Paecilomyces variotii[b]	17–50
Talaromyces leycettanus[b]	17–55
Rhizopus micro. var. *rhizopodiformis*[b]	15–55
Aspergillus quadrilineatus[b]	15–55
Aspergillus rugulosus[b]	15–55
Aspergillus fumigatus[a]	12–52
Mesophile	
Rhizopus nigricans[b]	8–30
Psychrophile	
Microdochium nivale	2–32

Based on [a]Cooney and Emerson (1964) and [b]Bokhary *et al.* (1984).

However, by far the majority of fungi known to develop in culture are mesophiles growing at between 5 and 37°C; the psychrophiles extend below that range of temperatures (Table 15.1).

As stressed by Hedger (1974), any discussion on thermophilic fungi must first underline the adopted definition of thermophilism. This is because a multiplicity of terminologies has been proposed. Apinis (1963) first classified fungi into three groups: microthermophilic species with an optimum temperature for growth between 25°C and 35°C and maximal temperature exceeding 40°C but not 48°C, e.g. *Byssochlamys nivea* and *Corynascus sepedonium*; psychrotolerant thermophiles with a wide temperature range but all growing well at 48°C, e.g. *Absidia corymbifera* and *Aspergillus fumigatus*; and thermophilic species with optimum temperature for growth between 40°C and 50°C and maxima up to 60°C but unable to grow at room temperature of about 20°C, e.g. *Rhizomucor pusillus* and *Thermomyces lanuginosus*.

Apinis' definition of thermophily is very close to that proposed by Cooney and Emerson (1964): a thermophilic fungus is one in which growth can take place at a maximum temperature at or above 50°C, and a minimum at or above 20°C; a thermotolerant fungus is one that has a thermal maximum near 50°C and a minimum below 20°C. Crisan (1973) suggested a wider definition, including as thermophiles all fungi with growth optima of 40°C or higher, thus including many fungi excluded by Cooney and Emerson. Craveri *et al.* (1964) suggested that a thermophilic fungus be regarded as one having a minimum growth temperature above 25°C. On the other hand Evans (1971) considers four groups: strong thermotolerants (which grow sparingly below 20°C and grow well at or above 50°C); thermotolerants in general (grow well at temperatures below 20°C and very slowly at or above 50°C); strong thermophiles (with minimum growth temperatures 25°C or above, and a high maximum temperature); weak thermophiles (minimum growth temperature of 20°C or just above, and a maximum at 50°C or above).

The commonly accepted classification scheme of thermophilic and thermotolerant fungi is that of Cooney and Emerson based on the minimum and maximum growth temperature values. This very simple scheme is sometimes difficult to apply. According to data published by Bokhary *et al.* (1984) *Melanocarpus albomyces*, a well-established thermophile, should be considered thermotolerant; the situation being the reverse for *Acremonium strictum*, a well-known mesophile.

Thermophilic and Thermotolerant Fungi

The first report of a thermophilic hyphomycete is often attributed to Tsiklinskaya (1899) who isolated a fungus from a potato inoculated with garden soil and grew it on bread kept at 52–53°C. She named the fungus *Thermomyces (Humicola) lanuginosus*. Although the Lindt description of *Rhizomucor pusillus* was published earlier in 1886, it was Tsiklinskaya who drew attention to thermophilism in fungi. Later (Miehe, 1905) serious investigation of self-heating hay produced the first extensive report on thermophily in fungi. Miehe isolated and studied a range of thermophiles including *Thermoascus aurantiacus* and *Malbranchea cinnamomea*. Griffon and Maublanc (1911) described the first thermophilic member of the genus *Penicillium*, *P. dupontii*, now *Talaromyces thermophilus* La Touche (1950) reported a new species of *Chaetomium*, *C. thermophile*, which generated much interest due to its cellulolytic nature.

The first comprehensive account on the morphology, taxonomy, biology and economic importance of thermophilic fungi was published by Cooney and Emerson (1964). Eleven thermophilic fungi were reported. Since then the number of taxa developing at high temperatures is expanding rapidly. In 1973, Crisan provided a list of 55 names of thermophilous fungi, i.e. thermophilic and thermotolerant ones. According to Hedger (1974), only 32 could be considered thermophilic *sensu* Cooney and Emerson's definition. Later, Samson and Tansey (1977) prepared a guide to species that can grow and sporulate at 45°C. Based on taxonomic characters, this list concerns eight mucorales, 19 ascomycetes, two basidiomycetes and 21 hyphomycetes, a total of 48 taxa when considering recent taxonomic decisions. The list prepared by Tansey and Brock (1978) reports 67 species or varieties of fungi growing at 50°C or above but a good proportion of these taxa still need to be defined at the species level. Finally, according to Abdullah and Al-Bader (1990), approximately 70 species detected in various substrates are now reported to be thermophilic or thermotolerant. Such an extended list awaits confirmation.

The number of fungi able to grow at high temperatures is evidently very limited in comparison to the number of fungi known to develop in culture. The latter represents a small part of the number of known fungal species, recently estimated by Hawksworth (1991) at about 69,000 species. The size of the thermophilous fungi will surely rapidly expand, partly as upper temperature growth limits for known species are determined and as new isolations are made by workers interested in these particular fungi. Such interest is enhanced by industrial microbiologists who are continuously using thermophilic microorganisms.

Habitats of Thermophilic Fungi

The first two thermophilic fungi to be identified were chance contaminants on organic substrates incubated at high temperatures. In 1905, Miehe investigated the spontaneous heating and ignition of hay. This was a pioneering detailed study of thermophilic fungi and of the role of mesophilic and thermophilic microorganisms in the thermogenic combustion of organic-rich substrates. This discovery aroused interest in the existence and the metabolic activities of this particular group of fungi. Cooney and Emerson (1964) and Emerson (1968) provided an account of the diversity of habitats where thermophiles were encountered. In 1973, Crisan reviewed current concepts of thermophilism and underlined points of future research. In 1978, Tansey and Brock examined reports regarding habitats in which thermophilic fungi occur; in most of these reports thermotolerant species were also found. Finally, a concise overview of the role of thermophilic fungi in agriculture has been published recently (Shekkar Sharma and Johri, 1992).

The occurrence of thermophilic and thermotolerant fungi in habitats rich in organic material is well documented (Tansey and Brock, 1978; Shekkar Sharma and Johri, 1992). Reports on less widespread similar habitats and habitats deserving future investigation were also considered by Tansey and Brock (1978). Thermophilic and thermotolerant fungi have also been isolated from industrial products (Apinis and Eggins, 1966), living animals (Tansey, 1984), plant debris, vegetation and soil (Gochenaur, 1975).

Thermophilic Fungi and Soil

Preliminary evidence on the occurrence of thermophiles in soil derives from studies conducted in temperate regions (Apinis, 1963; Eggins and Malik, 1969; Tansey and Jack, 1976). If the occurrence of fungi thriving at high temperatures in self-heated composts is not remarkable, the presence of thermophilic taxa in soils of particularly temperate regions was rather unexpected. Observed frequencies seemed too high to be explained by simple dispersal of spores from other sites. This favoured the hypothesis that thermophilic fungi are, at some periods, active in those soils from which they can be isolated. In 1972, Griffin stated that the ecology of thermophilic fungi in soil is a promising field for further study.

In temperate regions, the main source producing high temperature environments is solar heat radiation (Geiger, 1965). By direct solar heating the soil surface temperature frequently rises in summer to levels allowing the development of thermophilic fungi. Apinis (1963, 1972) reported on the occurrence of such fungi in alluvial grasslands in England and the discovery of some new taxa. Eggins and Malik (1969) also recorded some thermophiles from a British pasture soil. Using immersion tubes, Eggins *et al.* (1972) provided evidence that thermophilic fungi grow in sun-heated soils. The effect of sun-heating on soil thermophiles was further investigated in south-central Indiana soils (Tansey and Jack, 1976). Thermophilic and thermotolerant fungi were isolated most frequently from sun-heated soils, less so from grass-shaded and still less so from tree-shaded soil.

In tropical and subtropical zones the top soil layers are regularly warmed up to above 35°C, enough to enable the development of detectable populations of thermophiles (Hedger, 1974). In arid zones soil surface temperatures of 50–70°C are frequently recorded (Ranzoni, 1968).

Research Conducted in Arid Zones

Desert rocks and soils provide fascinating habitats for a variety of microorganisms (Friedmann and Galun, 1974). The most unusual habitat of such hot dry areas is the so-called desert rock varnish. According to Krumbein and Jens (1981) rock varnish or desert lacquer is mainly produced by the activity of often lichenized epi- and endolithic cyanobacteria, chemoorganotrophic bacteria and fungi, in association with the still debatable *Metallogenium symbioticum*. Fungi of this habitat have received little attention. Recently the new dematiaceous genus *Halsiomyces* was described from rock material collected in the Arizona Sonoran Desert (Simmons, 1981).

Serious study of desert soil microorganisms followed the pioneer work of Killian and Féher (1939) on Saharan soils. Killian and Féher stressed that 'l'exploration des phénomènes biologiques dans ces sols présente un interêt de tout premier ordre. Par suite des conditions climatiques extrêmes, on y trouve réalisé un milieu biologique qui entrave souvent la vie microbienne' and concluded 'les sols désertiques, considérés autrefois comme stériles, renferment des microorganismes à l'état de vie active'. In 1952, addressing the First Symposium on Desert Research, Thornton reviewed particular problems related to the study of arid soils micropopulations, characterized by high prevalent temperatures, lack of apparent water, and high salt levels, all major features of that environment.

Some significant preliminary contributions on fungi of desert soils appeared in 1960. Durrell and Shields (1960) investigated soils of the Navada Test Site. Forty mesophilic taxa were isolated with black-spored species predominating in isolation plates; this observation is however an artefact resulting from the high weight of soil inoculum employed. They suggested that in strongly insolated regions hyphal melanin pigment affords a degree of protection against ultraviolet light. Borut (1960) reported on the arid northern Negev soils; these revealed low population levels with a flora (60 taxa) composed mostly of cosmopolitan elements that need not be looked upon as specialized. A similar conclusion was also reached by Nicot (1960) in her analysis of Saharan desert sands.

Ranzoni (1968) surveyed the mycobiota of virgin Sonoran Desert soils (USA). He identified over 200 species from plates incubated at room temperature. In terms of species numbers, taxa were more numerous in soils with higher fungal density. He noted that 'no fungus grew on any of the isolation plates incubated at temperatures over 45°C, although *Rhizopus chinensis* was the only fungus that appeared on the 45°C plates'. According to Ranzoni, the Sonoran fungi are capable of surviving high soil temperatures although no thermophilic fungi in the Cooney and Emerson sense were found. A serious analysis of soil mycoflora of Peru was later published by Gochenaur (1970) which provided data on the distribution and floristic content of microfungi inhabiting natural plant communities inclusive of localities from the coastal desert, one of the driest regions on earth. The distribution of certain fungal taxa was found to be regulated by pH and temperature. *Penicillium* and *Trichoderma* species were prominent in cool, acid soils whereas aspergilli were abundant in warm dry alkaline soils.

Most of the subsequent major reports on fungi of arid zones deal with soils located in the Middle East, part of the greath North African and Arabian desert continuum. For soils submitted to arid stress conditions in other geographic areas, some reports of the check-list type have been produced. Reporting on the role of microfungi in ecosystems, Kjoller and Struwe (1982) stressed that in contrast to the other biomes only few international biology projects studied fungal occurrence in desert soils.

Work Done in the Middle East

For the Middle East, data on soil fungi have accumulated over the last two decades. However, analysis of published information emphasizes that further research is needed for a comprehensive account of the distribution of thermophilic and thermotolerant fungi in the diverse habitats of that region. For example for countries such as Syria, Jordan, Kuwait and Iraq only one published report on that topic could be located. For the Saudi Arabian Peninsula and Egypt only two for each are available. Regarding Egypt, this finding is surprising, as a large number of publications appear annually from that country. The soil mycobiota of North African countries other than Egypt still await exploration for their thermophilous and even for their mesophilic components.

The publication by Moustafa *et al.* (1976) on the thermophilous fungi of Kuwait, essentially a hot arid desert, is apparently the first comprehensive account. They provided compound data records for 40 taxa isolated from 100 desert but mostly salt-marsh soils. Analysis was by using both direct soil and dilution techniques with plates incubated at 43°C. Thermophilous fungi constitute a very small part of the soil population (in terms of cfu g^{-1} soil) as compared with the relatively higher counts of their mesophilous components, an observation in agreement with that reported by Apinis (1972) for temperate soils. No comparative data are, however, presented on the qualitative variation of the mycobiota and the population structure of Kuwaiti soils with the increase in incubation temperature. Such is the case for all examined Middle East reports.

For Syrian soils, similar compound data are reported by Abdel-Hafez *et al.* (1983) for 40 mostly cultivated soils. Using the dilution plate method with both glucose and cellulose media and a 45°C incubation temperature, data were collected on the total colony counts and number of cases of isolation of the 40 identified taxa. The same analytical procedure was applied by Moubasher et al. (1981) to Jordanian soils. Aggregate data from 100 mostly cultivated soils are reported for fungi developing at 45°C. Thermophilous population densities vary from 40 to 6000 cfu g^{-1} soil with aspergilli accounting for the largest number of colonies appearing in processed plates. More recently Abdullah and Al-Bader (1990) examined 200 Iraqi soils by various isolation techniques: soil plating and dilution, heat and alcohol soil treatment, three different media at 45°C. Compound data on the occurrence of the 35 identified species were provided. In addition, for each fungus production of amylase, cellulase, protease, and lipase enzymes was also assessed.

The thermophilous mycoflora of the Saudi Arabian Peninsula was first documented by Abdel-Hafez (1982). Desert soils (40) were examined with 18 plates each, six for each of the three media used, incubated at 45°C. A check-list of almost 50 species resulted from this analysis. Bokhary et al. (1984) also studied these fungi in that country and tried to compare their abundance in various substrata and find information on their seasonal fluctuations. Using the soil plate technique and a 46°C incubation temperature, 25 species were found to inhabit all sites examined with total abundance ranging from 670 cfu (flower garden) to 16 cfu for a bare desert soil. Abdulla and El-Gindy (1987) reported on soils of Jezan; at 45°C only three aspergilli were isolated. Three thermophilic species were recently isolated from soil underneath *Tirmania* native truffles in Saudi Arabia (Bokhary and Sarwat Parvez, 1992).

In Egypt, Sabet (1935) undertook a pioneering study of common fungi inhabiting some arid localities with the isolation of *Penicillium egyptiacum*. Numerous papers have since been published on crop pathogens and fungi active in agricultural soils as compared to soils of strictly arid areas in Egypt. The first account on thermophilous fungi relates to 74 salt-marsh soils along the Western Mediterranean coast and natural or cultivated sites in the Western Desert Oasis (Abdel-Fattah et al., 1977). Clay samples of the southerly oasis soils disclosed high fungal density values (200–300 cfu) as compared to the northern sandy salt-marshes. Compound data for the distribution of 20 identified taxa are provided. A short recent report deals with cultivated and bare areas exposed to cement dust particles near Assiut (Hemida, 1992). The analysis of 40 different sites provided data on a limited group of species. Data were not correlated with the impact of cement dust particles.

Thermophiles and Thermotolerants in the Middle East Soils

Work conducted in the Middle East suggests that such fungi inhabit all investigated biotopes from cultivated to bare land surfaces. In every report a limited number of thermophiles is usually observed with a comparatively larger number of thermotolerants. In quantitative terms of fungal density a significant reduction follows the shift from cultivated to soils devoid of natural plant cover. However, from such published compound data no ecological correlations can be extracted relating the fungi observed to features of the soil they inhabit.

For two countries both available reports were grouped to establish one list of fungi per country irrespective of soil nature. Overall, 19 thermophilic taxa were observed; these comprise elements from all major taxonomic groupings. Type material of *Thermophymatospora fibuligera* originates from Iraq where the isolation of *Thermoascus aegyptiacus* is the second record for that ascomycete (Table 15.2). The presence of these two recently described taxa favours the hypothesis that the Middle East thermophilic mycobiota comprises unknown interesting taxa. Their discovery is probably simply hindered by the lack of local taxonomic expertise as reflected by the absence of soil mycologists in Syria and Jordan. Tansey and Jack (1976) stressed that careful study of thermophilous fungi present in a particular heated habitat can usually be expected to yield undescribed taxa.

Four rather rare thermophiles were recorded in a single country. On the other hand, *Scytalidium thermophilum* (inclusive of *Humicola* spp.), *Thermomyces lanuginosus* (the first described thermophilic fungus) and *Myceliophthora thermophila* (anamorph of *Corynascus heterothallicus*) show the largest geographic distribution. *Malbranchea cinnamomea* produces distinctive yellow-coloured cultures. *Rhizomucor pusillus* exhibits a wider distribution than *R. miehei*; the first has often been confused with *R. miehei* which is regularly homothallic.

Talaromyces thermophilus, *Chaetomium thermophile*, *Melanocarpus albomyces* and *Thermoascus aurantiacus* form an intermediate group present in four or five countries. The first two ascomycetes proved to be rare taxa in all reports and such is the case with the majority of other thermophiles. For example *Thermoascus aurantiacus* was observed in only two out of 200 Iraqi soils sampled. Thermophilic fungi seem to represent elements having comparatively reduced numbers of propagules in soil. This would account for the restricted observed distribution in soils of every country. Huang and Schmitt (1975) also reported low concentrations in terms of cfu for the thermophiles of some Ohio soils.

In 1976, Tansey and Jack compared the thermophiles originating from sun-heated soils with species reported in 34 previous studies of temperate and tropical soils. No taxon was cited in more than 20 reports. A further 12

Table 15.2. Thermophilous fungi recorded in major studies on soil in the Middle East.

	SY	IR	KU	SA	JO	EG
Thermophiles						
Scytalidium thermophilum	ooo	o	oo	oo	oo	o
Thermomyces lanuginosus	o	oo	o	o	o	o
Myceliophthora thermophila	o	o	o	o	o	o
Malbranchea cinnamomea	o	o	–	o	oo	o
Chaetomium thermophile	o	–	o	o	o	o
Melanocarpus albomyces	o	–	–	o	oo	oo
Talaromyces thermophilus	o	–	–	o	o	o
Thermoascus aurantiacus	o	o	–	oo	o	–
Rhizomucor pusillus	o	–	–	oo	o	o
Myceliophthora fergusii (Corynascus thermophilus)	o	o	–	o	–	–
Myriococcum thermophilum	–	oo	–	o	–	–
Thielavia terrestris (Acremonium alabamense)	–	–	o	o	–	–
Rhizomucor miehei	–	oo	–	oo	–	–
Chaetomium thermophile var. coprophile–	–	–	–	oo	–	–
Chaetomium thermophile var. dissitum	–	–	–	oo	–	–
Talaromyces emersonii	–	–	–	oo	–	–
Thermoascus crustaceus	–	–	–	o	–	–
Thermoascus aegyptiacus	–	o	–	–	–	–
Thermophymatospora fibuligera	–	o	–	–	–	–
Thermotolerants						
Aspergillus fumigatus	ooo	ooo	ooo	ooo	ooo	ooo
A. terreus	oo	ooo	ooo	ooo	oo	oo
A. niger	oo	ooo	oo	oo	oo	oo
A. nidullelus	oo	oo	oo	oo	oo	oo
A. candidus	o	oo	o	o	–	o
A. flavus	o	–	oo	oo	–	o
A. nidulans var. latus	o	–	–	o	o	o
A. quadrilineatus	o	–	–	o	o	o
A. carneus	–	–	oo	o	o	o
A. alutaceus	o	–	o	o	–	o
A. egyptiacus	o	–	o	o	–	–
A. sydowii	o	–	o	o	–	–
A. flavipes	–	–	oo	oo	o	–
A. versicolor	oo	–	–	o	–	o
A. ustus	o	–	–	o	–	o
A. fischeri	–	–	–	oo	–	o
A. rugulosus	–	–	–	o	oo	–
A. violaceus	–	–	–	o	o	–
Paecilomyces variotii	ooo	o	o	o	o	o
Chaetomium olivaceum	o	–	o	o	–	o

Table 15.2. continued

	SY	IR	KU	SA	JO	EG
Chaetomium globosum	o	–	–	o	–	o
Acrophialophora fusispora	o	–	–	o	o	–
Sepedonium chrysospermum	o	–	–	o	–	o
Absidia corymbifera	–	o	o	o	–	–
Rhizopus stolonifer var. stolonifer	o	–	–	o	–	o
Chaetomium bostrychodes	o	–	–	o	–	–
Scopulariopsis brevicaulis	o	–	–	o	–	–
Chaetomium atrobrunneum	–	o	o	–	–	–
Corynascus sepedonium	–	o	–	–	o	–
Mucor circinelloides	–	–	o	o	–	–

Thermophilous fungi observed in only one country in the Middle East

SYRIA (SY): *Byssochlamys* sp., *Acremonium* sp., *Aspergillus terreus* var. *africa*, *Cladosporium herbarum*. Total: 4 taxa.

IRAQ (IR): *Cunninghamella echinulata, Mycotypha africana, Rhizopus* sp., *Byssochlamys verrucosa, Chaetomium subcurvisporum, Emericella similis, Talaromyces* sp., *Thielavia* sp., *Acrophialophora levis, Cladosporium* sp., *Gilmaniella macrospora, Myceliophthora* sp., *Penicillium* sp., *Torula terrestris, Trichoderma* sp. Total: 15 taxa.

KUWAIT (KU): *Actinomucor elegans, Cunninghamella phaeospora, Achaetomium* ? *strumarium, Arachniotus dankaliensis, Talaromyces flavus, Thielavia arenaria, T. microspora, T. terricola, Aspergillus glaucus* gp, *A. fischeri* var. *spinosus, Acremonium* sp., *Penicillium* ? *crustosum, Phoma* sp., Hyphomycete ×2. Total: 13 taxa.

SAUDI ARABIA (SA): *Rhizopus microsporus* var. *microsporus, R. microsporus* var. *rhizopodiformis, Achaetomium* sp., *Talaromyces leycettanus*, ? *Sphaerospora saccata, Acremonium strictum, Geotrichum* sp., *Humicola* sp., *Penicillium* ? *argillaceum, P. piceum, Phialophora* sp., *Scolecobasidium* sp., *Scopulariopsis brevicaulis, Sphaerosporium* sp., *Sporotrichum pruinosum* (= *S.* ? *pulverulentum*). Total: 15 taxa.

JORDAN (JO): *Syncephalastrum racemosum, Byssochlamys nivea, Eurotium* sp., *Neurospora crassa; Aspergillus nidulans* var. *echinulatus, A. nidulans* var. *dentatus, A. terreus* var. *aureus, Geotrichum candidum, Scopulariopsis candida, Sporotrichum carnis, Sporotrichum roseolum, Trimmatostroma* sp., *Torula herbarum, Macrophomina phaseoli*. Total: 14 taxa.

EGYPT (EG): *Aspergillus flavus* var. *columnaris, Emericella nivea, Humicola grisea*. Total: 3 taxa.

Frequency of occurrence: ooo, high; oo, moderate; o, rare.

or so thermophiles occur commonly in soil and another 12 or so are less common but can be detected by persistent searches. The 12 widespread species all figure among thermophiles recorded in the Middle East. A comparison of Tansey and Jack's observations with frequencies disclosed in the Middle East gives the following:

Thermomyces lanuginosus	18/34; 6 countries
Thermoascus aurantiacus	14/34; 4 countries
Talaromyces thermophilus	14/34; 4 countries
Rhizomucor pusillus	14/34; 4 countries
Malbranchea cinnamomea	12/34; 5 countries
Thielavia heterothallica	11/34; 6 countries
Thielavia terrestris	8/34; 2 countries
Melanocarpus albomyces	8/34; 4 countries
Scytalidium thermophilum	7/34; 6 countries
Rhizomucor miehei	5/34; 2 countries
Chaetomium thermophile var. *coprophile*	4/34; 1 country
Chaetomium thermophile var. *dissitum*	3/34; 1 country
Thermomyces stellatus	2/34; 0 country

A good agreement in species frequencies derived from both sources is apparent. The remaining five Middle East taxa are probably members of the less common group (Tansey and Jack, 1976). Such agreement suggests a cosmopolitan distribution of these specialized fungi. However, taxa unidentified to the species or even generic level are constantly reported in every published study suggesting that elements with limited distribution are not unlikely to occur. Progress in the taxonomy of these fungi is needed to ensure rapid and sound identification of thermophiles for a proper understanding of their ecological characteristics.

As expected, the thermotolerant group has a far larger number of taxa with most exhibiting limited distribution associated with rare frequencies of occurrences (Table 15.2). A major observation concerns the genus *Aspergillus* represented by 25 species, a size equivalent to that of reported ascomycetes and the latter would outnumber, when considering aspergilli that develop in culture, their *Emericella* teleomorphs. A second observation relates to the omnipresence of four aspergilli with moderate to high frequencies of occurrence in all reports, *A. fumigatus* being abundant everywhere, less so for *A. terreus*, *A. niger* and *A. nidulans*.

A. fumigatus was omnipresent in the 40 Saudi Arabian desert sites samples and in isolation plates examined its colonies amounted to half of all aspergilli or to 15.7% of total colonies counted (Abdel-Hafez, 1982). For Jordan, it appeared in 88/100 cultivated soils and its observed colonies attain 20.4% of the genus colonies or 13.9% of the total count (Moubasher et al., 1981). Such observations are in agreement with the biology of *A. fumigatus*, a well-known thermotolerant with a wide range of growth temperature for good growth. Although the fungus is not limited to habitats with relatively high temperatures, it is more commonly found in subtropical and tropical regions. The prevalence of *A. terreus*, *A. niger*, *A. nidulans* and *A. candidus* in the Middle East soils coincides with their physiological characteristics (Domsch *et al.*, 1980).

Paecilomyces variotii is a well-known thermotolerant with an obvious concentration in warm climates and arid regions; in scanned reports it proved to be abundant only in Syrian cultivated soils, suggesting it requires comparatively higher humidity levels. Among the 11 less widespread thermotolerants are five ascomycete species, four of *Chaetomium* and *Cornyascus sepedonium*. Finally the species observed in only one country (64 taxa) include 24 ascomycetes. Gochenaur (1975) noted that for the arid coconut grove investigated most abundant thermotolerant forms were of *Aspergillus* and *Chaetomium*. Additional knowledge of this group of fungi will accumulate when further soil thermophilous elements are investigated.

Characteristics of Arid Soil Fungal Communities

A thorough analysis of some communities of Kharga Oasis soils was recently undertaken. This oasis is the most important depression in the Western Desert of Egypt. Three sites without vegetation were sampled: E3 is a grey soil at 32 km south of Kharga City, E11 is a reddish ferruginous soil at 35 km and E1 a mobile sand dune on the eastern border of the depression. Using the direct soil plate method, large batches of inoculated Petri dishes were respectively incubated at 25°C, 35°C and 45°C. The weight of the soil inoculum was adjusted to give at maximum a mean number of 15 cfu plate^{-1} at 25°C. Values of relative abundances with regard to the fungal density (FD) were calculated; frequencies of occurrences were also noted. For each soil, developing taxa were grouped according to their presence at tested temperatures. A second grouping based on taxonomic characters was also realized. Selected levels were either generic or of higher taxonomic rank depending on the numerical importance of the total relative contributions (RC) of their components.

Table 15.3 groups data on the composition of the population observed at both high temperatures with relative abundance values; data for the 25°C level were omitted. For each soil, the first group of fungi comprises strong thermotolerants persisting along the temperature gradient 25°C–45°C. Less-tolerant fungi are again present only at 35°C (second group). Some species absent at 25°C appear at 35°C (third group) and sometimes also for some or all again at 45°C (fourth group). These mesophilic and thermophilic fungi add to the fungal spectrum observed at 25°C; their numbers vary from 9 (E1), 13 (E3) to 25 (E11). Finally in Table 15.3, the difference in the species numbering sequence represents mesophiles that have not developed above 25°C.

Table 15.3. Fungi of Kharga soils observed on plates incubated at high temperatures.

	35°C	45°C
Grey soil E3		
1 Aspergillus quadrilineatus	33.20	32.91
2 Thielaria subthermophila	13.88	15.88
3 Aspergillus niger	12.07	13.76
4 Rhexothecium globosum	10.27	4.14
5 Chaetomium jodhpurense	8.72	11.74
6 Chaetomium cymbiforme	6.46	2.51
7 Aspergillus fruticulosus	4.01	5.78
8 Emericella desertorum	1.48	3.66
9 Rhizopus stolonifer	1.42	0.96
10 Aspergillus nidulans	0.32	1.36
11 Aspergillus terreus	0.77	1.15
12 Aspergillus nidulans var. latus	0.26	0.19
13 Aspergillus fumigatus	0.26	0.48
14 Aspergillus flavus	0.06	0.19
15 Paecilomyces variotii	0.06	0.19
TOTAL Rel. Ab.	93.34	94.90
16 Chaetomium olivaceum	1.42	
17 Desertella globulifera	1.42	
18 Absidia heterospora	1.03	
19 Thielavia arenaria	0.71	
20 Chaetomium coarctatum	0.32	
21 Chaetomium globosum	0.26	
22 Chaetomium perlucidum	0.19	
23 Drechslera spicifera	0.19	
24 Podospora faurelii	0.06	
TOTAL Rel. Ab.	5.60	
SUM Rel. Ab.	98.94	
84 Sordaria fimicola	0.19	
85 Chaetomium aureum	0.06	
86 Chaetomium bostrychodes	0.06	
87 Chaetomium senegalensis	0.06	
88 Neotestudina rosatii	0.06	
89 Acremonium furcatum	0.06	
90 Trichothecium roseum	0.06	
91 Aspergillus egyptiacus	0.06	
92 Melanocarpus albomyces*	0.45	0.48
93 Rhizomucor miehei*		2.80
94 Rhizomucor pusillus*		0.67
95 Thermomyces lanuginosus*		0.67

83 Chaetomium thermophile var. coprophile*		0.48
TOTAL Rel. Ab.	1.06	5.10
SUM Rel. Ab.	100.00	100.00
No of replicates	146	84
No of species	33	20
Fungal density	1550	1040
Thermophilic Rel. Ab.	0.45	5.10

Red soil E11

1 Aspergillus flavus	18.90	2.05
2 Aspergillus niger	7.59	9.23
3 Aspergillus fumigatus	5.00	7.69
4 Sterile S1	3.89	1.08
5 Thielavia subthermophila	2.96	2.05
6 Chaetomium senegalense	1.85	2.05
7 Thielavia arenaria	1.48	2.05
TOTAL Rel. Ab.	41.67	26.15
8 Penicillium funiculosum	8.53	
9 Sterile S2	2.22	
10 Rhizopus stolonifer	1.11	
11 Rhexothecium globosum	1.11	
12 Thielavia sp.	0.74	
13 Aspergillus quadrilineatus	0.74	
14 Talaromyces flavus var. flavus	0.37	
15 Aspergillus nidullelus	0.37	
16 Saccobolus truncatus	0.37	
TOTAL Rel. Ab.	15.56	
SUM Rel. Ab.	57.23	
58 Aspergillus egyptiacus	0.74	
59 Corynascus setosus	0.74	
60 Cunninghamella echinulata	0.74	
61 Syncephalastrum racemosum	0.74	
62 Talaromyces trachyspermus	0.37	
63 Cladosporium herbarum	0.37	
64 Penicillium restrictum	0.37	
65 Aspergillus carneus	0.37	
66 Chaetomium indicum	0.37	
67 Wardomyces anomalus	0.37	
68 Doratomyces columnaris	0.37	
69 Thielavia microspora	0.37	
70 Paecilomyces variotii	7.96	1.03
71 Chaetomium jodhpurense	0.74	3.08
72 Chrysosporium luteum	0.74	2.05
73 Arachniotus ruber	0.37	1.03
74 Corynascus sepedonium	0.37	2.05
75 Chaetomium sp.	0.37	1.03

Table 15.3. continued

	35°C	45°C
76 Thermomyces lanuginosus*	2.22	
77 Thermoascus aurantiacus*	0.37	
78 Thielavia terrestris*	10.01	23.58
79 Melanocarpus albomyces*	5.00	17.95
80 Sterile S7*	5.00	7.69
81 Talaromyces byssochlamydioides*	2.59	2.05
82 Talaromyces emersonii*	0.74	5.64
83 Talaromyces thermophilus*	0.37	6.67
TOTAL Rel. Ab.	42.77	73.85
SUM Rel. Ab.	100.00	100.00
No of replicates	97	93
No of species	42	19
Fungal density	540	195
Thermophilic Rel. Ab.	24.06	55.89
Dune E1		
1 Aspergillus niger	15.62	3.20
2 Rhexothecium globosum	9.28	2.40
3 Paecilomyces variotii	7.56	7.60
4 Absidia heterospora	7.56	3.20
5 Aspergillus terreus	6.34	10.00
6 Sterile S2	3.90	1.60
7 Aspergillus quadrilineatus	3.90	5.20
8 Aspergillus nidulans	3.17	10.00
9 Aspergillus fumigatus	3.17	9.20
10 Thielavia subthermophila	3.17	2.40
11 Chrysosporium luteum	2.44	0.80
12 Thielavia arenaria	1.95	4.40
13 Chaetomium jodhpurense	1.22	0.80
TOTAL Rel. Ab.	69.28	60.80
14 Chaetomium cymbiforme	6.34	
15 Drechslera spicifera	5.61	
16 Drechslera halodes	1.22	
17 Chaetomium indicum	0.73	
18 Microascus cirrosus	0.73	
19 Chrysosporium indicum	0.73	
TOTAL Rel. Ab.	15.36	
SUM Rel. Ab.	84.64	
94 Petriella setifera	1.22	
95 Lasiobolidium orbiculoides	1.22	
96 Neotestudina rosatii	0.73	

97 *Thielavia microspora*	0.73	
98 Sterile S7*	1.22	
99 *Chaetomium thermophile**	0.73	
100 *Thermomyces lanuginosus**	5.12	10.80
101 *Rhizomucor pusillus**	1.95	4.40
102 *Melanocarpus albomyces**	1.22	3.20
103 *Rhizomucor miehei**	1.22	20.80
TOTAL Rel. Ab.	15.36	39.20
SUM Rel. Ab.	100.00	100.00
No of replicates	77	105
No of species	29	17
Fungal density	410	250
Thermophilic Rel. Ab.	11.46	39.20

Asterisks indicate thermophilic species.

With regard to estimated FD values, those for soil E3 are numerically more important. Based on the rate of decrease of this characteristic, soil E3 community exhibits higher thermoresistance capacities. Increase in temperature brings also a reduction in the number of species observed at each level. However, in this study high numbers of species recorded for each population temperature has to be correlated with the large numbers of plates processed: 100–150 plates.

For the grey soil E3, the relative abundance of the first group of fungi almost totally represents FD values at both temperatures; the five thermophilic fungi present at 45°C exhibit very low abundance values and frequencies of occurrence. The former thus accounts for the thermoresistance capacities of this soil community. For the red soil E11, the FD at 35°C results almost equally from the contributions of the first group and the 25 additional elements of the third group, of which eight are thermophiles; at 45°C, relative abundance of the latter accounts for a large part of the FD. Clearly soil E11 thermoresistance capacities derive mainly from the large spectrum of its thermophilic fungi with high abundance levels. The sand community compares with soil E11 regarding FD but is closer to soil E3 with respect to group contributions. Its thermoresistance capacities originate in equal proportions from the behaviour of some strong thermotolerants and a reduced spectrum of not very abundant thermophilic species.

The second step would be to underline elements accounting for the evident differential response of each soil community. This could be achieved by considering contributions of the taxonomic groupings. Eight such groupings have been established (Table 15.4). Soil E3 population at 25°C is clearly dominated by the relative importance of the genus *Aspergillus* and the ascomycetes followed by the *Penicillium* and dematiaceous species (Fig. 15.1); at higher temperatures, both dominant groupings accentuate their

Table 15.4. Thermotolerant and thermophilic fungi of Kharga soils abundant at 45°C and relative contributions of major taxonomic groupings.

Thermotolerant/thermophilic fungi abundant at 45°C				Relative contributions of major taxonomic groupings			
Soil	E1	E11	E3			25°C	45°C
Rhizomucor pusillus*	4.40			Aspergillus	E1	9.57	37.60
Rhizomucor miehei*	20.80				E11	35.27	18.97
Thermomyces lanuginosus*	10.80				E3	37.95	55.82
				Ascomycetes	E1	7.66	13.20
Aspergillus nidulans	10.00				E11	15.77	69.23
Aspergillus terreus	10.00				E3	31.00	38.89
Absidia heterospora	3.20						
Thielavia arenaria	4.40			Penicillium	E1	37.08	–
Paecilomyces variotii	7.60				E11	37.41	–
Aspergillus fumigatus	9.20	7.69			E3	20.71	–
Aspergillus niger	3.20	9.23	13.76				
Chaetomium jodhpurense	–	3.08	11.74	Dematiaceous	E1	37.90	10.80
Aspergillus quadrilineatus	5.20	–	32.91		E11	6.88	–
					E3	7.38	0.67
Melanocarpus albomyces*	3.20	17.95					
Sterile S7*		7.69					
Thielavia terrestris*		23.58					
Talaromyces emersonii*		5.64					
Talaromyces thermophilus*		6.67					
Thielavia subthermophila			15.88				
Rhexothecium globosum			4.14				
Emericella desertorum			3.66				
Aspergillus fruticulosus			5.78				
No of abundant species	12	8	7				
No of observed species	17	19	20				
Total Rel. Ab.	92.00	78.53	87.87				

Asterisks indicate thermophilic species.

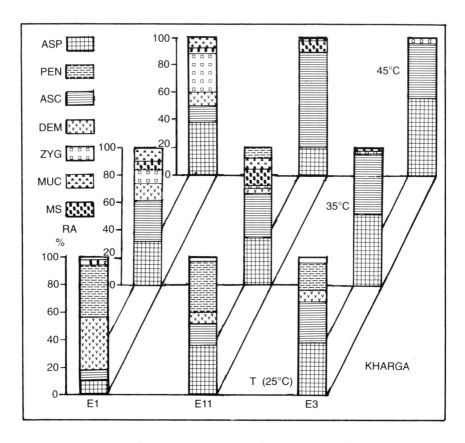

Fig. 15.1. Variation in the structure of Kharga soil community populations with the increase of plate incubation temperature. (ASP: *Aspergillus*; PEN: *Penicillium*; ASC: Ascomycetes; DEM: *dematiaceous*; ZYG: Zygomycetes; MUC: *mucedinaceous*; MS: *mycelia sterilia*).

respective contributions which finally represent the 45°C FD, with aspergilli being more preponderant. For soil E11, the penicillia and aspergilli dominate at 25°C; then come the ascomycetes and dematiaceous fungi. The genus *Penicillium* is absent at 45°C and the dominant position of the aspergilli is gradually reduced whereas that of the ascomycetes shows a reverse trend.

For the dune E1, penicillia and dematiaceous fungi are the two dominant basic groups; the former however disappear rapidly whereas the latter exhibit a gradual decrease in their contribution. The genus *Aspergillus* and some *Mucorales* then consolidate their relative positions, becoming dominant at 45°C. It is thus apparent that both soils have in common the high basic RC of the genus *Aspergillus*. Also in both cases the aspergilli and the ascomycetes (initially non-dominant in E11) accentuate their relative position. The sand dune is mainly characterized by the high RC of its dematiaceous species and penicillia but these fungi exhibit no thermoresistance abilities; such abilities are expressed by few aspergilli and *Mucorales*, which become relatively dominant at high temperature.

The final step would be to ascertain which species are responsible for the divergent behaviour of the two soil communities and their relationships with the sand dune community. Relative abundance data (Table 15.3) confirm that RC of groups of fungi derive mainly from corresponding values of a few members that proved to be abundant. By neglecting respective values below 3%, total contributions of species abundant at 45°C largely account for estimated FD values (Table 15.4). It is thus evident that the thermoresistance capacities of soil E3 result from the relative abundance in the sample analysed of propagules of four thermotolerant ascomycetes and three aspergilli. For soil E11, the same derives from the relative abundance of propagules of thermophilic ascomycetes and three high thermotolerants, among the latter only *Aspergillus niger* is also abundant in soil E3. In the dune, apart from *Melanocarpus albomyces*, three other distinctive thermophiles are abundant, with *A. fumigatus* being also abundant in E11 and *A. quadrilineatus* in E3.

Species abundant at high temperature form a group of 21 elements, the majority of which are ascomycetes (nine species) and aspergilli (six species); also thermotolerant species slightly outnumber thermophiles: respectively 13 and 8 species. No mesophilic penicillia or dematiaceous hyphomycetes are present. The distribution pattern of these fungi confirms that the thermoresistance capacities of soils E3 and E11 originate from the relative abundance of a few aspergilli, thermotolerant and thermophilic ascomycetes, not all simultaneously abundant in both soils. The relative abundance of thermophilic ascomycetes and *A. fumigatus* in soil E11 might be related to the ferruginous nature of that soil where the heating effect would be too severe for mesophilic thermotolerant taxa. In the dune community a balanced contribution is evident among its 12 abundant fungi, distinctive by the presence of mucedinaecous, dematiacous and mucoralean fungi.

Soils Submitted to Arid Stress Conditions: a Model for Mycological Analysis

The distribution patterns of species abundant at high temperatures stress that soils from the same locality differing in their physical and chemical characteristics possess different spectra of abundant fungi. The response of this small number of taxa clearly accounts for the different behaviour disclosed by each community in our analysis of the effect of increase in incubation temperature. Significant differences in quantitative and qualitative characteristics describing soil fungal communities can thus be seen if an appropriate analytical procedure is applied. This should of course involve the examination of an adequate sample of soil.

Kjoller and Struwe (1982), reporting on the occurrence and the activity of fungi in litter and soil, stressed that arid regions should provide excellent conditions for providing information on the difference between fungi in rhizosphere and non-rhizosphere soils. In arid regions all biotopes are submitted to climatic conditions generally unfavourable for the development of all plant groups. Under such conditions the impact of the microclimate is intensified and this sharply contrasts neighbouring dissimilar habitats. Arid soil fungal communities are mainly characterized by low fungal density values coupled with a high species biodiversity. They thus represent a simplified model of the complex soil fungal community system. Study of fungi inabiting arid zones should provide a better understanding of the ecology of these microscopic soil cummunities.

Biotechnological Potential of Thermophilic Fungi

Pure culture studies of thermophilic fungi have provided clear evidence that they possess a battery of extracellular enzymes able to hydrolyse polymers such as starch, protein, pectin, hemicellulose and cellulose. They have also been reported to produce among others a number of antibacterial and antifungal antibiotics, extracellular phenolic compounds and organic acids (Satyanarayana *et al.*, 1992). Some thermophilic fungi are already used in industries involving food processing and bioconversion of organic materials.

Fungi developing only at high temperatures offer several advantages for their use in industry. This is mainly due to their production of unusually thermostable enzymes. Such enzymes have a high degree of stability in the presence of detergents, aqueous organic solvents and reduced loss of activity during storage and purification. Mass cultivation of thermophilic fungi precludes contamination problems encountered with mesophiles with a concomitant reduction in investment costs. For such reasons the discovery of new thermophilic species from habitats submitted to arid stress conditions is promising for exploring future industrial applications.

References

Abdel-Fattah, M., Moubasher, A.H. and Abdel-Hafez, S.I. (1977) Studies on mycoflora of salt marshes in Egypt. III. Thermophilic fungi. *Bulletin Faculty of Sciences, Assiut University* 6, 225–235.

Abdel-Hafez, S.I.I. (1982) Thermophilic and thermotolerant fungi in the desert soils of Saudi Arabia. *Mycopathologia* 80, 15–20.
Abdel-Hafez, S.I.I., Abdel-Hafez, A.I.I. and Abdel-Kader, M.I.A. (1983) Composition of the fungal flora of Syrian soils. IV. Thermophilic fungi. *Mycopathologia* 81, 77–181.
Abdulla, M. El-S. and El-Gindy, A.A. (1987) Mesophilic and thermotolerant fungi in soil of Jezan, Saudi Arabia. *Zentrallblatt für Mikrobiologie* 142, 143–147.
Abdullah, S.K. and Al-Bader, S.M. (1990) On the thermophilic and thermotolerant mycoflora of Iraqi soils. *Sydowia* 42, 1–7.
Apinis, A.E. (1963) Occurrence of thermophilous microfungi in certain alluvial soils near Nottingham. *Nova Hedwigia* 5, 57–78.
Apinis, A.E. (1972) Thermophilous fungi in certain grasslands. *Mycopathologia et Mmycologia Applicata* 48, 63–74.
Apinis, A.E. and Eggins, H.O.W. (1966) *Thermomyces ibadanensis* sp. nov. from oil palm kernel stacks. *Transactions of the British Mycological Society* 49, 629–632.
Bokhary, H.A. and Sarwat Parvez (1992) Soil mycoflora from truffle native areas of Saudi Arabia. *Mycopathologia et Mycologia Applicata* 118, 103–107.
Bokhary, H.A., Sabek, A.M., Abu-Zinada, A.H. and Fallatah Jow, M.O. (1984) Thermophilic and thermotolerant fungi of arid regions of Saudi Arabia: occurrence, seasonal variation and temperature relationships. *Journal of Arid Environments* 7, 263–274.
Borut, S.H. (1960) An ecological and physiological study of soil fungi of the Northern Negev (Israel). *Bulletin of the Research Council of Israel* 8D, 65–80.
Brock, T.D. (1969) Microbial growth under extreme conditions. *Symposium of the Society of General Microbiology, London* 19, 15–41.
Brock, T.D. (1970) High temperatures systems. *Annual Review of Ecology and Systematics* 1, 191–220.
Cooney, D.G. and Emerson, R. (1964) *Thermophilic Fungi. An Account of Their Biology, Activities and Classification*. W.H. Freeman, San Francisco, London, 188 pp.
Craveri, R., Manachini, P.L. and Craveri, A. (1964) Eumiceti termofili presenti nel suols. *Annali di Microbiologia ed Enzimologia* 14, 13–26.
Crisan, E.V. (1973) Current concepts of thermophilism and the thermophilic fungi. *Mycologia* 65, 1171–1198.
Domsch, K.H., Gams, W. and Anderson, T.H. (1980) *Compendium of Soil Fungi*, Vols 1 and 2. Academic Press, London, 405 pp.
Durrell, L.W. and Shields, L.M. (1960) Fungi isolated from soils of the Nevada Test Site. *Mycologia* 52, 636–641.
Eggins, H.O.W. and Malik, K.A. (1969) The occurrence of thermophilic cellulolytic fungi in a pasture land soil. *Antonie van Leeuwenhoek* 35, 178–184.
Eggins, H.O.W., von Szilvinyi, A. and Allsopp, D. (1972) The isolation of actively growing thermophilic fungi from insolated soils. *International Biodeterioration Bulletin* 8, 53–58.
Emerson, R. (1968) Thermophiles. In: Ainsworth, G.C. and Sussman, A.S. (eds) *The Fungi*, Vol. 3. Academic Press, New York, pp. 105–128.
Evans, H.C. (1971) Thermophilous fungi of coal spoil tips. I. Taxonomy. *Transactions of the British Mycological Society* 57, 241–254.
Friedmann, I. and Galun, M. (1974) Desert algae, lichens and fungi. In: Brown,

G.W. Jr (ed.) *Desert Biology 2*. Academic Press, New York, pp. 166–213.
Geiger, R. (1965) *The Climate near the Ground*, 4th edn. Harvard University Press, Cambridge, MA.
Gochenaur, S.E. (1970) Soil mycoflora of Peru. *Mycopathologia et Mycologia Applicata* 42, 259–272.
Gochenaur, S.E. (1975) Distributional patterns of mesophilous and thermophilous microfungi in two Bahamian soils. *Mycopathologia et Mycologia Applicata* 57, 155–164.
Griffin, D.M. (1972) *Ecology of Soil Fungi*. Chapman & Hall, London.
Griffon, E. and Maublanc, A. (1911) Deux moisissures thermophiles. *Bulletin de la Société Mycologique de France* 27, 68–74.
Hawksworth, D.L. (1991) The fungal dimension of biodiversity: magnitude, significance, and conservation. *Mycological Research* 95, 641–655.
Hedger, J.H. (1974) The ecology of thermophilic fungi in Indonesia. In: Kilbertus, G., Reisinger, O., Mourey, A. and Cancela da Fonseca, J.A. (eds) *Biodegradation et Humification I*. Pierron, Sarreguemines, pp. 59–65.
Hemida, S.K. (1992) Thermophilic and thermotolerant fungi isolated from cultivated and desert soils exposed continuously to cement dust particles in Egypt. *Zentralblatt für Mikrobiologie* 147(3–4), 277–281.
Huang, L.H. and Schmitt, J.A. (1975) Soil microfungi of central and southern Ohio. *Mycotaxon* 3, 55–80.
Killian, C. and Féher, D. (1939) Recherches sur la microbiologie des sols désertiques. *Encylopédie Biologique* 21.
Kjoller, A. and Struwe, S. (1982) Microfungi in ecosystems: fungal occurrence and activity in litter and soil. *Oikos* 39, 391–422.
Krumbein, W.E. and Jens, K. (1981) Biogenic rock varnishes of the Negev Desert (Israel): an ecological study of iron and manganese transformation by cyanobacteria and fungi. *Oecologia* 50, 25–38.
La Touche, C.J. (1950) On a thermophilic species of *Chaetomium*. *Transactions of the British Mycological Society* 33, 94–104.
Miehe, H. (1905) Uber die Selbsterhitzung der Heues. *Arbeiten der Deutschen Landwirtschaftsgesellschaft* 111, 76–91.
Moubasher, A.H., Mazen, M.B. and Abdel-Hafez, A.I.I. (1981) Some ecological studies on Jordanian soil fungi. V. Thermophilic fungi. *Naturalia Monspeliensa* 45, 1–8.
Moustafa, A.F., Sharkas, M.S. and Kamel, S.M. (1976) Thermophilic and thermotolerant fungi in the desert and salt-marsh soil of Kuwait. *Norwegian Journal of Botany* 23, 213–220.
Nicot, J. (1960) Some characteristics of the microflora in desert sands. In: Parkinson, D. and Waid, J.S. (eds) *The Ecology of Soil Fungi*. Liverpool University Press. Liverpool, pp. 94–97.
Ranzoni, F.V. (1968) Fungi isolated in culture from soils of the Sonoran Desert. *Mycologia* 60, 356–371.
Sabet, Y. (1935) A preliminary study of the Egyptian soil fungi. *Bulletin of the Faculty of Science, Egyptian University* 5, 1–29.
Samson, R.A. and Tansey, M.R. (1977) Guide to thermophilic and thermotolerant fungi. *Abstracts, Second International Mycological Congress*, Tampa, Florida, p. 5.
Satyanarayana, T., Johri, B.N. and Klein, J. (1992) Biotechnological potential of

thermophilic fungi. In: Arora, D.K., Elander, R.P. and Mukerjii, K.G. (eds) *Handbook of Applied Mycology*, Vol. 4. Marcel Dekker, New York, pp. 729–761.

Shekkar Sharma, S.H. and Johri, B.N. (1992) The role of thermophilic fungi in agriculture. In: Arora, D.K., Elander, R.P. and Mukerjii, K.G. (eds) *Handbook of Applied Mycology*, Vol. 4. Marcel Dekker, New York, pp. 707–728.

Simmons, E.G. (1981) *Halsiomyces*, a new dematiaceous genus from Arizona's Sonoran desert. *Mycotaxon* 13, 407–411.

Tansey, M.R. (1984) Effective isolation of thermophilic and thermotolerant mucoralean fungi. *Mycopathologia et Mycologia Applicata* 85, 31–42.

Tansey, M.R. and Brock, T.D. (1978) Microbial life at high temperatures: ecological aspects. In: Kushner, D.J. (ed.) *Microbial Life in Extreme Environments*. Academic Press, London, pp. 159–216.

Tansey, M.R. and Jack, M.A. (1976) Thermophilic fungi in sun-heated soils. *Mycologia* 68, 1061–1075.

Thornton, H.G. (1952) Some problems presented by the microbiology of arid soils. *Desert Research, Proceedings of the First International Symposium*, Jerusalem. Weizmann Science Press, Jerusalem, pp. 295–300.

Tsiklinskaya, P. (1899) Sur les mucédinées thermophiles. *Annales de l'Institut Pasteur* 13, 500–505.

Biodiversity of the Rock Inhabiting Microbiota with Special Reference to Black Fungi and Black Yeasts

16

C. Urzì[1], U. Wollenzien[2], G. Criseo[1] and W.E. Krumbein[2]

[1] *Istituto di Microbiologia, Facoltà di Scienze, Università di Messina, Italy;* [2] *AG Geomikrobiologie, ICBM, Carl von Ossietzky Universität, Oldenburg, Germany.*

Introduction

Following the definition of Whittaker (1975) the term diversity is defined by the number of species in a standard size sample. This simple definition causes some problems concerning the microbial ecology of many habitats, including rocks, which are not easily accessible to a multitude of techniques used in other fields of ecology. From the ecological point of view, the habitat 'monument' can be considered as a particular one that offers many unsettled problems of microbial ecology. In biodeterioration studies many of the approaches commonly used in ecological studies (wrongly or rightly as sorted by Brock, 1987) are used.

The simultaneous establishment and presence of phototrophic and chemotrophic microbiotas demonstrates, however, that rock surfaces are a good substrate and offer many possibilities of interactions among several types of microbiota in an environment which is seemingly quite uniform.

Biodeterioration of rocks with its solubilization, mobilization, immobilization and reprecipitation processes plays a very important role in nature in the recycling of the elements, but is without doubt undesirable when the rocks subjected to it are considered works of art and thus something unique that needs to be protected and maintained for the coming generations.

Interactions between microorganisms and stone monuments lead to different patterns of surface alterations. Aesthetic, but also chemical and physical, damage can occur and has been observed. Climatic and environmental factors deeply affect the extent and type of colonization as well as the spatial orientation of the material and its physicochemical status. Previous conservation procedures, with addition of nutrients, may further complicate the situation.

For the colonization and stabilization of microbial communities, water availability and sun irradiation and shading seem to be the main factors that can affect the extent and the type of colonizers in and on the rock surfaces. Formerly it was thought that, as in the case of plant communities the nutrient content of the rock would be of importance. Today it is thought that energy sources and nutrients may originate from the immediate and far environment of the rock and that the rock itself perhaps more or less just physically structures the habitat of the microbial community.

Several detailed studies have been done concerning special rock environments in terms of algae and cyanobacteria, lichens and chemolithotrophic and chemoorganotrophic bacteria. Very little is known, however, about fungi inhabiting rock surfaces and their deeper layers, about their ecology and taxonomy in relation to the rock substrate, the immediate environment and general climate conditions. This is witnessed by the recent discovery of a group of fungi so far unknown for the rock environment but detected in many geographically different locations.

Some general concepts that are commonly used in ecological research on soil, aquatic habitats, the phylloplane, phyllosphere, etc. can also be applied to the rock substrate.

Some parameters to study a fungal community should be taken into consideration in order to avoid confusing data.

- Standardization of the sampling procedure – inadequate sampling procedures can affect enormously the community pattern in a given environment. Media composition and fast-growing species can give an erroneous idea of the size of the population and diversity of species present.
- Presence of nutritional factors – readily assimilable carbon sources can be used by fast-growing fungi, whereas the 'difficult' carbon sources can be degraded and assimilated only by specialists, generally with a slow-growing pattern; the extent of a microbial community belonging to the same group or to different ones can affect the diversity referring to number of individual organisms and of species present; distribution can differ widely between the surface and the deeper layers of the rock analysed.
- Survival strategies – the possibility for some fungal species to form specific survival structures can give them more chances in a given environment with respect to those species unable to do so.
- Substrate specificity – this is related in particular to the possibility of

attacking (degrading) an organic substrate by fungi. Concerning the rock substrate (and thus on first view an inorganic one) the different composition of rock material (pH, mineral composition, porosity) and the presence of man-applied chemical products can greatly affect the colonization of these surfaces as well as the amount of organic pollution or eutrophication of the air and particulate deposits in the surroundings of the rock surface.

On the basis of these preliminary definitions the following can be assumed.

1. Rocks situated in places in which water is available and sun exposure is not too strong should be considered as 'physically favourable environments' and thus the colonization and establishment of a polymicrobial biota is possible in which all the bionts are in equilibrium among themselves. In this case the biodiversity of the total microbiota is guaranteed, because the flow of nutrients is maintained by the succession of organisms and the deposition of organic and inorganic particles. Also the utilization of readily degradable and more refractory carbon sources is possible by the cooperation of the microbial community.

2. Rock exposed to a high solar exposure and with low humidity allows the environment to be defined as 'extreme', being generally characterized by the presence of only a few species and also a low number of individual organisms. Microorganisms, and especially fungi, living in such environments are characterized by a high frequency of survival strategies (spores, sclerotia, reduced metabolic rates, etc.).

The factors affecting microbial communities are summarized in Table 16.1. A favourable rock environment can be transformed into a disturbed one when the intervention of humans modifies established equilibria.

With such premises it is clear that the ecological study of the microbiota

Table 16.1. Factors affecting microbial communities.

Favourable environment	Equilibrium among microbiota living in the microniche; the growth rate is controlled by temperature, carbon sources, pH.
Disturbed environment	Sudden events modify equilibria existing (e.g. cleaning methods, application of consolidants, biocides, etc.). These events can lead to: (a) destruction of all the microbiota; (b) favouring one or a few species by enrichment (eutrophication)
Extreme environment	Selection of species which are stress-tolerant and capable of producing different kinds of survival strategies

inhabiting rock surfaces and the application of such acquired knowledge to conservation of the stone monuments are very intriguing and interesting. In particular, in recent years, our attention was attracted to some specific alteration patterns, such as macropitting and pseudo- or biokarst caused by dematiaceous and black yeast-like fungi without any apparent evidence of lichen growth.

In this chapter we give an outline of the fungal colonization of rock monuments, particularly calcareous ones, the problems concerning the ecology and taxonomy and their role in the decay of rock monuments related to climate and conservation status.

Materials and Methods

Samples were taken from different statues and rock materials situated in outdoor environments in various geographical and climatic areas in Europe (Fig. 16.1).

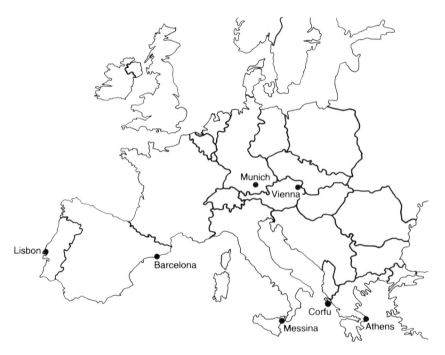

Fig. 16.1. Geographical distribution of samples. Two main areas can be more or less delineated by the Alps: Central Europe (Vienna and Munich samples) and the Mediterranean Basin (Lisbon, Barcelona, Messina, Corfu, Athens). Lisbon does not strictly belong to the Mediterranean Basin but it is included with this group for this review.

Analyses

Microbiological analyses were carried out according to the Italian (Normal 9/88; Commissione Normal, 1990) and German standard methodologies as well as stereomicroscopic and scanning electron microscopic (SEM) observations. The preparation of SEM samples was carried out by the usual techniques (glutaraldehyde fixation, dehydration in ethanol) reported also in the Normal 8/81 (Commissione Normal, 1981). For most samples a staining procedure was also used (periodic acid Schiff, PAS) in order to show the extent of microbial colonization directly on the rock samples by staining specific extracellular polymeric substances (polysaccharides, etc.) produced by these microorganisms (Warscheid, 1990).

For enumeration, isolation of fungi specific media were used: DRBC (King et al., 1979) and Czapek-Dox Agar diluted 1:50 (CZDd). Such media were found useful for the isolation of slow-growing fungi because of the presence of inhibitors (dichlorane and rose bengal in DRBC) (Urzì et al., 1992a) and oligotrophic conditions (CZDd). Morphological features were taken into consideration for the identification of most of the genera (Barnett and Hunter, 1972) isolated whereas the species identifications were checked in specialized laboratories (Centraalbureau voor Schimmelcultures (CBS) in Baarn, The Netherlands) and in the collaborative work by one of us (U. Wollenzien). Special attention was given to those colonies belonging to dematiaceous fungi and especially black yeasts with a very small size of colony and slow growth pattern.

Some tests to assess their role in the biodeterioration of rock monuments were also carried out: (i) solubilization of $CaCO_3$ in Petri dishes of calcium carbonate powder-enriched agar and (ii) tests on marble slabs inoculated with fungal spore suspensions and incubated for different time intervals (4, 8 and 12 months) in a wet chamber. Stereomicroscopic and SEM observations were then carried out on the marble slabs.

Results

The microbiological analyses and examinations of the original rock samples by SEM microscopy show that many different microniches and microhabitats exist in rock materials. In suitable environmental conditions and in relation with the state of conservation of the stone, dense polymicrobial communities were found to live at different depths in the stone and with different relationships between the microorganisms. One common characteristic of these communities was the production of prolific amounts of extracellular polymeric substances (EPS). This production is not only due to the phototrophic biota but also to the other microorganisms (bacteria and fungi, in some cases also actinomycetes). In general, the presence of

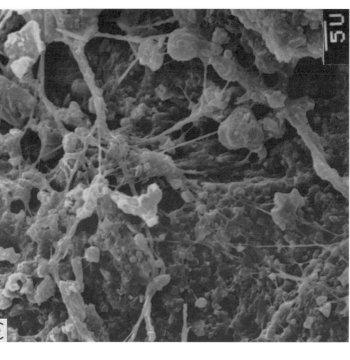

Fig. 16.2. (and opposite) (a) SEM observations of a sample taken from a Carrara marble statue exposed in the old cemetery in Munich. The sample surface is widely colonized by a polymicrobial population. Algae and bacteria embedded in common extracellular polymeric substance (EPS) are clearly visible. (b) SEM observations of a sample taken from a calcareous obelisk in Corfu. In this case the polymicrobial colonization is mainly found in the inner layer of the stone beneath the surface. Algae, fungi and bacteria are present. (c) Thin section from a Pentelic marble block view under light microscopy. Magnification × 640. Clusters of dark cells are clearly visible inside calcite crystals. (d) SEM observations of a sample taken from the same marble block described above. Also in this case clusters of cells are situated among marble crystals.

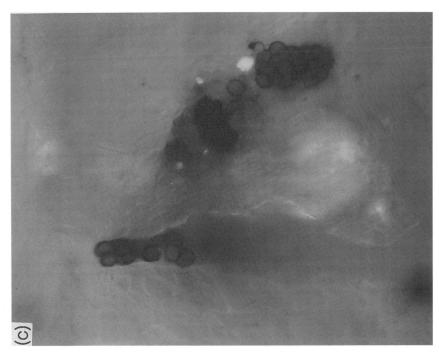
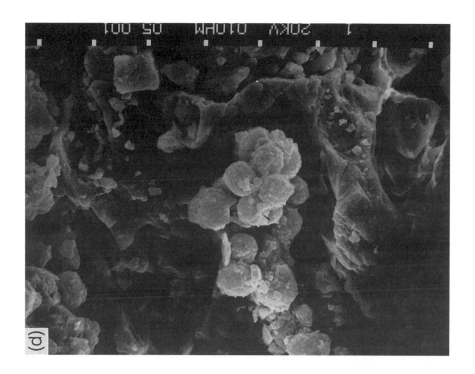

dematiaceous fungi was higher than that of moniliaceous ones. The latter were found in higher percentage when samples were taken in specific areas more protected from sun irradiation (e.g. dress folds in a marble statue).

On the basis of our results it is possible to divide the samples into two main groups.

1. Samples taken from monuments situated in climatic areas with high humidity and lower exposure to sun irradiation (Vienna and Munich samples from marble statues). Such conditions are characterized by a heavy colonization and establishment of a polymicrobial microbiota both on the surface and in the deeper layers. The presence of fungi is in some cases affected by the algal colonization that supports or is supported by a dense bacterial colonization (e.g. samples 17 and 18 of Munich, where no fungi were detected) (Fig. 16.2a) but in Vienna samples where a similar situation was found (algae and bacteria closely connected) fungi (*Exophiala jeanselmeii*) were also found. SEM observations show clearly the massive production of EPS that embeds all the microbial community. The surface appearance (dark green to grey and black) is due to degradation of chlorophyll and carotenoids as reported by Urzì *et al.* (1992b). The fungal strains isolated belong mainly to the following genera: *Alternaria, Ulocladium, Cladosporium*, and *Phoma*. It is noteworthy that such fungi, although not able to produce organic acids in laboratory tests, are able to spread rapidly over the surface in search of nutrients or to catch them from the air. This fact, ecologically important for the fungi, has as a consequence, in rock monuments, the formation of a black patina. Among the black yeasts, *Aureobasidium pullulans* was quite common, but more interesting was the presence of colonies of *Exophiala jeanselmeii*, found in a very high amount in 50% of the samples from Vienna (from 70 to 96.5% of fungal colonies isolated). This species, often related with some human pathologies (chromoblastomycoses and phaeohyphomycoses), seems to have a good natural genetic reservoir in the rock habitats of Central Europe as first stated by Braams (1992). In fact, until now it was never isolated from samples derived from the Mediterranean Basin. On the contrary, *Aureobasidium pullulans* was never found in high numbers, but is quite ubiquitous and did not exhibit specific geographical correlations.

2. Samples taken from rocks of the Mediterranean Basin: different situations can be observed depending on water availability and shadowing possibilities. When these factors are not limiting the growth of microorganisms, their colonization is comparable with that described above (Fig. 16.2b). Higher sun exposure and lower water availability drastically change the typology of alterations and the colonization patterns. Phototrophic microorganisms, when present, are only in the deep layer, whereas on the surface lichens, mainly endolithic, and dematiaceous fungi are commonly found. These extreme conditions seem particularly interesting

for the particular interactions existing among the microorganisms and for the different strategies operated by the dematiaceous fungi to colonize those substrates.

In these conditions only a few filamentous fungal species were found, sometimes associated with yeasts and bacterial strains producing slime. The extracellular polymeric substances protect the embedded cells against the strong solar irradiation. Algae, when present, are confined only to the deeper layer but not to the surface. The fungi grow in a very characteristic intercrystalline pattern, similar to a lichen colonization that was excluded by the cultural isolation and microscopical observations (Fig. 16.2c, d).

This phenomenon was observed at various times in some marble samples from the Parthenon (Athens) and in Carrara marble statues exposed for 80 years in an outdoor environment in the garden of the Messina Museum. The deteriorative pattern seems to be due to the mechanical action exerted by the meristematic cells and not by acid production as observed in laboratory study with marble slabs. It is a matter of discussion whether such structures, considered to have a very low metabolism, can produce acid in natural conditions.

Solar energy is necessary for the growth of phototrophic microorganisms, but direct sun irradiation can be fatal for microorganisms that live on the surface unless they possess particular protective pigments. As a consequence of direct sun exposure we can observe on rock surface the growth of dark-pigmented microorganisms (epilithic organisms) or the remains of old algal colonization, whereas in deeper layers the active growth of light-sensitive organisms will prevail. The deep colonization can be due to an active drilling activity (endolithic organisms) or to fissure and crack colonization (chasmolithic organisms). On the other hand, in light-protected situations the colonization of microorganisms is mostly on the surface with formation of microbial films and mats, a 'biological patina', or 'pellicola' (Urzì et al., 1993). How can this be transferred to questions of biodiversity of the rock microbiota?

Under high light conditions at the immediate surface only a few microorganisms can survive. It can be considered an extreme environment where only a few selected microorganisms can grow in unequilibrated conditions, sometimes leading to misleading high numbers in plate counts. As stated by Gochenaur (1984), radiation (gamma, UV) should not affect biodiversity in terms of numbers of species present, but can result in the selection of the most resistant ones (e.g. melanin producers and ones with protective morphology). Some lichenized and dark fungi belonging to the dematiaceous group, however, can exert their attacking strategies on the rock. In the same environment the deep layers are colonized by an equilibrated and active biota of algae, cyanobacteria, filamentous fungi, streptomycetes and bacteria. The biodeteriorative activity of single microorganisms, shown by a higher or lower capacity to produce organic acids and the capability to use a wide range of organic compounds, is much less relevant than the physical and chemical activity of the polymicrobial community. In each microniche there is always potential to produce many trigger compounds that in the appropriate concentration will revive the dormant community. This can explain the large fluctuations of chemo-organotrophic bacteria found in the rock. It is interesting to note that the microbiota is characterized by a predominance of the Gram-positive group (spore-forming bacteria, cocci and coryneform rods).

In monuments exposed alternately to sun and shadow the surface growth is related to the season and the hours of shade (including water availability). A wide active biological growth is found on the surface appearing as a green patina. After the death of the algae the colour turns black and abundant growth in the form of chemoorganotrophic microorganisms is a consequence. This can be called aesthetic deterioration. In the deeper layers, microbial activity also continues during unfavourable conditions at the surface (rock and microfilm buffering effects). The spatial relationship between green algae and chemoorganotrophic bacteria seems to be so close that mutual advantages (uptake, storage and release of water, gases and metabolites) are evident. Due to a large amount of organic compounds released by the superficial dead layer it may no longer be necessary for the chemoorganotrophic microorganisms to be in close contact with phototrophic ones. Thus the repeated finding of algae that are able to grow in chemoorganotrophic media even in the dark seems to demonstrate a particular need of the alga for substances excreted by the bacteria rather than the bacteria themselves.

Discussion

First, rock is practically the only natural habitat devoid of noteworthy macroorganisms. Second, it exhibits drastic local disparities of climatic, physical and nutritive conditions. Third, the fact of exposure without macroorganismic 'buffering systems' makes these ecosystems very vulnerable to sudden or periodic changes in environmental stresses. From these general remarks we can derive several observations and comments on the potential for biodiversity. As stated by Widden (1981) the concepts of phenotype, ecotype and genotype are on one hand important for ecological approaches to microorganisms. On the other hand, the concept of phenology also relates to biological periodicity in relation to the seasonal sequence of climatic factors. This concept, which was originally applied to plants, can be useful in the study of fungal rock communities. Several survival strategies may be applied within the framework of phenology. Shape and form of the microorganisms are related to the phenological responses (e.g. clamydospore formation, sclerotia, etc.). Also the energy dispersion in building aerial mycelia plays a role as a fast response to environmental changes (growth close to the surface is a less energy-consuming process, but can be too slow for population maintenance). The responses are usually genetically programmed when changes are fundamental for the survival and adaptive strategies. According to Mueller-Dombois and Ellenberg (1984) the cyclic appearance and disappearance of microfungi in a given habitat can also be an important aspect of the phenology of fungal populations. Direct and indirect field studies and laboratory model studies can permit a better understanding of these events.

Interesting observations also exist on the seasonality of fungi: November, for example, is the time when the highest number of spores is found in the atmosphere, but summer time is the period of maximum activity of mycelia (Widden, 1981). Light intensity and wavelength influence sporulation in many fungi. Thus there is a great deal of potential and actually analysed factors influencing the distribution and diversity of fungi in the rock environment.

If we look, however, at the totality of microorganisms living on and in rocks, some older concepts should at least be briefly mentioned. The theory of 'autotrophic' microorganisms being the pioneer organisms in rock decay environments, which would be followed by the less important chemoorganotrophic ones, seems to be outdated and obsolete in view of the great number of rock samples that have been analysed in the past years which always exhibited a great variety of chemoorganotrophic species with few, if any, algae, cyanobacteria and/or chemolithoautotrophs associated.

Fungi, in particular, seem to be very well adapted to the living conditions in rock crevices and cavities. In all the texts on the ecology of fungi the bulk of information is reported on soil and aquatic fungi, on those colonizing phyllosphere, phyllophane, rhizosphere, rhizoplane, etc. A first detailed study was done by Braams (1992) on the ecology of fungal microbiota inhabiting sandstone monuments of Middle Europe, but little more is available from the literature. The taxonomical literature on the dematiaceous fungi (e.g. Ellis, 1971, 1976) reports on hundreds of species of this group typically occurring on and in plants, animals and humans, fewer reports exist on soil isolates and practically none on rock habitats. Some attention was given to black fungi attack on wallpapers and paints in damp buildings but little consideration has been given so far to the natural and monument rock-inhabiting black yeasts and fungi so frequently isolated from rock environments in our study. In two cases, a Pentelic marble block from the Dionysos theatre in Athens and a dolomitic marble from Namibia, other black yeasts and fungi were practically the exclusive inhabitants of the rock surfaces and crevices. It seems evident that their resistance to dryness, to extensive and intensive sunlight and UV irradiation enables them to conquer this niche almost exclusively. Another type of fungi conquering these habitats was described by Krumbein and Jens (1981) and Grote (1992). These are fungi of different groups, which are capable of enriching iron and manganese and causing very dense black layers on the rocks, classically called rock or desert varnish. Both groups, the melanin-forming and the black manganese-forming fungal biotas may have found two different lines of adaptation to extended periods of dryness and very high irradiation doses.

Conclusions

The study of the effect of the colonization and subsequent deterioration of rock surfaces can provide much information concerning not only problems related to restoration and conservation of stone monuments, but also the possibility to study the ecology of this particular habitat as one of the many possible in natural environments and barren rock outcrops.

Very little is known about the ecology of fungal communities and our results stress some points, in particular:

- the need for better communication between ecologists and taxonomists;
- a revision of taxonomic concepts used to classify black fungi and in particular black yeasts: in fact, extreme environments seem to select a special fungal biota that has developed specialized survival strategies, including the shifting from mycelial form to the unicellular or budding one. This has a very important relevance in saving energy because aerial mycelia and sexually or otherwise reproductive ones are more energy-consuming; phenotypic rearrangements can also be accompanied by genetic ones and thus the ecological and taxonomic aspects of these fungi should be considered as a whole entity;
- techniques of molecular biology can usefully be employed to compare deteriogens and pathogens;
- the need to study the distribution and spread of human pathogenic species like *Exophiala jeanselmeii*.

Acknowledgements

This work has been partially supported by a CNR contribution for a bilateral project Italy/Germany 1992/1994 and a Vigoni-travel support programme.

We wish to thank Elaine Johnston for the revision of the English text. We also wish to thank Renate Kort of the Geomicrobiology Division of Oldenburg for her precious technical assistance in SEM observations and all the collaborators in Messina and Oldenburg who helped us in the examination and analyses of samples, culture and maintenance of strains. Some of the samples (Munich and Vienna) were taken in the frame of the EUREKA project EU 496 Euromarble, Athens samples were taken under the kind but close control of Parthenon conservators; we wish to thank Professor T. Moropolou of the NTU of Athens for the Corfu samples, and for the Messina ones the Director of the Museum, Dr F. Campagna for her open hospitality.

References

Barnett, H.L. and Hunter, B.B. (1972) *Illustrated Genera of Imperfecti Fungi*. Burgess, Minneapolis.
Braams, J. (1992) Ecological studies on the fungal microflora inhabiting historical sandstone monuments. Doctorate Thesis, Oldenburg.
Brock, T.D. (1987) The study of microorganisms *in situ*: progress and problems. In: Fletcher, M., Gray, T.R.G. and Jones, J.G. (eds) *Ecology of Microbial Communities*. Cambridge University Press, Cambridge, pp. 1–17.
Commissione Normal (1981) Esame delle caratteristiche morfologiche al microscopio eletronico a scansione (SEM). Doc. 8/81, ICR, CNR, Rome, 12 pp.
Commissione Normal (1990) Raccomandazioni Normal: 9/88 Microflora autotrofa ad eterotrofa: tecniche di isolamento in cultura. CNR – ICR, Rome.
Ellis, M.B. (1971) *Dematiaceous Hyphomycetes*. Commonwealth Mycological Institute, Kew, Surrey, UK.
Ellis, M.B. (1976) *More Dematiaceous Hyphomycetes*. Commonwealth Mycological Institute, Kew, Surrey, UK.
Gochenaur, S.E. (1984) Response of soil fungal communities to disturbance. In: Wicklow, D.T. and Carroll, G.C. (eds) *The Fungal Community, its Organization and Role in the Ecosystem*. Marcel Dekker, New York, pp. 459–479.
Grote, G. (1992) Mikrobieller Mangan- und Eisentransfer an Rock Varnish und Petroglyphen arider Gebiete. Dissertation, Oldenburg, 168 pp.
King, A.D. Jr, Hocking, A.D. and Pitt, J.I. (1979) Dichloran rose bengal medium for enumeration and isolation of moulds from foods. *Applied and Environmental Microbiology* 37, 959–964.
Krumbein, W.E. and Jens, K. (1981) Biogenic rock varnishes of the Negev Desert (Israel), an ecological study of iron and manganese transformation by cyanobacteria and fungi. *Oecologia* 50, 25–38.
Mueller-Dombois, D. and Ellenberg, H. (1984) *Aims and Methods of Vegetation Ecology*. Wiley, New York.
Urzì, C., Lisi, S., Criseo, G. and Zagari, M. (1992a) Comparazione di terreni per l'enumerazione e l'isolamento di funghi deteriogeni isolati da materiali naturali. *Annals of Microbiology and Enzymology* 42(1), 1–9.
Urzì, C., Krumbein, W.E. and Warscheid, T. (1992b) On the question of biogenic colour changes of marbles on Mediterranean monuments (coating – crust – microstomatolite – patina – scialbatura – skin – rock varnish). In: Decrouez, D., Chamay, J. and Zezza, F. (eds) *The Conservation of Monuments in the*

Mediterranean Basin. Proceedings of the 2nd International Symposium, Musée d'Art et Histoire de Genève, Geneva, 19–21 November, pp. 397–420.

Urzì, C., Krumbein, W.E., Criseo, G., Gorbushina, A.A. and Wollenzien, U. (1993) Are colour changes of rocks caused by climate, pollution, biological growth or by interactions of the three? *21 RILEM Proceedings of the International Congress on the Conservation of Stone and Other Materials*, Paris, 29 June–1 July, Vol. 1, pp. 279–286.

Warscheid, T. (1990) Untersuchungen zur Biodeterioration von Sandsteinen unter besonderer Berücksichtigung der chemo-organotrophen Bakterien. Dissertation, Oldenburg, 147 pp.

Whittaker, R.H. (1975) *Communities and Ecosystems*, 2nd edn. Macmillan, New York.

Widden, P. (1981) Patterns of phenology among fungal populations. In: Wicklow, D.T. and Carroll, G.C. (eds) *The Fungal Community, its Organization and Role in the Ecosystem*. Marcel Dekker, New York, pp. 387–401.

VI INVENTORYING AND MONITORING MICROORGANISMS

Statistics, Biodiversity and Microorganisms 17

E. RUSSEK-COHEN AND D. JACOBS

Department of Animal Sciences, University of Maryland, College Park, Maryland 20742-2311, USA

Introduction

Microorganisms are a part of every ecosystem. If we are to understand biodiversity at all, we need to understand which microorganisms are present and the role they play. Statistical methodology, when properly used, leads to sound scientific conclusions. This implies that statistics and statisticians are important in the biodiversity question.

This chapter highlights several aspects of statistics as it relates to our understanding of microbial diversity, and some of the challenges of big science, namely coordinating long-term studies involving multiple laboratories and/or investigators. Since systematics is such a key component of this area, comment is made also on some of the techniques in this area from a statistical perspective. This chapter cannot be regarded as comprehensive since the field is so broad. It is hoped that this discussion will stimulate collaborative efforts of microbiologists and statisticians.

The Challenges of Laboratory Data

Overview

Statistics can be used to evaluate laboratory methods and can assist in improving quality of the results. Statistics can also be used to predict properties of restriction enzymes when processing long molecular sequence data, so as to improve laboratory efficiency (Karlin and Macken, 1991).

This chapter is mainly confined to binary data, molecular sequence data and most probable numbers. A more extensive discussion of quantitative variables can be found elsewhere (Hertzberg and Russek-Cohen, 1993). However, when general principles are described they apply to quantitative variables as well.

In the evaluation of laboratory methods, one needs to identify systematic and random sources of variation in measurements. For example, if two technicians differ in their reading of a binary measurement, this would be a systematic source of variation one would like to eliminate. So would a situation in which a reading was more likely to be negative over time, as when a reagent or medium deteriorates over time. Random sources of variations are those that are not systematic. These are typically harder to eliminate. However, different sources of random variation may differ in relative magnitude. For example, samples of water collected from different ponds in an area may differ in microbe counts, but multiple plate counts collected within each pond may differ considerably less (i.e. environmental variation is often more than analytical variation). If one wants to establish prevalence of organisms in the area, sampling more ponds rather than plate counts per pond is in order. A study of random variation is often done to minimize variability subject to cost constraints (Snedecor and Cochran, 1967). These same issues occur whether sampling in the laboratory, sampling water, soil or a host organism.

Binary Data: Measurement Error, Sensitivity, and Specificity

Many areas of microbiology record characters or attributes as presence or absence. Traditional taxonomic methods make heavy use of such characters corresponding to whether or not an organism grows in a particular medium. Such phenotypic data are often numerous and as varied as possible to obtain a 'representative' sampling of phenetic relatedness (Sneath and Sokal, 1973; Sneath, 1984). Newer molecular probes are used to discern whether or not a particular organism is present in a sample. These characters are considered binary characters. However, they can be somewhat subjective in some cases or sensitive to the particular conditions in which the character is measured (e.g. plankton or water samples). In this case measurement error can arise. This can be particularly troublesome when more than one laboratory is involved in study. Furthermore, probes are often developed for identifying a class of organisms. For example, one may have a fluorescent antibody procedure to detect the presence of *Vibrio cholerae*. If samples are collected from new habitats, the probe may not always function as intended. There are two kinds of problem that can result. Statistically, sensitivity is defined as the probability that a cholera organism is detected, given that it is present. Specificity is defined as the probability that the probe does not detect a non-cholera organism, given that it is present. Ideally, a probe will have high specificity and high sensitivity. It is important to define at what limits of detection sensitivity and specificity are evaluated, i.e. what level of organisms.

In order to practically evaluate the sensitivity and specificity of a probe, there needs to be a set of reference strains. There needs to be a sufficient number of cholera and non-cholera (but closely related) strains. Too many collections have breadth and not depth. That is, a collection often consists of a few (<10) strains of a 'species'. Furthermore, if some strains are more prevalent than others, weighting all strains equally in assessing sensitivity or specificity may be inappropriate. Finally, adding depth to these databases will also aid in probabilistic identification of microorganisms (Russek-Cohen and Colwell, 1986).

Most Probable Numbers (MPN)

As molecular techniques continue to develop, there will be less use of MPN. However, in field situations in which samples are often messy with lots of 'junk', MPN still continue to have value. For example, the US Department of Agriculture Food Safety and Inspection Service continues to use MPN values for *Salmonella* in field samples. The problem with MPN values is that it provides an estimate of actual counts and not the counts themselves (Russek and Colwell, 1983). With a nine tube MPN (i.e. three tubes each of three dilutions), the error may be off by an order of magnitude. For some purposes, this may be adequate. However, in order to understand how organisms respond to their environment, more tubes and possibly more dilutions are needed. Furthermore, subsequent analyses need to incorporate the variability of the MPN estimate into the analysis (Fuller, 1987). This is currently almost never done.

Molecular Data

By far the most exciting progress in the biodiversity question has come from progress in molecular biology. With phenotypic characters, establishing a 'gold standard' for a species or a species definition has been almost impossible. Although there is a whole wealth of literature on using sequence data in the context of inferring evolutionary trees, there is a lesser known set of literature for using statistics to improve laboratory efficiency.

Statistical methodology can be used (i) to identify coding and non-coding regions of DNA (Michel, 1986), (ii) to decide how long a sequence to measure for purposes of phylogenetic analysis (Weir, 1990), (iii) for estimating fragment lengths when using restriction enzymes, for (iv) for assessing outliers in aligned databases (Russek-Cohen and Jacobs, 1989). There is also the possibility of using statistics to predict the properties of some restriction enzymes prior to their use (Karlin and Macken, 1991).

Recently there have been some efforts to model error rates in DNA sequences (Churchill and Waterman, 1992). Such models may become necessary if one hopes to distinguish intraspecific variation from errors in sequencing, and to have a measure of sequence quality that agrees with a given sequence in a database.

Monitoring Quality in Multilaboratory Studies

There are two issues that must be considered in the context of multi-laboratory or multi-investigator studies. There is the question of whether a laboratory running an established procedure achieves a correct response more than a certain percentage of the time (i.e. comparing a laboratory to a gold standard). There is also the issue of whether the laboratories achieve comparable results, perhaps in the absence of a gold standard. It is important that each laboratory in a multi-investigator study processes a set of samples that is identical. It also makes sense that the investigator is blind to the true identity of the sample, so as not to bias the results. Statistical methods for comparing laboratories is discussed below.

Monitoring Quality Over Time

Control charts are a simple mechanism for monitoring the quality of data collected over a long period of time. For example, reagents, antibodies, and other factors may vary during the course of a long-term study (one that takes longer than 6 months to complete). A simple mechanism is to take a preselected group of strains/specimens and run them through tests at periodic intervals (e.g. once a month). If one agrees on a correct response for each test run on each specimen, then one can compute the percentage correct for each attribute recorded. This can be plotted over time using a P-chart (Ryan, 1989) for binary data. An X-bar chart can be similarly constructed for continuous data.

The Challenge of Sampling the Environment

Overview

In order to understand the types of organisms present as well as their relative importance, key attention needs to be paid to the sampling schemes used to extract the organisms from their habitat. Statisticians strongly recommend some variation on a random sample rather than a representative sampling scheme. The latter tends to impose too much personal bias on the selection scheme. Most statistical techniques are in turn not valid unless some effort is made to follow this random sampling scheme.

As an aside, when specimens are inventoried via an environmental sampling scheme, the conditions under which the samples are collected need to be included. This will facilitate developing estimates of global biodiversity.

Confounding of Space and Time Effects

Sampling for convenience can be a serious problem. For example, marine scientists often rely on cruises to collect data. So if a cruise collects data in Puerto Rico in January and in the mid-Atlantic area in July, it becomes difficult to compare samples from these two environments. When one cannot distinguish temporal variation from location difference this is called 'confounding'. If methodology varied in any way between the two cruises, further confounding may result. Thus, if one is going to compare two sites one needs to find a way to make them comparable in time. Perhaps the following year, one can sample north in January and south in July. Thus it is important to balance space and time when designing a sampling scheme. Furthermore, one needs to sample in all seasons if one does not wish systematically to omit a group of organisms.

Validation of Methodology

Techniques should be carefully tested and validated prior to a final study. Statistics relies very heavily on the repeatability of a procedure. Procedures which are refined along the way may mean earlier collected data are the result of a different method from later data. This can be difficult in field situations, but still must be done to get accurate conclusions.

Units of Replication

One needs to understand the sources of variability and how to sample. An experiment which requires sampling from their field stations, with three gallon samples per station, three subsamples per gallon sample and two plates per subsample is not the same as a random sample of $4 \times 3 \times 3 \times 2 = 72$ samples, even though there may be 72 plates with data recorded. These samples are 'nested' and any analysis should reflect the kind of nesting that occurs. If one plans to use these samples in a subsequent systematic study, sampling four field stations may not yield the variability in six field stations even though the latter may also include 72 samples. A general rule of thumb is that organisms closer in space tend to be more similar. This means if you want to see what exists in an estuary you will need to adequately sample different areas of the estuary. One possibility is to divide the area into quadrats and select n quadrats at random using a random number table. Too many ecological studies tend to sample a small area intensively and then make very sweeping conclusions (Hurlbert, 1984). Good science is expensive and may mean collecting samples from distant locations.

Stratification

Environments are often heterogeneous. For example, levels of bacteria may be sensitive to salinity in an estuary and stratifying an estuary into areas with 'comparable salinity values' may be done prior to sampling. Then one can carry out a random sampling scheme within each stratum (Green, 1979). There are techniques that allow one to combine results across strata, but these vary with the objective (e.g. see Cochran, 1973).

Some Common Themes

Systematic and Random Sources of Variation

When the response variable is continuous (e.g. length or volume data), systematic and random components of variability are analysed by a mixed model analysis of variance (Sokal and Rohlf, 1981). In the laboratory, systematic components may include laboratory or investigator, time spent growing the organisms, storage time, or components of the media. Random effects may include batch to batch variation of the same media, or subsamples of the same sample. In the environment, systematic effects often include location and season, but may also include depth and type of sampling equipment used. A similar set of procedures called generalized mixed models (Schall, 1991) are available for binary data. Unfortunately, although commonly available packages (e.g. SAS, BMDP, GENSTAT) have mixed model procedures for continuous data, there do not appear to be preprepared procedures for binary data. When all the factors are systemic (i.e. fixed), the model for binary data is just a logistic regression model (Hosmer and Lemeshow, 1987) and again the packages do have easy to use procedures. The drawback to most binary models is that they rely on large sample theory, though for some special cases software for exact methods exists (e.g. STATXACT, TESTIMATE, LOGXACT). However, in general, complex models with insufficient data will perform poorly.

There has been some effort made to construct similar models for sequence data (Excoffier *et al.*, 1992). However, these are not as general or as well established as those for continuous or binary data.

Selection Biases

Whether in the laboratory or in the field, potential selection biases exist. In the environment, it may occur by virtue of when, where and how we sample. In the laboratory, it may occur by how we choose to store or process the samples collected. A particular medium may selectively exclude slow-growing organisms or non-culturables. Statistical methods cannot solve these problems.

Numerical Taxonomy

Overview

Over the last several decades, numerical taxonomic methods have been widely used in microbial taxonomy studies (MacDonell and Colwell, 1985). One aim of bacterial numerical taxonomy is to classify strains into related groups based on characteristics or traits of each strain (Sneath and Sokal, 1973; Austin and Priest, 1986). The typical procedure is to gather phenotypic and/or genotypic data for each operational taxonomic unit (OTU, e.g. organism, strain, isolate) of interest, determine the relationships between the OTUs by computing a measure of similarity or dissimilarity, examine the interrelationships between the OTUs using a variety of analytical techniques (e.g., cluster and principal component analyses), and show the relationships in dendrograms or other such graphical displays (Sneath and Sokal, 1973; Austin and Priest, 1986; Digby and Kempton, 1987).

Measures of Relatedness

For binary data, taxonomic relatedness between each pair of OTUs is expressed in terms of the number of characters in common relative to the total number of characters examined. Two of the most commonly used measures of similarity used in bacterial numerical taxonomy are the simple match coefficient (S_{sm}) and the Jaccard similarity coefficient (S_j) (Austin and Priest, 1986). S_{sm} is the percentage of positive characters in common relative to the total number of characters whereas S_j ignores the double negatives in the total number of characters (Jones and Sackin, 1980; Sneath, 1984; Austin and Priest, 1986). Dissimilarity measures and/or distance measures may also be used to measure OTU relatedness.

Similar measures of relatedness may be computed using quantitative data. For example, Sorenson similarity is the ratio of the sum of the minimal amount measured for each test (e.g., organic compounds from gas chromatography) relative to the total amount of material for all tests for the pair of OTUs (Magurran, 1988). If each test represents a dimension in

n-dimensional space, Euclidean, Manhattan and other such distance measures can be computed for each pair of OTUs (Digby and Kempton, 1987; Magurran, 1988). Phenotypic data (binary, quantitative and multistate) can also be used in combination to compute one measure of association (e.g., Gower's general similarity measure) (Gower, 1971; Sneath and Sokal, 1973; Digby and Kempton, 1987).

Phenotypic data, as described above, may not provide sufficient information to differentiate bacterial strains since only 5–20% of the genome is expressed (Bradley, 1980; Austin and Priest, 1986) or alternatively the characters selected may not be as representative as they seem (Russek-Cohen and Colwell, 1986). It is generally recognized that relationships deduced from nucleotide sequences are more accurate than those deduced from polypeptide sequences (Holmquist et al., 1972).

The sources of genetic data commonly used today in numerical taxonomy are nucleic acid sequences (e.g. DNA, 5S and 16S ribosomal RNA or rRNA) and hybridization results (e.g. DNA/DNA and DNA/rDNA) (De Ley, 1971; Krieg, 1988; Johnson, 1989). DNA sequences and DNA/DNA hybridization data delineate the OTUs at the lowest taxonomic levels (i.e., genus, species and subspecies) (De Ley, 1971; Bradley, 1980; Grimont, 1984; Owen and Pitcher, 1985). The homology values (DNA/DNA and DNA/rRNA) are only relative measures of sequence similarity and are not equivalent to the actual amount of similarity between two sequences (Austin and Priest, 1986; Steven, 1990). This is due to the conditions under which the nucleic acid sequences are isolated, partitioned and hybridized (Grimont, 1984; Steven, 1990). Related to this is that the resulting $n \times n$ matrix is not symmetric – hybridization values depend on which of each pair of OTUs was denatured and allowed to anneal to the other sequence which was radioactively labelled (Steven, 1990). Often an average of the two values is used to construct an overall similarity value.

Prokaryotic rRNA is primarily composed of 5S, 16S and 23S subunits and function, in part, as a structural component in protein synthesis (Erdmann et al., 1986). These rRNA sequences can be used to delineate the OTU relationships spanning the levels of kingdom to subgenus because these sequences contain both conserved and variable regions (Hori and Osawa, 1979, 1987; Woese, 1987, 1991). After the bases for each OTU sequence are determined, the primary structure (i.e. linear) of the rRNA sequences are aligned using one of the many alignment schemes (e.g. Erdmann et al., 1986; MacDonell et al., 1987). The aligned sequences for each pair of OTUs are compared position by position, to determine a measure of relatedness. Analogous to S_{sm} computed for binary data, a simple match coefficient can be computed as the ratio of the number of positions in agreement (i.e. same base) to the total number of aligned positions (Austin and Priest, 1986). Kimura (1980) derived a measure

(K_{nuc}) that estimates the evolutionary distance between two sequences in terms of the number of nucleotide substitutions per site (i.e., number of transitions and transversions). When a gap (i.e., a space inserted due to the alignment procedure) is paired with a base, it is usually counted as a transversion. The K_{nuc} values computed for each pair of OTUs form a dissimilarity (i.e. evolutionary distance) matrix.

One potential problem with these measures of relatedness is that they are influenced by how the alignment is done. Different alignment schemes may insert spaces along the sequence at different locations and may bias the coefficient. This may be due to underlying assumptions of the method as well as if the alignment is done pair-wise or over all sequences (i.e. multiple alignment) (Felsenstein, 1988). A related problem is if all of the sequences under consideration have been partially or completely sequenced. For example, 5S rRNAs typically have a length of 120 bases and are relatively easy to sequence completely. Larger rRNAs (e.g. 16S, *ca.* 1540 bases, 23S 2900 bases), although containing more information due to their length, are frequently only partially sequenced due to the effort involved in determining the bases of long sequences (MacDonell *et al.*, 1987). These measures also do not incorporate the effect that the secondary and tertiary structure may have on restricting what nucleotides may occur at a given position (Hori and Osawa, 1979).

Cluster Analyses and Ordination Methods

Once a similarity or dissimilarity measure has been computed for each pair of OTUs, the $n \times n$ symmetric matrix is analysed to determine the taxonomic relationships (Sneath and Sokal, 1973). The relationships between OTUs can be examined by ordination methods, such as principal component analysis (PCA) (Grimont and Popoff, 1980; MacDonell *et al.*, 1986) and multidimensional scaling (Kruskal and Wish, 1978). These techniques provide an alternative to dendrograms in visualising complex data and may assist in determining how to group a number of vibrio–enteric bacteria (MacDonell *et al.*, 1986), though some cautions are given about PCA in the context of closely related species (Sneath and Sokal, 1973).

There are a number of hierarchical clustering methods used to place the OTUs into groups (e.g. single-linkage, complete-linkage, centroid and average-linkage) (Sneath and Sokal, 1973). The process starts with each OTU in its own group. The groups are merged together in a pair-wise fashion, using the criteria of the chosen clustering method. The joining of the groups continues until all OTUs are in one group. The most common method of measuring distances among clusters is average linkage, or unweighted pair-group method, using arithmetic averages (UPMGA) (Sneath and Sokal, 1973; MacDonell and Colwell, 1985; Austin and Priest, 1986; MacDonell *et al.*, 1986). Simulations and other work suggest that this may be the most robust among single linkage, complete linkage and UPGMA for phenotypic data. However, Felsenstein (1988) cautions against using it when comparing taxa with different rates of evolution. It is unclear which algorithm is best for rRNA data and more work needs to be done.

A common way to display the results of the cluster analysis is as a dendrogram, phenogram or tree (Pielou, 1984; Austin and Priest, 1986). The termini of the branches in the tree represent the OTUs and the axis show the relatedness values at which the groups form (Sneath and Sokal, 1973; Jones and Sackin, 1980). A cophenetic correlation coefficient can be computed and is a measure of how well the relationships between OTUs depicted in the tree reflect those in the relatedness matrix (i.e. the correlation between the tree-determined relatedness to the matrix values) (Farris, 1969; Sneath and Sokal, 1973). However, this cophenetic correlation coefficient is based on non-independent observations, so the usual table values for correlation coefficients (e.g. Sokal and Rohlf, 1981) do not apply.

Sometimes it is desired to determine group membership at a given level of relatedness (Austin and Priest, 1986). For example, Steven (1990) suggests using 95% similarity to delineate genera when using 5S rRNA data. Once the major groups are determined, the original data (phenotypic, genotypic) can be reexamined to determine which tests are unique to each group (i.e. determine why the groups diverge). One possible method proposed by Krichevsky and Walczak (1982) is to compute the group partitioning index. This is a measure of how effectively a particular test divides a group in two and ranges between 0.0 and 0.5 (the test occurs only in one group). Other statistics, such as inter- and intragroup means relatedness and variability, can also be computed.

Comparing Trees

Statisticians regard the results of a cluster analysis as an exploratory tool rather than a confirmatory tool. The latter would typically involve formal tests of hypotheses. Using the types of tests described above following a cluster analysis is helpful but they are not confirmatory tests. However, one can begin to confirm the results of a 'dendrogram' or cluster analysis in a variety of ways. Felsenstein (1985) suggests the use of a bootstrap method to repeatedly resample characters, as if characters were selected independently and at random from a bigger population of characters. Each time a sample of characters is selected, a tree is constructed. If the same organisms continue to group together then there is a high level of confidence in the node that contains them. An alternative approach is based on the comparison of two or more trees. Trees can be constructed using more than one kind of information, e.g. DNA/DNA hybridization and comparing rRNA, using K_{nuc} distances. Consensus indices can be defined as the percentage of nodes that two trees have in common. This has typically arisen in cladistic analyses (see below). However, there have been randomization tests to see if the consensus index is higher than would be expected with a random pair of trees (Shao and Rohlf, 1983). Fowlkes and Mallows (1983) present an approach that does something similar. However, a separate index (B_k) is computed to describe overlap for the data divided into k clusters ($k = 1,...n$).

A third approach can be to compare the distance matrices for each pair of trees. Although the usual table for comparing correlation coefficients (Sokal and Rohlf, 1981) does not apply, an alternative randomization test by Manly (1991) does.

Cladistics

Another possible approach to classifying prokaryotes is cladistic or phylogenetic analysis (Hennig, 1966; Woese, 1987). The goals of cladistic analysis are to identify natural (i.e. monophyletic) groups of organisms and to produce the most parsimonious tree(s) possible. Parsimony methods try to find the shortest branching tree that minimizes the amount of evolutionary change. For bacterial classification, the phenotypic and sequence data (described above) are converted into cladistic character matrices and outgroup analysis is used to polarize the character states (the value that represents the ancestral state), with *Escherichia coli* often chosen as the functional outgroup (Watrous and Wheeler, 1981). More than one tree may be found to be the most parsimonious and these can be compared to one another using various consensus methods (e.g. Adams, Nelson or strict, majority rule) (Swofford, 1990). The hierarchy of this tree reflects the hierarchy of the shared derived characters (synapomorphies) which unite the groups (Nelson and Plantick, 1981). A consistency index and retention index can be computed to measure the amount of homoplasy (convergence) for a given data set (Farris, 1989a,b). Using these index values and the tree the characters can be traced to, one can find which are reliable indicators of phylogeny (similar to the group partition index described above) (Swofford, 1990).

The numerical taxonomic and cladistic methods described above can yield incongruent results due to the very different underlying assumptions. Farris (1981, 1985) argues for the cladistic approach due to the information content of parsimony. Felsenstein (1984) and Nei (1987) defend bacterial numerical taxonomy methods and state that they are more appropriate. Depending on the source of prokaryotic data, both methods may or may not be appropriate. For example, hybridization data (DNA/DNA, DNA/rRNA) can only be analysed using one or more of the methods available in numerical taxonomy. Phenotypic and sequence data, on the other hand, are amenable to both approaches.

Likelihood Approaches

The most statistical of all the approaches to establishing phylogenies is the likelihood-based procedures (Felsenstein, 1988). Here explicit models are used to estimate branch lengths in evolutionary trees. There are cases in which likelihood methods can be shown to give results similar to UPGMA or to parsimony methods. The real strength of this set of procedures is that all the model assumptions need to be defined explicitly prior to solving a likelihood. It also helps in conceptualizing some of the strengths and/or weaknesses of the other procedures described above.

Conclusions

The biodiversity question will remain unanswered without careful planning and collaboration of multiple investigators. Resources will need to be invested in order to assure data quality and comparability of data combined from many sources. Too often, we see journal articles with a careful description of the molecular techniques used but very little attention paid to the methodology used in data analysis. Efforts need to be made to understand the limitations of all aspects of methodology used. More interdisciplinary research among statisticians and biologists is needed. Clearly statistics has a role to play in every facet of data analysis in the context of biodiversity.

References

Austin, B. and Priest, F. (1986) *Modern Bacterial Taxonomy.* Van Nostrand Reinhold, Wokingham, UK, 145 pp.

Bradley, S.G. (1980) DNA reassociation and base composition. In: Goodfellow, M. and Board, R.G. (eds) *Microbial Classification and Identification.* Academic Press, London, pp. 11–26.

Churchill, G.A. and Waterman, M.S. (1992) The accuracy of DNA sequences: estimating sequence quality. *Genomics* 14, 89–98.

Cochran, W.G. (1973) *Sampling Techniques,* 3rd edn. Wiley, New York.

De Ley, J. (1971) Hybridization of DNA. In: Norris, J.R. and Ribbons, D.W. (eds) *Methods in Microbiology,* Vol. 5A. Springer-Verlag, New York, pp. 311–329.

Digby, P.G.N. and Kempton, R.A. (1987) *Multivariate Analysis of Ecological Communities.* Chapman & Hall, New York, 206 pp.

Erdmann, V.A., Pieler, T., Wolters, J., Digweed, M., Vogel, D. and Hartmann, R. (1986) Comparative structural and functional studies on small ribosomal RNAs. In: Hardesty, B. and Kramer, G. (eds) *Structure, Function, and Genetics of Ribosomes.* Springer-Verlag, New York, pp. 164–183.

Excoffier, L., Smouse, P.E. and Quattro, J.M. (1992) Analysis of molecular variance inferred from metric distances among DNA haplotypes: application to human mitochondrial DNA restriction data. *Genetics* 131, 479–491.

Farris, J.S. (1969) On the cophenetic correlation coefficient. *Systematic Zoology* 18, 279–285.

Farris, J.S. (1981) Distance data in phylogenetic analysis. In: Funk, V.A. and Brooks, D.R. (eds) *Advances in Cladistics. Proceedings of the First Meeting of the Willi Hennig Society.* New York Botanical Garden, Bronx, New York.

Farris, J.S. (1985) Distance data revisited. *Cladistics* 1, 67–85.

Farris, J.S. (1989a) The retention index and rescaled consistency index. *Cladistics* 5, 417–419.

Farris, J.S. (1989b) The retention index and homoplasy excess ratio. *Systematic Zoology* 38, 406–407.

Felsenstein, J. (1984) Distance methods for inferring phylogenies: a justification. *Evolution* 39, 783–791.

Felsenstein, J. (1985) Confidence limits on phylogenies: an approach using the bootstrap. *Evolution* 39, 783–791.

Felsenstein, J. (1988) Phylogenies from molecular sequences: inference and reliability. *Annual Review of Genetics* 22, 521–565.

Fowlkes, E.B. and Mallows, C.I. (1983) A method for comparing two hierarchical clusterings. *Journal of the American Statistical Association* 78, 553–584.

Fuller, W.A. (1987) *Measurement Error Models.* Wiley, New York.

Gower, J.C. (1971) A general coefficient of similarity and some of its properties. *Biometrics* 27, 857–872,

Green, R.H. (1979) *Sampling Design and Statistical Methods for Environmental Biologists.* Wiley Interscience, New York.

Grimont, P.A.D. (1984) DNA/DNA hybridization in bacterial taxonomy. In: Sanna, A. and Morace, G. (eds) *New Horizons in Microbiology.* Elsevier Science Publishers, Amsterdam, pp. 11–19.

Grimont, P.A.D. and Popoff, M.V. (1980) Use of principal component analysis in interpretation of deoxyribonucleic acid relatedness. *Current Microbiology* 4, 337–342.

Hennig, W. (1966) *Phylogenetic Systematics.* University of Illinois Press, Urbana, 263 pp.

Hertzberg, V. and Russek-Cohen, E. (1993) Statistical methods in molecular epidemiology. In: *Molecular Epidemiology: Principles and Practices.* Academic Press, New York, 588 pp.

Holmquist, R., Cantor, C. and Jukes, T. (1972) Improved procedures for comparing homologous sequences in molecules of proteins and nucleic acids. *Journal of Molecular Biology* 64, 145–161.

Hori, H. and Osawa, S. (1979) Evolutionary change in 5S rRNA secondary structure and a phylogenic tree of 352 5S rRNA species. *BioSystems* 19, 163–172.

Hori, H. and Osawa, S. (1987) Origin and evolution of organisms as deduced from 5S ribosomal RNA sequences. *Molecular Biology of Evolution* 4, 445–472.

Hosmer, D. and Lemeshow, S. (1987) *Applied Logistic Regression.* Wiley Interscience, New York, 307 pp.

Hurlbert, R. (1984) Pseudoreplication and the design of ecological field experiments. *Ecological Monographs* 54, 187–211.

Johnson, J.L. (1989) Bacterial classification III – nucleic acids in bacterial classification. In: Williams, S.T., Sharpe, M.E. and Holt, J.G. (eds) *Bergey's Manual of Systematic Bacteriology*, Vol. 4. Williams & Wilkins, Baltimore, pp. 2306–2309.

Jones, D. and Sackin, M.J. (1980) Numerical methods in the classification and identification of bacteria with special reference to the Enterobacteriaceae. In: Goodfellow, M. and Board, R.G. (eds) *Microbiological Classification and Identification*. Academic Press, London, pp. 73–106.

Karlin, S. and Macken, C. (1991) Some statistical problems in the assessment of inhomogeneities of DNA sequence data. *Journal of American Statistical Association* 86, 27–35.

Kimura, M. (1980) A simple method for estimating evolutionary rates of base substitutions through comparative studies of nucleotide sequences. *Journal of Molecular Evolution* 16, 111–120.

Krichevsky, M.I. and Walczak, C.A. (1982) Computer-aided selection of group descriptors as exemplified by data on *Capnocytophaga* species. *Current Microbiology* 7(4), 199–204.

Krieg, N.R. (1988) Bacterial classification: an overview. *Canadian Journal of Microbiology*, 34, 536–540.

Kruskal, J. and Wish, M. (1978) *Multidimensional Scaling*. Sage Publications, Beverley Hills.

MacDonell, M.T. and Colwell, R.R. (1985) The contribution of numerical taxonomy to the systematics of Gram-negative bacteria. In: Goodfellow, M., Jones, D. and Priest, F.G. (eds) *Computer-assisted Bacterial Systematics*. Academic Press, New York, pp. 107–135.

MacDonell, M.T., Swartz, D.G., Ortiz-Conde, B.A., Last, G.A. and Colwell, R.R. (1986) Ribosomal RNA phylogenies for the vibrio–enteric group of eubacteria. *Microbiological Science* 3, 172–178.

MacDonell, M.T., Hansen, J.N. and Ortiz-Conde, B.A. (1987) Isolation, purification and enzymatic sequencing of RNA. In: Colwell, R.R. and Grigorova, R. (eds) *Methods in Microbiology*, Vol. 19. *Current Methods for Classification and Identification of Microorganisms*. Academic Press, London, pp. 357–404.

Magurran, A.E. (1988) *Ecological Diversity and Its Measure*. Princeton University Press, Princeton, NJ, 179 pp.

Manly, B.F. (1991) *Randomization and Monte Carlo Methods in Biology*. Chapman & Hall, New York, 281 pp.

Michel, C.J. (1986) New statistical approach to discriminate between protein coding and nonprotein coding regions in DNA sequences and its evaluation. *Journal of Theoretical Biology* 120, 223–236.

Nei, M. (1987) *Molecular Evolutionary Genetics*. Columbia University Press, New York, 610 pp.

Nelson, G. and Platnick, N. (1981) *Systematics and Biogeography: Cladistics and Vicariance*. Columbia University Press, New York, 567 pp.

Owen, R.J. and Pitcher, D. (1985) Current methods for estimating DNA base composition and levels of DNA–DNA hybridization. In: Goodfellow, M. and Minnikin, D.E. (eds) *Chemical Methods in Bacterial Systematics*. Academic Press, New York, pp. 67–93.

Pielou, E.C. (1984) *The Interpretation of Ecological Data: A Primer on Classification and Ordination*. Wiley, New York, 263 pp.

Russek, E. and Colwell, R.R. (1983) The analysis of most probable numbers. *Applied and Environmental Microbiology* 45, 1146–1150.

Russek-Cohen, E. and Colwell, R.R. (1986) Application of numerical taxonomy procedures in microbial ecology. In: Tate III, R.L. (ed.) *Microbial Autoecology: a Method for Environmental Studies*. Wiley, New York, pp. 133–146.

Russek-Cohen, E. and Jacobs, D. (1989) Detecting outliers in a 5S rRNA database. *Binary* 1, 115–125.

Ryan, T.P. (1989) *Statistical Methods for Quality Improvement*. Wiley, New York, 446 pp.

Schall, R. (1991) Estimation in generalized linear models with random effects. *Biometrika* 78, 719–727.

Shao, K. and Rohlf, F.J. (1983) Sampling distribution of consensus indices when all bifurcating trees are equally likely. In: Felsenstein, J. (ed.) *Numerical Taxonomy*. Springer-Verlag, New York.

Sneath, P.H.A. (1984) Numerical taxonomy. In: Krieg, N.R. and Holt, J.G. (eds) *Bergey's Manual of Systematic Bacteriology*, Vol. 1. Williams & Wilkins, Baltimore, pp. 5–7.

Sneath, P.H.A. and Sokal, R.R. (1973) *Numerical Taxonomy: The Principle and Practice of Numerical Classification*. W.H. Freeman, San Francisco, 573 pp.

Snedecor, G. and Cochran, W. (1967) *Statistical Methods*. Iowa State Press, Ames, 593 pp.

Sokal, R.R. and Rohlf, F.J. (1981) *Biometry*, 2nd edn. Freeman, New York, 859 pp.

Steven, S.E. (1990) Molecular systematics of *Vibrio* and *Photobacterium*. PhD dissertation, University of Maryland, College Park, 251 pp.

Swofford, D.L. (1990) *PAUP: Phylogenetic Analysis Using Parsimony, Ver. 3.0*. Illinois Natural History Survey, Champaign, IL, 134 pp.

Watrous, L.E. and Wheeler, Q.D. (1981) The outgroup comparison method of character analysis. *Systematic Zoology* 30, 1–11.

Weir, B. (1990) *Genetic Data Analysis*. Sinauer Associates, Sunderland, MA, 377 pp.

Woese, C.R. (1987) Bacterial evolution. *Microbiology Review* 51, 221–271.

Woese, C.R. (1991) Prokaryote systematics: the evolution of a science. In: Balows, A., Truper, H.G., Dworkin, M., Harder, W. and Schleifer, K.H. (eds) *The Prokaryotes*, 2nd edn, Vol. 1. Springer-Verlag, New York, pp. 3–18.

Traditional Methods of Detecting and Selecting Functionally Important Microorganisms from the Soil and the Rhizosphere

F.A.A.M. DE LEIJ[1], J.M. WHIPPS[2] AND J.M. LYNCH[1]

[1]*School of Biological Sciences, University of Surrey, Guildford, Surrey GU2 5XH, UK.;* [2]*Department of Microbial Biotechnology, Horticulture Research International, Worthing Road, Littlehampton, West Sussex BN17 6LP, UK.*

Introduction

The interactions that occur between the soil biota, plant roots and physical and chemical properties of the soil are of major importance to plant productivity. Accurate methods are needed to quantify the presence and activity of microorganisms in soil/root ecosystems and the role these organisms have in sustaining plant productivity. Plant–microbe interactions can be of a negative nature as is the case with fungal and bacterial pathogens, or positive, as with symbionts that supplement the plant with nutrients (i.e. nitrogen-fixing bacteria and many mycorrhizal fungi), as well as organisms that reduce the occurrence of plant diseases, produce growth-stimulating substances, break down organic components into nutrients that can be taken up by the plant, improve soil structure and improve the water-holding capacity of the soil.

In modern agricultural practices the activity of the soil biota has been largely marginalized by the use of agrochemicals, such as pesticides, which suppress damaging organisms or which, through application of chemical fertilizers, bypass nutrient cycles. However, with increasing pressure on farmers and governments to reduce the use of polluting chemicals, the role of the soil biota on plant productivity has gained increasing attention.

Despite this, scientists involved in soil ecology are faced with an almost unending array of species and interactions that are all likely to have an effect on the way a soil ecosystem behaves. Although the recent developments of molecular techniques have enabled scientists to look at specific aspects of microbial biology and microbial genetics (Pickup, 1991) it is unlikely that these techniques will solve the more holistic issues of microbial/soil/root interactions and their implications for plant productivity. It is therefore not surprising that the study of microbial ecology still heavily relies on the use of the more traditional methods such as bioassays, dilution plating on agars, most probable number (MPN) assays, respiration measurements and acetylene reduction tests (nitrogen fixation). Many of these techniques (like any technique) have obvious limitations in that they often rely on the culturability of the sampled species (Colwell *et al.*, 1985) or give a fairly imprecise and coarse answer to biological activity of the sampled microbial community. Despite these problems, traditional methods can, if used with appropriate care, help to elucidate some of the questions concerning microbial interactions in the rhizosphere. This, however, does not mean that there is no need to improve or adapt those methods to suit the problems that need investigation. In this chapter the use and limitations of some traditional methods are discussed, using the development of microbial biological control agents against nematode pests and the monitoring of environmental risks associated with the release of genetically modified microorganisms (GEMMOs) into the environment as two examples.

Development of Biological Control Agents

There are three questions that have to be considered in any research programme. First, why is the research done (motivation); second, what areas are of most importance (goals); and third, how can we best reach the set goals (methods). In the case of biological control, it is not difficult to answer the first question as effective biological control agents would reduce the use of environmentally harmful pesticides. However, a range of promising research areas is possible. For example, biological control may be achieved by either developing agricultural practices that help promote natural suppressive soils, or selecting one promising organism and developing this further as a biological pesticide. In the following sections, the latter approach, normally consisting of several steps, will be considered.

Selection of Effective Biological Control Agents

The first step in the development of a biological control agent normally involves selecting an effective organism against the target pest. Frequently, biological control agents are selected from soils that are suppressive to the target pest (Stirling, 1979; Crump, 1987; Kerry, 1990; Deacon, 1991). However, the isolation procedure itself determines to a large extent which organisms are to be found and there is no guarantee that the organisms isolated are the ones that are responsible for the disease/pest suppressiveness of the soil. For instance, observation chambers as described by Crump (1987) can be used to select fungal parasites of cyst and root knot nematodes. This method is very useful for the selection of (fungal) parasites of cyst and root knot nematodes as well as for the selection of (fungal) parasites that kill females and/or eggs; however, parasites of juveniles go undetected. In comparison, agar plates sprinkled with soil and baited with nematodes (Barron, 1977) will only detect fungi that destroy juveniles or free-living nematodes.

The isolation of fungi from diseased nematodes poses another problem. If a nematode female dies, and if a microorganism can be isolated from the cadaver, there is no guarantee that the isolated organism is a true nematode parasite. Plant pathogens in particular are known to invade and destroy the giant cells on which the nematode feeds (Fattah and Webster, 1983; Moussa and Hague, 1988) and so deprive the nematode of nutrients; therefore, plant pathogenic fungi might well be isolated from nematodes that are killed in this way. If selection is aimed at true nematode parasites, promising organisms that prevent nematodes from invading plant roots, such as mycorrhizas (Marx, 1972) and rhizosphere bacteria (Oostendorp and Sikora, 1989) are left undetected. Nevertheless, whatever method(s) are employed to select promising biological control agents, it is likely that a large collection of microbial isolates that might be involved in nematode suppression will be obtained.

Screening

The next task is to select from the collection of isolates that may be involved in nematode suppression, the one or two organisms that are particularly effective as biological control agents and which can be easily mass produced. If large numbers of organisms have to be screened there is a tendency to develop an *in vitro* screening programme. Although quick and simple, the disadvantage of such a programme is that it tests organisms out of their ecological context; effectiveness *in vitro* is no guarantee for effectiveness in the field (Deacon, 1991). A better approach is, therefore, to screen organisms in a more realistic way. The most commonly used strategy is the empirical approach whereby candidate organisms are introduced into field soil or onto seeds that are subsequently planted in field soil. Effects are compared with an appropriate control treatment. The control treatment in such experiments is crucial, since the biomass or the food base, which is often used to help establishment of the biocontrol agents, can by itself have a dramatic effect on the target organism by stimulation of indigenous microbial activity (de Leij and Kerry, 1991). Sadly, from 25 experiments purporting nematode control with the fungus *Paecilomyces lilacinus*, only half were conducted with a suitable control treatment (Kerry, 1990). Although quick and more reliable than the *in vitro* screening procedure, empirical screening is very much a hit or miss approach. The disadvantage is that organisms with great potential are easily missed, because the organism is applied either in the wrong form or in a situation that does not offer the right window of opportunity for effective biological control. For example the method of mass production (liquid versus solid culture) can have a significant effect on how the organism behaves in soil (McQuilken *et al.*, 1992). In natural suppressive soils, different nematophagous fungi succeed each other (Crump, 1987). If an organism is selected that dominates late in the season, it is quite useless to test its efficacy on young plants. Ecological understanding of the organisms that are to be tested is therefore of great importance if unnecessary failure is to be avoided.

Gaining Insight in the Ecology of the Biological Control Agent

To illustrate the importance of ecological understanding in the development of biological control agents the ecology of the nematophagous fungus *Verticillium chlamydosporium* will be used as an example. This fungus is a facultative parasite of cyst and root knot nematodes (Willcox and Tribe, 1974; Morgan-Jones *et al.*, 1981; Freire and Bridge, 1985) and the ease with which it can be mass produced on a range of substrates makes it an ideal organism to be used for augmentive purposes (Tribe, 1980). Since *V. chlamydosporium* is a facultative parasite, its interactions with its environment are of great importance for its efficacy as a biological control agent. The interactions of this fungus with its environment can be visualized as follows (Fig. 18.1; Kerry and de Leij, 1991).

In order to study these interactions a range of criteria must be met. The fungus must be accurately monitored in non-sterile soil and on the roots of the host plant; its activity must be measured in the environment; the nematode population must be accurately monitored to determine the effect of the fungus; and, preferably, it must be possible to reisolate the fungus from dead nematodes so that effects can be directly related to the activity of the fungus. The development of a (semi) selective medium for *V. chlamydosporium* (de Leij and Kerry, 1991) made it possible to monitor the fungus in non-sterile environments. The advantage of such a medium is that it is easy to use and it allows processing of many samples taken in time or from different parts of the soil or the root system. The disadvantage is that it only provides information on the number of propagules that are present which may be in the form of hyphae, conidia or resting spores and not on the activity of these propagules. This is a common problem with the use of

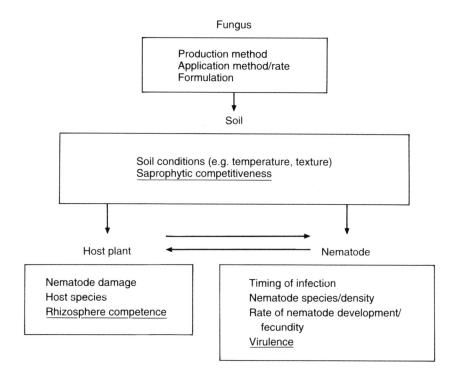

Fig. 18.1. Factors affecting the establishment of *Verticillium chlamydosporium* in soil and on the roots and the numbers of nematodes infected.

culture media. Lumsden *et al.* (1990) found that in the case of biological control activity of *Trichoderma harzianum* against the damping-off disease *Pythium ultimum*, colony forming unit (cfu) counts were inversely correlated with the fungal activity, which suggested that the fungus produces large amounts of non-active conidia under conditions that affect fungal activity adversely. Similar findings apply to *V. chlamydosporium* (de Leij *et al.*, 1992b). It is clear, therefore, that cfu counts have to be supplemented with other measurements to establish whether the fungus is active or not. In the case of *T. harzanium*, ATP measurements and chitin utilization provided the necessary complementary measures (Lumsden *et al.*, 1990) whereas fungal cfu formation *in vitro* provided the clues which were necessary for the interpretation of cfu counts from environmental samples (de Leij *et al.*, 1992b).

Reisolation of the biological control agent from the host is seen as an important prerequisite to establish biological control activity (Stirling, 1988) and indeed if the organism can be found to be associated with the

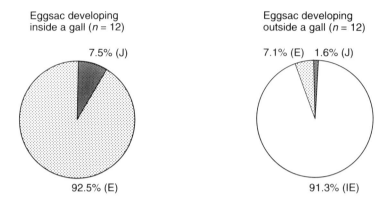

Fig. 18.2. Availability of *Meloidogyne incognita* eggs developing inside and outside the root gall to *Verticillium chlamydosporium* (E = healthy eggs; IE = infected eggs; J = juveniles).

target pest, it is a clear sign that the organism is active and effective. This, however, is only reasonable if the biological control agent is a true parasite (and not a competitor or has inhibitory properties) and the pest organism can be clearly identified and isolated (as is the case with cyst and root knot nematodes, but not with most fungal and bacterial pathogens). Although, when reisolation or, even better, direct quantification of parasitized pest organisms is possible, it will establish activity, it seldom establishes efficacy. Measuring efficacy requires some understanding of the life cycle and the damage caused by the pest. In the case of measuring biological control activity of *V. chlamydosporium*, several parameters that are related to biological control activity can be measured. First, the proportion of infected eggs can be quantified. This parameter is related to the time of sampling. The fungus needs time to infect the eggs in an egg mass; the later in the life cycle of the nematode that sampling takes place, the greater the proportion of infected eggs that are found. Because infected eggs are not extracted using the extraction procedures described by Coolen and d'Herde (1973), egg masses have to be picked off individually from the root system and examined using a microscope. Egg masses that are not exposed on the root surface are in general missed, and an overestimation of fungal efficacy is the likely result (Fig. 18.2; de Leij *et al.*, 1992a). The second possibility is to use the total number of eggs that are produced by the nematode as a measure for efficacy. Although the extraction procedures described by Coolen and d'Herde (1973) can be employed, this method suffers from the same disadvantage as the previously described procedure in that the measurement is dependent on the time of sampling. An underestimation of efficacy is the likely result. The only measurement that reliably estimates nematode control, independent of the time of sampling, is the measurement of the proportion of fully embryonated eggs and juveniles in a sample, since those have escaped fungal infection (Irving and Kerry, 1986).

Another possibility is the use of bioassays in which plant damage is directly measured (Bridge and Page, 1980). Reliable bioassays normally involve the use of the biological control agent against a range of pest densities. Because a biological control agent is seldom 100% effective, high levels of the pest can give considerable damage resulting in a failure to detect biological control activity using bioassays, despite the possibility of considerable activity of the biological control agent (de Leij *et al.*, 1992c). This brings us full circle back to the empirical approach which is aimed at measuring the potential of biological control agents against certain pests and diseases. The ecological understanding gained in this type of approach can now be used to devise more sophisticated and reliable tests (de Leij *et al.*, 1993a).

Other Approaches

If reliable biological activity can be induced using the programme described above, there are several possibilities to progress the work. To develop a commercially viable product, programmes should be started that aim to mass-produce, formulate and market the product and further test the product's reliability to control pests or diseases. To develop further our ecological understanding of the biological control agent, it is possible to develop mathematical models in which key parameters can be used to predict the activity and effect of the organism under different sets of circumstances. In addition, to extend the study of the modes of action of the organism, traditional methodology has to be combined with enzyme assays, DNA probes and other molecular methods. Although these molecular methods will provide scientifically interesting information, they are unlikely to increase the efficacy or extend the range of applications of biological control agents.

Traditional Methods to Monitor Survival, Spread and Impact of Genetically Modified Microorganisms (GEMMOs)

Genetically modified microorganisms (GEMMOs) have potential for crop protection, improvement of soil structure and nutrient status, bioremediation and mineral leaching. However, before release of such functionally

Table 18.1. Indigenous soil-inhabiting bacteria (all species and fluorescent pseudomonads) resistant to 100 ppm kanamycin and/or able to utilize lactose as a carbon source ($n = 3$).

Component	Medium	Cells g^{-1} soil	SE	Percentage of total population
Total bacteria g^{-1} soil	0.1 strength TSA	2.5×10^7	2.6×10^6	
Catechol utilization	0.1 strength TSA	0	0	0
Kanamycin (km) resistance	0.1 strength TSA, km (100 ppm)	8.8×10^5	4.4×10^4	3.5
Lactose utilization	M9, lactose (1%), Xgal (50 ppm)	2.3×10^5	3.3×10^4	0.93
Lactose utilization + kanamycin (km) resistance	M9, lactose (1%), Xgal (50 ppm), km (100 ppm)	8.5×10^4	1.2×10^3	0.03
				Percentage pseudomonad population
Total pseudomads g^{-1} soil	P-1	8.9×10^5	8.8×10^4	
Catechol utilization	P-1	0	0	0
Kanamycin (km) resistance	P-1, km (100 ppm)	15	15	0.0017
Lactose utilization	P-1, Xgal (50 ppm)	0	0	0
Lactose utilization + kanamycin (km) resistance	P-1, Xgal (50 ppm), km (100 ppm)	0	0	0

Media used: TSA, triptic soy agar; P-1, selective agar for pseudomonads; M9, minimal salts agar.

active GEMMOs into the environment can take place, a better understanding of behaviour of microbes released into the environment and their effects on indigenous microbial populations is needed (Tiedje *et al.*, 1989; Smit *et al.*, 1992). To achieve this goal, three main areas connected with GEMMO introductions need to be addressed. First, more information is needed on the survival and dispersal of GEMMOs in the environment. Second, information is needed on possible interactions of GEMMOs with other species and/or biological systems in the environment where introduction will take place, and third, information is needed on the transfer of introduced genetic material to indigenous organisms.

Spread and Survival in the Environment

Monitoring of spread and survival of a GEMMO in the environment is probably the easiest of the three areas mentioned above to solve. The methods used to monitor spread and survival may involve the insertion of marker genes into the genome of innocuous bacteria. These markers can encode for antibiotic resistance and/or consist of metabolic markers that enable the GEMMO to utilize specific substrates (Drahos *et al.*, 1988). Media with the appropriate antibiotics and substrates incorporated into them can be used to isolate viable propagules from environmental samples. Such a system was employed to study the behaviour of the common rhizosphere bacterium *Pseudomonas aureofaciens* (de Leij *et al.*, 1993b). This bacterium was chromosomally marked with the gene cassettes *lacZY* and km^r-*xylE* (Bailey, unpublished). The *lacZY* system (Drahos *et al.*, 1986, 1988; Barry, 1988) is one of the most widely used metabolic markers (Kluepfel and Tonkyn, 1992). In this system the *Escherichia coli* genes *lacZ* (β-galactosidase) and *lacY* (lactose permease) enables the modified organism to utilize lactose as a sole carbon source. When organisms with the *lacZY* genes are plated out on media containing the chromogenic substrate X-gal, the cleavage of this substrate results in easily identifiable blue colonies. The km^r gene encodes for resistance to the antibiotic kanamycin and the *xylE* gene enables the conversion of the substrate catechol to 2-hydroxymuconic semi-aldehyde, a yellow product. However, natural field soil can harbour a large proportion of organisms that can utilize lactose, are kanamycin-resistant or are both kanamycin-resistant and able to utilize lactose. For example, with a minimal medium (Sambrook *et al.*, 1989) that was amended with 100 ppm kanamycin, 50 ppm X-gal and 1% lactose, 0.03% of the total culturable microbial population from a brick-earth soil formed colonies (Table 18.1). This meant that, with dilution-plating, recombinant cells could only be detected if the concentration of GEMMOs was greater than 300–1000 cells g^{-1} soil. On roots where bacterial numbers are normally one to two log units higher than in soil, detection would only be possible if the recombinant was present in concentrations $> 10^4$ cells g^{-1} root. Consequently, a more sensitive method to recover recombinant bacteria from the environment was required. The use of a selective medium for fluorescent *Pseudomonas* spp. that is based on the capacity of fluorescent pseudomonads to utilize betaine as a sole carbon source (Katoh and Itoh, 1983) proved to be a better basal medium than the one described by

Sambrook *et al.* (1989). Incorporation of 100 ppm kanamycin and 50 ppm Xgal into this medium for the selection and identification of the recombinant provided conditions where only the recombinant could grow (Table 18.1). Since only 1% of a 1-g soil or root sample is normally plated out onto a 9 cm diam. Petri dish, recovery of recombinant cells using dilution-plating is possible if the concentration of recombinants exceeds 100 cfu g^{-1} sample. By using an MPN assay which gives reliable results if the concentration of soil or roots does not exceed 1% in the selection broth, this detection level could be improved. By using bottles that contain 100 ml selection broth, recombinants can routinely be found when only one cfu g^{-1} soil or roots is present (de Leij *et al.*, 1993b). This method is currently used successfully for monitoring the above-described GEMMO in a field release study. Although the method is extremely sensitive, some care has to be taken that no alternative carbon sources are added to the medium. To minimize this kind of contamination, root and leaf samples are sonicated to dislodge bacterial cells and any plant material is removed before the bacterial suspension can be processed with this MPN method.

Quantification of Interactions of GEMMOs with Other Species and/or Biological Systems in the Environment

An ecosystem that provides the right conditions for plant growth and crop productivity might have the following characteristics:

1. Suppressive to a range of pests and diseases.
2. Good functioning carbon and nitrogen cycles.
3. A soil structure that is optimal for root growth.
4. Good water retention characteristics.

The introduction of organisms that multiply and spread through the environment and have a negative effect on such ecosystems should be avoided, since, once introduced, such organisms cannot be removed from the environment without great difficulty. Thus negative effects on ecosystem functioning that might result from GEMMO introductions must be addressed for risk assessment, but can be very difficult to quantify (Smit *et al.*, 1992). A useful strategy to assess environmental risks resulting from GEMMO introductions might be to compare the above-mentioned characteristics in soil ecosystems where GEMMOs are introduced with control ecosystems without GEMMOs. This approach is similar to the empirical approach described in testing biological control agents; it will give a crude indication if introductions of a particular GEMMO have an effect on ecosystem functioning. However, without an understanding of the ecological interactions involved, this is likely to be a hit or miss approach. As with the development of biological control agents, it is important to develop an ecological understanding of the interactions that might occur between the GEMMO and the environment. To develop this kind of understanding, the biological characteristics of the GEMMO (the wild type organism from which it was derived as well as the novel genes and their function) as well as the ecological niche in which the GEMMO can be active have to be taken into account. Only if both these aspects can be quantified accurately can the risk associated with the release of a GEMMO be assessed accurately (Straus et al., 1985).

In the case of the double-marked *P. aureofaciens* described earlier, the marking system in combination with the use of suitable selection procedures can provide the tools to define the ecological niche that this organism occupies. Although this organism can colonize a wide variety of host plants it seems to do particularly well in habitats that are uncrowded and provide readily available nutrients. In situations where microbial competition is great, growth and survival are severely limited (de Leij et al., 1994). In ecological terms this strain of *P. aureofaciens* can be described as a typical *r*-strategist (MacArthur and Wilson, 1967; Andrews and Harris, 1986). The concept of *r*/*K* strategy can also be used to describe possible effects this bacterium might have on other organisms that share its environmental niche. When the microbial community that inhabits the rhizosphere is plated onto a general bacterial medium, colonies will appear after different time intervals. The distribution of fast and slow developing colonies seems to fit an *r*/*K* continuum and can be used to quantify disturbances of the indigenous microbial community after introductions of this rhizosphere-competent GEMMO (de Leij et al., 1993c). Although useful, this technique is purely ecological and does not give any information about the functional properties of the microbial community that might be affected. Furthermore, characterization of a microbial community on an agar plate does not tell anything about the activity of the isolated bacteria *in situ* (Lumsden et al., 1990). However, on the *Pseudomonas* medium described by Katoh and Itoh (1983) it appears that the time taken for a bacterial colony to appear is related to the physiological state of the bacterium; colonies that appear quickly might be in a state of growth *in situ*, whereas those that appear after a long incubation period might come from starved or non-active cells (F.A.A.M. de Leij, unpublished). This would support some of the findings of Hattori (1982, 1983) who used colony appearance on agar to describe the physiological state of the bacteria that were isolated from different environmental samples. Further development of these traditional plate methods might therefore be of use to describe some important features of microbial ecology that cannot be described using modern molecular techniques, or assessments that use either total culturable bacterial counts on agars, direct counting methods of bacterial cells or direct viable counting methods (Kogure et al., 1979).

Gene Transfer to Indigenous Microbial Populations

Although the use of enzyme polymorphisms has indicated that some bacterial populations inhabiting the environment have a clonal structure (Milkman, 1973; Selander and Levin, 1980), there is ample evidence that bacteria can exchange genetic information by a variety of mechanisms both within and between different species as well as different genera (Reanney *et al.*, 1983; Smit *et al.*, 1991). Bacteria may harbour a great variety of mobile genetic elements, such as plasmids, transposons and insertion elements that have at least the potential for considerable genetic exchange and also rearrangement of chromosomal genes (Syvanen, 1984; Bale *et al.*, 1987; Klingmüller *et al.*, 1990; Dykhuizen and Green, 1991; Istock *et al.*, 1992).

Establishment of new genetic elements in a microbial community is determined by the mobility of the genetic information and by the selective advantage that acquisition of the novel genes gives to the recipient organism. The latter is probably of greater importance than the former. Therefore, it is of importance that inserted genetic information either does not give a selective advantage to the GEMMO (Smit *et al.*, 1992) or gives only a temporary advantage as is the case with GEMMOs that are designed for bioremediation (Brokamp and Schmidt, 1991) or specific biological control purposes. When the polluting compounds are degraded or the pest is reduced, the novel genes will be a burden, and recombinants as well as possible transconjugants are likely to disappear from the environment (Tiedje *et al.*, 1989; Lenski, 1991). If proper care is taken that the inserted genes offer no, or only a temporary, selective advantage, monitoring gene transfer to other members of the microbial community is more of scientific interest than of environmental concern. Efforts to monitor the transfer of genes that have no selective advantage in the natural environment are likely to be very time consuming and with an extremely small chance of success. Monitoring transfer of genes that give a temporary advantage can be done by sampling the target habitat and subsequent dilution-plating (as in the case of improved biological control agents) or by plating samples from the polluted site onto media that have the polluting agent incorporated (as in the case of bioremediation). Bacteria, other than the recombinant, can then be isolated and tested to determine if the novel genetic elements are present using methods such as DNA hybridization and polymerase chain reactions.

Conclusions

Traditional methods are important to study and select functionally important microorganisms in the environment. Although empirical methods normally give quick answers under realistic circumstances, they provide little ecological information and are therefore very much a hit or miss approach. Ecological understanding of biological processes is essential in the development of biological control agents and assessment of risks associated with the release of GEMMOs. Traditional methods can be used to elucidate many aspects of microbial ecology and there is scope to develop these methods further to answer some of the more holistic issues related to soil microbial ecology.

Acknowledgements

Parts of the work presented in this paper were funded by the Agricultural Genetics Company and the Department of the Environment. The views expressed in this chapter are those of the authors and not necessarily those of the Department of the Environment.

References

Andrews, J.H. and Harris, R.F. (1986) r- and K-selection and microbial ecology. *Advances in Microbial Ecology* 9, 99–147.
Bale, M.J., Fry, J.C. and Day, M.J. (1987) Plasmid transfer between strains of *Pseudomonas aeruginosa* on membrane filters attached to river stones. *Journal of General Microbiology* 133, 3099–3107.
Barron, G.L. (1977) *The Nematode Destroying Fungi*. Canadian Biological Publications Ltd, Guelph, 140 pp.
Barry, G.F. (1988) A broad host range shuttle system for gene insertion into chromosomes of Gram-negative bacteria. *Gene* 71, 75–84.
Bridge, J. and Page, S.L.J. (1980) Estimation of root knot nematode infestation levels using a rating chart. *Tropical Pest Management* 26, 296–298.
Brokamp, A. and Schmidt, F.R.J. (1991) Survival of *Alcaligenes xylosoxidans* degrading 2,2-dichloropropionate and horizontal transfer of its halidohydrolase gene in a soil microcosm. *Current Microbiology* 22, 229–306.
Colwell, R.R., Brayton, P.R., Grimes, D.J., Roszak, D.B., Huq, S.A. and Palmer, L.M. (1985) Viable but non-culturable *Vibrio cholerae* and related pathogens in the environment: implications for release of genetically engineered microorganisms. *Biotechnology* 3, 817–820.
Coolen, W.A. and d'Herde, C.J. (1973) *A Method for the Quantitative Extraction of Nematodes from Plant Tissue*. Ghent State Agriculture Centre, 77 pp.
Crump, D.H. (1987) A method for assessing the natural control of cyst nematode populations. *Nematologica* 33, 232–243.
Deacon, J.W. (1991) Significance of ecology in the development of biocontrol agents against soil-borne plant pathogens. *Biocontrol Science and Technology* 1, 5–20.
de Leij, F.A.A.M. and Kerry, B.R. (1991) The nematophagous fungus *Verticillium chlamydosporium* as a potential biological control agent for *Meloidogyne arenaria*. *Revue de Nématologie* 14, 157–164.
de Leij, F.A.A.M., Davies, K.G. and Kerry, B.R. (1992a) The use of *Verticillium*

chlamydosporium Goddard and *Pasteuria penetrans* (Thorne) Sayre & Starr alone and in combination to control *Meloidogyne incognita* (Kofoid & White) Chitwood on tomato plants. *Fundamental and Applied Nematology* 15, 235–242.

de Leij, F.A.A.M., Dennehy, J.A. and Kerry, B.R. (1992b) The effect of temperature and nematode species on interactions between the nematophagous fungus *Verticillium chlamydosporium* and root-knot nematodes (*Meloidogyne* spp.). *Nematologica* 38, 65–79.

de Leij, F.A.A.M., Kerry, B.R. and Dennehy, J.A. (1992c) The effect of fungal application rate and nematode density on the effectiveness of *Verticillium chlamydosporium* as a biological control agent for *Meloidogyne incognita*. *Nematologica* 38, 112–122.

de Leij, F.A.A.M., Dennehy, J.A. and Kerry, B.R. (1993a) *Verticillium chlamydosporium* as a biological control agent for *Meloidogyne incognita* and *M. hapla* in pot and micro-plot tests. *Nematologica* 39, 115–126.

de Leij, F.A.A.M., Bailey, M.J., Whipps, J.M. and Lynch, J.M. (1993b) A simple most probable number technique for the sensitive recovery of a genetically modified *Pseudomonas aureofaciens* from soil. *Letters in Applied Microbiology* 16, 307–310.

de Leij, F.A.A.M., Whipps, J.M. and Lynch, J.M. (1993c) The use of colony development for the characterisation of bacterial communities in soil and on roots. *Microbial Ecology* 27, 81–97.

de Leij, F.A.A.M., Sutton, E.J., Whipps, J.M. and Lynch, J.M. (1994) Spread and survival of a genetically modified *Pseudomonas aureofaciens* in the phytosphere of wheat. *Applied Soil Ecology* 1, 207–218.

Drahos, D.J., Hemming, B.C. and McPherson, S. (1986) Tracking recombinant organisms in the environment: β-galactosidase as a selectable non-antibiotic marker for fluorescent pseudomonads. *Bio/Technology* 4, 439–444.

Drahos, D.J., Barry, G.F. and Hemming, B.C. (1988) Pre-release testing procedures: US field test of a *lac*ZY-engineered bacterium. In: Sussman, M., Collins, G.H., Skinner, F.A. and Steward-Tull, D.E. (eds) *The Release of Genetically Engineered Micro-organisms*. Academic Press, San Diego, pp. 181–192.

Dykhuizen, D.E. and Green, L. (1991) Recombination in *Escherichia coli* and the definition of biological species. *Journal of Bacteriology* 173, 7257–7268.

Fattah, F.A. and Webster, J.M. (1983) Ultrastructural changes caused by *Fusarium oxysporum* f.sp. *lyscoperici* in *Meloidogyne javanica* induced giant cells in *Fusarium* resistant and susceptible tomato cultivars. *Journal of Nematology* 15, 128–155.

Freire, F.C.O. and Bridge, J. (1985) Parasitism of eggs, females and juveniles of *Meloidogyne incognita* by *Paecilomyces lilacinus* and *Verticillium chlamydosporium*. *Fitopathologi Brasileira* 10, 577–596.

Hattori, T. (1982) Analysis of plate count data of bacteria in natural environments. *Journal of General and Applied Microbiology* 28, 13–22.

Hattori, T. (1983) Further analysis of plate count data of bacteria. *Journal of General and Applied Microbiology* 29, 9–16.

Irving, F. and Kerry, B.R. (1986) Variation between strains of the nematophagous fungus *Verticillium chlamydosporium* Goddard. II. Factors affecting parasitism of cyst-nematodes. *Nematologica* 32, 474–485.

Istock, C.A., Duncan, K.E., Ferguson, N. and Zhou, X. (1992) Sexuality in a natural population of bacteria – *Bacillus subtilis* challenges the clonal paradigm. *Molecular Ecology* 1, 95–103.

Katoh, K. and Itoh, K. (1983) New selective media for *Pseudomonas* strains producing fluorescent pigment. *Soil Science and Plant Nutrition* 29, 525–532.

Kerry, B.R. (1990) An assessment of progress toward microbial control of plant-parasitic nematodes. *Journal of Nematology* 22, 621–631.

Kerry, B.R. and de Leij, F.A.A.M. (1991) Key factors in the development of fungal agents for the control of cyst and root-knot nematodes. In: *Biological Control of Plant Diseases: Progress and Challenges for the Future*. NATO, Athens.

Klingmüller, W., Dally, A., Fentner, C. and Steinlein, M. (1990) Plasmid transfer between soil bacteria. In: Fry, J.C. and Day, M. (eds) *Bacterial Genetics in Natural Environments*. Chapman & Hall, London, pp. 133–151.

Kluepfel, D.A. and Tonkyn, D.W. (1992) The ecology of genetically altered bacteria in the rhizosphere. In: Tjamos, E.S. (ed.) *Biological Control of Plant Diseases*. Plenum Press, New York, London, pp. 407–413.

Kogure, K., Simidu, U. and Taga, N. (1979) A tentative direct microscopic method for counting live marine bacteria. *Canadian Journal of Microbiology* 25, 415–420.

Lenski, R.E. (1991) Quantifying fitness and gene stability in microorganisms. In: Ginzburg, L.R. (ed.) *Assessing Ecological Risks of Biotechnology*. Butterworth-Heinemann, Boston, Massachusetts, pp. 173–192.

Lumsden, R.D., Carter, J.P., Whipps, J.M. and Lynch, J.M. (1990) Comparison of biomass and viable propagule measurements in the antagonism of *Trichodermia harzianum* against *Pythium ultimum*. *Soil Biology and Biochemistry* 22, 187–194.

MacArthur, R.H. and Wilson, E.O. (1967) *The Theory of Island Biography*. Princeton University Press, Princeton, 203 pp.

Marx, D.H. (1972) Ectomycorrhizae as biological deterrents to pathogenic root infections. *Annual Review of Phytopathology* 10, 429–454.

McQuilken, M.P., Whipps, J.M. and Cooke, R.C. (1992) Use of oospore formulations of *Pythium oligandrum* for biological control of *Pythium* damping-off in cress. *Journal of Phytopathology* 135, 125–134.

Milkman, R. (1973) Electrophoretic variation in *Escherichia coli* from natural sources. *Science* 182, 1024–1026.

Morgan-Jones, G., Godoy, G. and Rodriguez-Kabana, R. (1981) *Verticillium chlamydosporium* fungal parasite of *Meloidogyne arenaria* female. *Nematropica* 11, 115–119.

Moussa, E.M. and Hague, N.G.M. (1988) Influence of *Fusarium oxysporum* f.sp. *glycines* on the invasion and development of *Meloidogyne incognita* on soybean. *Revue de Nématologie* 11, 437–439.

Oostendorp, M. and Sikora, R.A. (1989) Seed treatments with antagonistic rhizobacteria for the suppression of *Heterodera schachtii* early root infection of sugar beet. *Revue de Nématologie* 12, 77–83.

Pickup, R.W. (1991) Development of molecular methods for the detection of specific bacteria in the environment. *Journal of General Microbiology* 137, 1009–1019.

Reanney, D.C., Gowland, P.C. and Slater, J.H. (1983) Genetic interactions among microbial communities: In: Slater, J.H., Whittenbury, R. and Wimpenny, J.W.T. (eds) *Microbes in their Natural Environment*. Cambridge University Press, Cambridge, pp. 379–421.

Sambrook, J., Fritsch, E.F. and Maniatis, I. (1989) *Molecular Cloning: a Laboratory Manual*, Vol. 3. Cold Spring Harbor Laboratory Press, Cold Spring Harbor, New York.

Selander, R.K. and Levin, B.R. (1980) Genetic diversity and structure in *Escherichia coli* populations. *Science* 210, 545–547.

Smit, E., van Elsas, J.D., van Veen, J.A. and de Vos, W. (1991) Detection of plasmid transfer from *Pseudomonas fluorescens* to indigenous bacteria in soil by using bacteriophage φR2f for donor counterselection. *Applied and Environmental Microbiology* 57, 3482–3488.

Smit, E., van Elsas, J.D. and van Veen, J.A. (1992) Risk associated with the application of genetically modified microorganisms in terrestrial ecosystems. *Microbial Review* 88, 263–278.

Stirling, G.R. (1979) Techniques for detecting *Dactylella oviparasitica* and evaluating its significance in field soils. *Journal of Nematology* 11, 99–100.

Stirling, G.R. (1988) Biological control of plant-parasitic nematodes. In: Poinar, G.O. and Jansson, H.B. (eds) *Diseases of Nematodes*, Vol. 2. CRC Press, Boca Raton, FL, pp. 93–139.

Straus, H.S., Hattis, D., Page, G., Harrison, K., Vogel, S. and Caldart, C. (1985) Direct release of genetically engineered microorganisms: a preliminary framework for risk evaluation under TSCA. *Report no. CTPID 85-3*, Centre for Technology, Policy and Industrial Development, Massachusetts Institute of Technology, Cambridge, MA.

Syvanen, M. (1984) The evolutionary implications of mobile genetic elements. *Annual Review of Genetics* 18, 271–293.

Tiedje, J.M., Colwell, R.K., Grossman, Y.L., Hodson, R.E., Lenski, R.E., Mack, R.N. and Regal, P.J. (1989) The planned introduction of genetically modified organisms: ecological considerations and recommendations. *Ecology* 70, 298–315.

Tribe, H.T. (1980) Prospects for the biological control of plant parasitic nematodes. *Parasitology* 81, 619–639.

Willcox, J. and Tribe, H.T. (1974) Fungal parasitism in cysts of *Heterodera* I: Preliminary investigations. *Transactions of the British Mycological Society* 62, 585–594.

Problems in Measurements of Species Diversity of Macrofungi 19

E. ARNOLDS

Biological Station, Centre for Soil Ecology, Wageningen Agricultural University, Kampsweg 27, 9418 PD Wijster, The Netherlands.[1]

Introduction

Fungi are essential components of terrestrial ecosystems by their contribution to the decomposition and mineralization of organic compounds and by the ability of some species to form mutual symbioses with algae (lichens) and higher plants (mycorrhizas). This chapter is restricted to the methodological aspects of research on diversity of macrofungi, an artificial group based on the size of the reproductive structures (sporocarps, sporophores, carpophores), which are clearly visible to the naked eye, i.e. larger than 1 mm. The division between macrofungi and microfungi is neither sharp, nor fundamental in taxonomic or ecological respect. Therefore it seems meaningful to spend a few words on the reasons for a separate treatment of these groups. The division is particularly based on practical differences in methodology of field studies: research on microfungi is related to microbiological studies and mainly carried out with various isolation techniques and subsequent culturing of the isolates on artificial media in the laboratory (Gams, 1992). Research on macrofungi is mainly carried out using sporocarps in the field as indicators of the presence of certain species. This approach resembles methods used in the floristics and biogeography of green plants and, on the level of ecosystems, methods used in phytocoenology. The analysis of communities of macrofungi is called mycocoenology, the methodology of which is described by Winterhoff (1984), Barkman (1987) and Arnolds (1992b), among others. Macrofungi are relatively rarely isolated using microbiological techniques due to the fragmentation of the

---
[1] Comm. no. 507 of the Biological Station Wÿster.
© 1995 CAB INTERNATIONAL. *Microbial Diversity and Ecosystem Function*
(eds D. Allsopp, R.R. Colwell and D.L. Hawksworth)

mycelia and their relatively slow growth rate, so that they are usually not able to compete for carbon with microfungi and bacteria from the same substrate.

Fortunately, the division into micro- and macrofungi roughly coincides with some broad taxonomic and ecological groups. The majority of macrofungi, in The Netherlands represented by ±3500 species, belong to the Holobasidiomycetes, and the rest to the Heterobasidiomycetes and some orders of Ascomycetes. Microfungi belong mainly to the Ascomycetes (where the borderline between the two groups is very arbitrary indeed) and Deuteromycetes. Most macrofungi are involved in the decomposition of complex organic molecules, such as cellulose, hemicellulose and lignin, the latter substance being almost exclusively degraded by larger basidiomycetes. Most microfungi utilize more simple organic compounds or special substances such as keratin or chitin. In symbiotic relationships macrofungi are mainly involved in ectomycorrhizas with a limited number of woody plants, whereas microfungi predominantly participate in other types of mycorrhizas (Harley and Smith, 1983). Lichenized fungi are not considered in this chapter. In addition, most soil-inhabiting microfungi appear to be almost cosmopolitan and possess a very wide ecological range, whereas macrofungi tend to be more specialized and restricted in their distribution (Gams, 1992).

As stated above, macrofungi are to date, with few exceptions, only recognizable by their reproductive structures. This presents some fundamental problems. The occurrence of sporocarps does with certainty indicate the presence of a mycelium in the substrate, but the absence of sporocarps does not necessarily mean that the fungus is absent. The numbers of sporocarps do not necessarily reflect the abundance of mycelia. In addition, sporocarps of most species of macrofungi show strong temporal dynamics due to their limited longevity, pronounced periodicity and strong fluctuations. These factors complicate studies on the diversity of macrofungi considerably. This chapter will treat the connected methodological problems in some detail, exemplified by the results of mycocoenological monitoring in a grassland plot during seven years.

Longevity of Sporocarps

Sporocarps of macrofungi are more or less ephemeral, but detailed data on the longevity of sporocarps of various species are surprisingly scarce in the mycological literature. One of the noticeable exceptions is the study by Richardson (1970) in a Scots pine plantation in Scotland, who found for 11 more frequent species with larger sporocarps mean lifetimes between 4 and 19 days. The longevity of sporocarps differs from species to species and may also be strongly influenced by (i) predation, in particular by insects and snails, but also by mammals such as squirrels and roe deer (Richardson, 1970), (ii) incidence of parasitic moulds, e.g. the ascomycete *Apiocrea chrysospermus* on *Boletales* and (iii) weather conditions. Almost all fleshy sporocarps are sensitive to drought, especially effective in summer, and most of them cannot stand temperatures below $-5°C$, which usually terminate the fruiting season at the end of autumn. Sporocarps of a single species are usually more persistent in cool periods in late autumn than in early autumn or summer, when the decay rate, predation and risk of desiccation are larger.

On the basis of potential longevity (i.e. without taking into account premature dying due to external factors) the following schematic division into types of sporocarps may be made.

1. Ephemerals: sporocarps complete their development within one or two days, sometimes even within a few hours. Examples: small species of *Coprinus*; an estimated ± 2% of the West European macrofungi belong here.

2. Short-living: sporocarps last two days to one week. Many tiny saprophytes belong here, e.g. many species of *Galerina, Mycena, Conocybe,* and *Omphalina*, together about 20% of West European macrofungi.

3. Moderates: sporocarps may live between one and four weeks. This group includes most larger saprotrophic agarics such as species of *Clitocybe* and *Collybia*, as well as most ectomycorrhizal species, e.g. boletes, *Amanita, Russula*. Approximately 50% of the West European macrofungi.

4. Durables: sporocarps last between one month and a year, for instance many *Gasteromycetes* and wood-inhabiting *Aphyllophorales*. A special subgroup are the revivers, agarics which may dry out completely during dry weather, but are able to resume sporulation when the weather conditions have become favourable again, e.g. *Marasmius*. About 25% of the West European mycoflora.

5. Perennials: sporocarps which can live for more than one year by formation of a new hymenial layer each year. A few species of polypores, e.g. species of *Phellinus, Fomes* and *Ganoderma*, ± 1% of the West European mycoflora.

Periodicity

Most macrofungi show a pronounced periodicity or seasonality, which means that potential sporocarp formation is restricted to a certain proportion of the year. Periodicity is strongly related to climatological conditions and most pronounced in areas with alternating dry and moist and/or warm and cold episodes. Most species in the northern temperate region fruit in late summer and autumn, but many fungi have a more restricted optimum

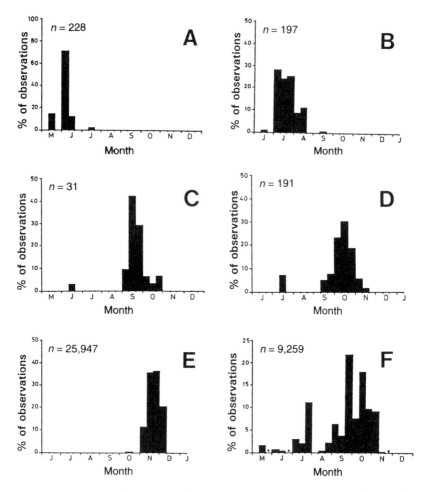

Fig. 19.1. Periodicity of sporocarp formation in various species of macrofungi in The Netherlands, based on added abundance values per decade (days) in the period 1974–1977 in 64 permanent plots (after Arnolds, 1982). n, total number of records of a species. (a) Vernal fruiting in *Agrocybe paludosa*; (b) summer fruiting in *A. pediades*, (c) early autumn fruiting in *Bovista nigrescens* Pers., (d) mid autumn fruiting in *Calocybe carnea*, (e) late autumn fruiting in *Mycena cinerella* P. Karst., (f) vernal until autumnal fruiting in *Marasmius oreades*.

within this period. In addition, some species produce sporocarps mainly or exclusively in different seasons. Again, relatively few accurate data are available on fruiting rhythms of individual species. For instance, Krieglsteiner (1977) gave detailed data on the periodicity of macrofungi in mixed

Abies forests in Germany, based on 7 years of observations. Arnolds (1982) produced diagrams of periodicity for species in grass- and heathlands in The Netherlands, based on data from three successive years. The diagrams of six species, showing different types of periodicity, are reproduced in Fig. 19.1.

As noticed by other authors (e.g. Ohenoja, 1993) the periodicity of a species may vary in different habitats in view of different microclimatological conditions. Exceptional weather conditions may also influence normal fructification patterns, for instance earlier fruiting of normally autumnal species in cool, wet summers. In addition, periodicity types differ from region to region according to climatological conditions. In boreal Scandinavian forests most ectomycorrhizal species fruit in July or August, whereas the same species in northern Germany and The Netherlands fruit in September and in Mediterranean areas in December. Regional differences are sometimes less easy to explain. For instance, most species of the large saprobic agaric genus *Hygrocybe* belong in northwestern Europe to the mid- or late autumnal fruiters, but in the northeastern United States the same species are predominantly aestival (Hesler and Smith, 1963; Arnolds, 1990). This striking difference cannot simply be explained by climatological or other environmental factors and is not yet understood.

The following schematic classification of periodicity types in northwestern Europe can be made (after Arnolds, 1981).

1. Hibernal fruiters, mainly producing sporocarps from December to March, e.g. the agarics *Flammulina velutipes*, *Mycena tintinnabulum* and *Pleurotus ostreatus* and the ascomycete *Sarcoscypha coccinea*.
2. Vernal fruiters, mainly present from April until June, e.g. the agarics *Entoloma clypeatum* and *Calocybe gambosa*, many species of the ascomycete genera *Morchella* and *Helvella* (Fig. 19.1A).
3. Aestival fruiters, mainly fruiting between June and August, for instance the agarics *Hygrocybe helobia*, *Agrocybe pediades* and several small *Coprinus* species on stems of herbaceous plants (Fig. 19.1B).
4. Early autumnal fruiters, producing the bulk of sporocarps in August and September, e.g. many ectomycorrhizal agarics (*Boletus* spp.), saprobic fungi such as many *Entoloma* species (Fig. 19.1C).
5. Mid-autumnal fruiters, with maximum fruiting capacity early or mid-October, including many species of all ecological and taxonomic groups (Fig. 19.1D).
6. Late autumnal fruiters with main sporocarp productivity at the end of October or in November, e.g. many saprotrophic species of *Hygrocybe* and *Mycena* (Fig. 19.1E).

Numerous species do not exactly fit into one of these categories, but have a broader range, for instance fruiting during the entire autumn or from spring onwards (Fig. 19.1F).

Table 19.1. Annual abundance of sporocarps of macrofungi in a permanent plot (400 m²) in a grassland near Beilersluis, Beilen, The Netherlands, in the period 1974–1980.

Species indicated with * are characteristic of dung or places where excrement have been deposited.

Abbreviations in front of each species indicate the estimated average potential lifetime of individual sporocarps (see par. on longevity); e = ephemerals (1–2 days); s = short-living (2–6 days), m = moderates (7–28 days), d = durables (1–12 months).

Numbers in front of each species indicate the optimum fruiting period (see para. on periodicity): 1. Hibernal. 2. Vernal. 3. Aestival. 4. Early autumnal. 5. Mid-autumnal. 6. Late autumnal. 7. Early until late autumnal. 8. Vernal or aestival until autumnal.

FF = Fluctuation factor, i.e. quotient between highest and lowest abundance in any year; FF with †: lowest abundance (0) adjusted to 1.

	1974	1975	1976	1977	1978	1979	1980	Average	FF
Total no. of species	47	32	43	43	45	42	42	42	
Total no. of sporocarps	7457	4547	10,013	9187	10,029	1735	6635	7086	6
Species producing sporocarps in 7 years:									
5,m *Clavulinopsis helveola*	2501	1225	163	2292	1923	274	2373	1536	15
7,s *Rickenella fibula*	294	454	3538	2654	2502	115	540	1442	31
6,s *Mycena cinerella*	1342	175	1885	788	1624	421	829	1009	11
6,m *Hygrocybe virginea*	642	433	314	503	250	213	430	398	3
7,m *Mycena sepia*	174	246	815	321	699	27	308	370	30
7,s *Rickenella setipes*	90	49	960	447	544	20	102	317	48
6,m *Hygrocybe laeta*	852	184	42	203	77	61	391	259	20
7,m *Hygrocybe psittacina*	302	268	141	281	162	36	307	214	9
5,m *Entoloma papillatum*	193	76	171	174	290	45	182	161	6
7,m *Mycena avenacea* var. *avenacea*	2	53	401	55	201	21	165	128	100

5,m	Clavulinopsis luteoalba	164	90	—	82	59	52	161	88	33
7,m	Entoloma sericeum	15	5	214	93	107	10	35	68	43
7,s	Galerina hypnorum	4	1	37	138	132	4	6	46	138
7,m	Hygrocybe pratensis	73	17	9	83	15	25	56	40	8
7,s	Galerina atkinsoniana	61	11	49	63	36	2	19	34	31
5,m	Mycena leptocephala	3	1	48	44	47	14	57	31	57
7,s	Mycena galopus	32	14	18	21	39	3	3	19	13
7,m	Cordyceps militaris	6	11	12	9	17	13	28	14	5
5,m	Hygrocybe conica	28	32	9	7	12	2	3	13	16
5,m	Entoloma sericellum	9	2	2	2	3	4	17	6	8
Species producing sporocarps in 6 years:										
6,m	Mycena pelliculosa	303	937	863	685	607	—	145	506	937[†]
8,s	Mycena sanguinolenta	37	148	152	62	160	—	38	85	160[†]
7,m	Panaeolus fimicola*	—	7	8	2	114	67	2	29	114[†]
6,s	Galerina unicolor	11	2	5	18	2	8	—	7	18[†]
Species producing sporocarps in 5 years:										
5,m	Entoloma chalybaeum	—	—	47	28	121	5	61	38	121[†]
7,s	Tephrocybe tylicolor*	45	54	27	—	25	9	—	23	54[†]
8,m	Entoloma conferendum	67	1	—	13	—	4	32	17	67[†]
7,s	Galerina vittaeformis	6	2	25	—	20	—	4	8	25[†]

Table 19.1. continued.

		1974	1975	1976	1977	1978	1979	1980	Average	FF
Species producing sporocarps in 4 years:										
7,s	Psathyrella romagnesii*	67	41	—	—	267	69	—	63	267†
6,m	Collybia butyracea	—	—	—	1	11	78	55	21	78†
6,m	Geoglossum nigritum	40	1	—	17	—	—	2	9	40†
7,s	Psilocybe montana	—	—	3	27	1	3	—	5	27†
6,m	Hygrocybe ceracea	15	2	4	2	—	—	—	3	15†
6,s	Galerina pumila	10	—	1	5	4	—	—	3	10†
7,m	Panaeolus acuminatus	1	—	1	—	1	—	13	2	13†
8,s	Bolbitius vitellinus*	—	—	6	7	1	—	2	2	7†
7,s	Mycena stylobates	—	—	1	7	2	—	5	2	7†
8,m	Stropharia semiglobata*	—	3	—	1	2	3	—	1	3†
Species producing sporocarps in 3 years:										
4,m	Entoloma caesiocinctum	2	—	—	—	9	—	30	6	
6,m	Geoglossum glutinosum	16	—	—	8	4	—	—	4	
7,e	Coprinus patouillardii*	—	—	1	—	7	16	—	3	
3,e	Coprinus friesii	4	—	—	2	—	—	8	2	

7,m	Mycena avenacea var. roseofusca	–	–	8	–	4	1	–	2
5,m	Calocybe carnea	3	–	3	8	–	–	–	2
5,s	Clavaria vermicularis	2	–	2	3	–	–	–	1
	Species producing sporocarps in 2 years:								
7,e	Coprinus ephemeroides*	–	–	–	–	3	17	–	3
6,s	Mycena flavoalba	–	–	9	12	–	–	–	3
7,s	Conocybe tenera	2	–	–	3	–	–	–	1
7,s	Conocybe rickeniana	–	–	7	1	–	–	–	1
7,s	Conocybe siliginea*	–	–	–	–	–	–	2	1
3,m	Agrocybe pediades	3	–	–	–	–	3	1	–
5,m	Ramariopsis tenuiramosa	–	–	–	1	2	–	–	0
6,s	Hemimycena delectabilis	–	–	–	–	2	–	1	0

Species producing sporocarps in one year only: 1974, Marasmius oreades [8,d] 1, Mycena epipterygia [5,m] 6, Entoloma juncinum [5,m] 1, Psilocybe semilanceata [7,m] 1, Galerina mniophila [6,s] 4, Hygrocybe miniata [7,m] 2, Stropharia albocyanea [5,m] 1, Psilocybe inquilina [7,s] 11, Tubaria furfuracea [6,m] 9; 1975, Typhula incarnata [7,s] 2; 1976, Hygrophoropsis macrospora [5,m] 2, Agaricus cupreobrunneus [5,m] 2, Clavulinopsis corniculata [5,m] 2, Sclerotinia trifoliorum [6,s] 1; 1977, Geoglossum fallax [6,m] 14; 1978, Coprinus bisporus* [8,e] 2, Entoloma hispidulum [6,m] 6, Coprinus echinosporus [7,e] 1; 1979, Panaeolus sphinctrinus* [8,m] 15, Coprinus miser P.Karst.* [7,e] 34, Peziza cf. cerea Sow.* [7,s] 2, Coprinus curtus Kalchbr.* [7,e] 7, Coprinus stercoreus (Scop.)Fr.* [7,e] 5, Coprinus radiatus* [7,e] 6, Sphaerobolus stellatus [7,s] 12, Iodophanus carneus* [7,s] 8; 1980, Entoloma infula [3,m] 183, Marasmius graminum [3,m] 6, Entoloma serrulatum [4m] 6, Lepista nuda [6,m] 3, Clitocybe amarescens* [7,m] 18, Mycena pura [7,m] 1, Lepista sordida* [6,m] 1.

The mechanisms responsible for differences in periodicity are still insufficiently understood. It seems likely that these differences are caused in part by genetic factors, in part by climatological factors and the availability of adequate sources for sufficient mycelial growth (Ohenoja, 1993). It is to be expected that most ectomycorrhizal fungi will fruit in summer and autumn, when sufficient mycorrhizal rootlets have been formed and sufficient carbohydrates have been supplied to the roots as a result of photosynthesis. The end of fructification of ectomycorrhizal symbionts of deciduous trees coincides roughly with the end of leaf fall (Mason et al., 1982). It has been demonstrated that artificial defoliation of a tree can cause immediate inhibition of sporocarp formation of associated ectomycorrhizal fungi (Last et al., 1979), showing the dependence of fruiting on sufficient carbon supply. Field observations indicate that fruiting of ectomycorrhizal fungi in evergreen coniferous forests may last longer, at least when weather conditions are appropriate. Nevertheless, some supposedly ectomycorrhizal fungi are able to produce sporocarps in spring, e.g. *Hygrophorus marzuolus* and several species of *Entoloma*, associated with *Rosaceae*.

An explanation of a distinct periodicity in saprobic, litter-inhabiting fungi is less obvious. Most species produce sporocarps in autumn before leaf fall, so that they are apparently not dependent on fresh substrates. It is clear that winter and summer offer usually suboptimal conditions for fruiting, but the weather conditions in spring and autumn are comparable in many respects. Maybe most species are dependent on extension of mycelia in spring and summer before fruiting in autumn is possible. However, this hypothesis is frustrated by the existence of several species, often with large sporocarps, which exclusively fruit in spring.

Fluctuations

Fluctuations are changes in sporocarp productivity from year to year. We may distinguish between quantitative fluctuations, in which a species is present in a certain area each year, but with different numbers of sporocarps, and qualitative fluctuations, in which a species is completely absent in certain years.

Fluctuations have been studied on an ecosystem level by monitoring mycocoenological plots during several successive years. Again, few studies have been published over an adequate period to allow some conclusions on the degree and causes of fluctuations, e.g. Arnolds (1988b) during 7 years in heathlands in The Netherlands, Krieglsteiner (1977) during 7 years in mixed *Abies* forest in Germany, Ohenoja (1993) during 13 years in two conifer forests in Northern Finland and Eveling et al. (1990) during 14 years in a coniferous forest in Northern Ireland. An example is given in Table 19.1, concerning a plot of 400 m^2 in a regularly mown grassland on poor, acid,

Measurements of Species Diversity in Macrofungi

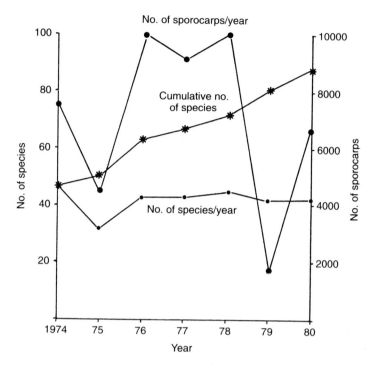

Fig. 19.2. Annual numbers of species (•—•) and sporocarps (●—●) of macrofungi and cumulative number of species (*—*) in a grassland plot (400 m²) in the northeastern Netherlands during the period 1974–1980.

sandy soil in The Netherlands, incidentally grazed by a horse. For details on the topography, plant community and soil conditions refer to Arnolds (1981: 361, plot 15), who published the results for the years 1974–1976. The observations for later years have not been published previously.

In the plot all sporocarps were counted at two-week intervals during potential fruiting periods in the period 1974–1980. For each species the total number of sporocarps in each year is indicated, together with the average number over the seven years. Altogether 86 species of macrofungi were observed. The annual number of species varied between 32 (37%) and 47 (55%) (Table 19.1). During one plot inspection a maximum number of 32 species (37%) was found on 3 October 1978. The increase of the number of species varied from 1 in 1975 to 13 in 1976, but there was no distinct trend for this factor to level off in later years (Fig. 19.2). Consequently, the number of species present in reality may exceed the total number after 7 years.

Qualitative fluctuations of species were considerable: only 20 (23%) of them were found fruiting each year and no less than 33 species (38%) were observed in one year only (Fig. 19.3). As to be expected, there was a positive correlation between the abundance of a species and its ability to produce sporocarps each year. However, there are some noticeable exceptions. For instance *Mycena pelliculosa*, one of the dominant species in the plot, was completely absent during one year. On the other hand, *Entoloma sericellum* was found each year but with only a few sporocarps.

One might expect that species with (moderately) durable sporocarps are less subject to qualitative fluctuations than those with short-living sporocarps, due to less sensitivity to unfavourable weather conditions and a larger chance to be found with the method applied here (Fig. 19.3A). However, no distinct trend could be found. Only the nine species with very short-living, ephemeral sporocarps were found consistently less regularly, between one and three years. Seven of these species are restricted to excrements of grazing animals, a substrate of irregular occurrence in the plot. Also coprophilic species with more durable sporocarps are found irregularly throughout the years, so that the strong fluctuations of ephemeral species are explained rather by the availability of a suitable substrate than by sporocarp characteristics.

It can be hypothesized that species fruiting during a long period or late in autumn are subject to less qualitative fluctuations than species with a

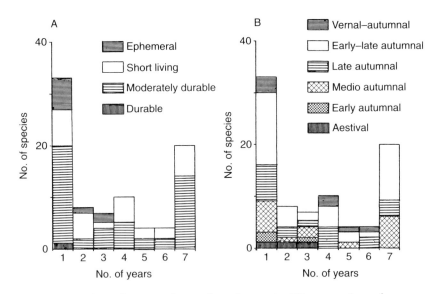

Fig. 19.3. Numbers of species of macrofungi fruiting in different numbers of years within a seven-year period (1974–1980) of monitoring in a grassland plot (400 m^2) in the northeastern Netherlands, in relation to (A) the average longevity of sporocarps and (B) the average periodicity type.

short fruiting period or with sporocarp formation in summer, the weather conditions in summer being more often unfavourable than in late autumn. In the investigated plot this hypothesis could not be confirmed (Fig. 19.3B). It is true that summer fruiters are only found in a few years, but these species are at the same time scarce in the plot and produce often ephemeral sporocarps.

Next to qualitative fluctuations considerable quantitative fluctuations occurred. The total annual number of sporocarps varied from 1735 to 10,029. Interestingly these numbers are not correlated with the annual numbers of species (Table 19.1). The degree of quantitative fluctuations for a species can be expressed in the fluctuation factor (FF), being the quotient between the highest and the lowest annual numbers of sporocarps (Arnolds, 1988b). FF values for all species present in at least four years are given in Table 19.1. For species lacking in one or more years the minimum number of sporocarps was adjusted to 1 since otherwise all FF values of these species would be infinite. Fluctuation factors in the grassland plot vary, for species present in all years, from 3 to 138, up to 937 for species lacking in some years. For six dominant and regular species in heathlands, Arnolds (1988b) found FF values between 8 and 87, consequently in the same order of magnitude. There is no relation between the FF values and characteristics of the individual species, such as average abundance, longevity of sporocarps and characteristic fruiting period.

The fluctuations in this plot concern exclusively saprobic litter decomposers. However, it is evident that quantitative and qualitative fluctuations among wood-inhabiting and ectomycorrhizal fungi in forests are also very strong (Krieglsteiner, 1977; Ohenoja, 1993). Ohenoja (1993) found in two coniferous forests studied during 13 years between 300 and 433,700 sporocarps ha^{-1} yr^{-1}, producing between 0.3 and 31 kg dry weight, the majority being ectomycorrhizal fungi.

Factors Causing Fluctuations

Fluctuations may be caused by environmental factors inducing sporocarp formation or by population dynamics of mycelia, including local extinctions and colonizations. In addition methodological factors could produce inaccurate results.

The current practice of analysing mycocoenological plots at intervals of one to three weeks may lead to an overestimation of quantitative and qualitative fluctuations. In view of the short lifetime of many sporocarps, a large proportion of them may be missed, and some species may not be noticed in a certain year. However, during mycocoenological research in grasslands there were no indications that ephemeral species had stronger fluctuations than other species (Fig. 19.3). Possibly the short lifetime of small sporocarps is compensated by the production of more sporocarps during a prolonged period. Also small sporocarps are easily overlooked, in particular in dense vegetation. Some species with hypogeous sporocarps demand special techniques and will certainly be underestimated.

It is difficult to discriminate between fluctuations in fruiting merely caused by environmental factors and those caused by fluctuations in mycelium density, as long as it is impossible to estimate the abundance of vegetative tissue of different species in the substrate. It may be that in the future molecular-biological techniques will enable us to carry out such analyses. At present it is possible to identify fungi in ectomycorrhizal roots by comparison with molecular-biological characteristics of the sporocarps.

Circumstantial evidence from mycocoenological observations suggest that under stable conditions most mycelia are permanently present during the period of research. Sporocarps are often found on nearly the same spot in a plot each year. Examples are known of ectomycorrhizal species fruiting with intervals over 20 years in exactly the same place. Sometimes indications are found of true local extinctions or new colonizations. In the grassland study presented here (Table 19.1) one could think of local extinction of, for example, *Hygrocybe ceracea* with its highest abundance in the first years and absent in the last three years. However, *Entoloma caesiocinctum* was also absent during three years, but again produced sporocarps afterwards. *Agrocybe pediades* even fruited with an interval of five years, but in this case recolonization is not unlikely.

On the other hand, most or all coprophilic species must have been recolonizing the plot repeatedly since the demanded substrate, dung, is not permanently available. New establishment of mycelia is also probable in the case of *Collybia butyracea*, which appeared in 1977 and spread strongly in subsequent years from a central point to a widening fairy ring. A spectacular phenomenon was the appearance of numerous sporocarps of *Entoloma infula* in 1980, which was never found before. For this species new colonization seems to be more unlikely since the pattern of sporocarps suggested the presence of an old mycelium.

The age of some fairy rings in grasslands has been estimated on the basis of the extension rate of the rings of sporocarps at 160–700 years (Kreisel and Ritter, 1985). The size and age of individual mycelia (genets) of several macrofungi have been determined using somatic incompatibility tests (Rayner and Todd, 1979). For instance, Dahlberg and Stenlid (1990) found for the ectomycorrhizal species *Suillus bovinus* that in pine forests large and smaller genets are present next to each other. The number of young, small genets was significantly larger in young forests than in old stands, suggesting a recent establishment of mycelia through spore dispersal in young forests.

In more or less stable environments the turnover of mycelia seems to be relatively unimportant (except for species growing on ephemeral substrates). The considerable qualitative and quantitative fluctuations are mainly caused by environmental factors, in particular weather conditions. It is a common belief that high amounts of precipitation in summer and autumn are correlated with a high sporocarp production. The correlations between weather conditions and sporocarp yields in boreal coniferous forests in Finland were extensively analysed by Ohenoja (1993). These investigations indicated a more complicated relationship between weather and fruiting. For instance, ectomycorrhizal fungi behaved differently from saprobic fungi on soil and those on wood. Low temperatures in the preceding winter period appeared to be unfavourable and, surprisingly, high temperatures in late autumn were found to be detrimental to the fruiting of ectomycorrhizal fungi in the next year. The complicated interactions between weather conditions and fruiting hamper simple predictions on the productivity of sporocarps in a certain year.

Conclusions

The combination of (i) generally short-living sporocarps, (ii) pronounced periodicity and (iii) strong quantitative and qualitative fluctuations complicate attempts to analyse the diversity of macrofungi in certain areas or biocoenoses. A mycological survey during one visit or one year will always be very incomplete, a situation different from floristic studies on, for example, phanerograms. For a more or less complete survey of an area, repeated monitoring throughout the fruiting season during several successive years is necessary. Estimations of the proportions of species observed in grassland plots (500 m^2), with different frequencies of observation, are presented in Table 19.2. The total number of species after seven years is

Table 19.2. Estimation of numbers of fruiting species of macrofungi observed in grasslands, following different sampling procedures, expressed in the expected percentage of the total number of species after sampling procedures, expressed in the expected percentage of the total number of species after sampling with fortnight intervals during seven successive years (=100%).

	1	2	3	4	5	6	7
Frequency of sampling							
1 × per two weeks	50	65	75	85	90	95	100
1 × per four weeks	40	55	65	75	85	90	95
1 × per eight weeks (during fruiting season)	25	40	55	65	75	80	85

considered to be 100%, but the maximum number is most probably not yet reached after this period. In other environments or with a different plot size these proportions might be different.

In view of the methodological problems, it is even more difficult to establish in a reliable way possible changes in frequency of species over longer periods, which is of crucial importance to show long-term changes in species performance. Nevertheless some methods have been used successfully for an analysis of both mycofloristical and mycocoenological data (e.g. Derbsch and Schmitt, 1987; Arnolds, 1988a, 1992a; Arnolds and Jansen, 1992). It was demonstrated that some ecological and taxonomic groups of macrofungi are strongly decreasing in Western Europe and that several regional extinctions have taken place in recent decades, with possibly considerable consequences for the functioning of ecosystems (e.g. Jansen and Dighton, 1990; Termorshuizen and Schaffers, 1991). A discussion of the declining diversity of macrofungi falls outside the scope of this chapter. More long-term studies on the population dynamics of macrofungi are urgently needed, both in undisturbed and disturbed environments, in order to confirm any serious disturbance of fungal communities while the evidence is still available.

References

Arnolds, E. (1981) *Ecology and Coenology of Macrofungi in Grasslands and Moist Heathlands in Drenthe, The Netherlands*, Vol. 1. J. Cramer, Vaduz.

Arnolds, E. (1982) *Ecology and Coenology of Macrofungi in Grasslands and Moist Heathlands in Drenthe, The Netherlands*, Vol. 2. J. Cramer, Vaduz.

Arnolds, E. (1988a) The changing macromycete flora in the Netherlands. *Transactions of the British Mycological Society* 90, 391–406.

Arnolds, E. (1988b) Dynamics of macrofungi in two moist heathlands in Drenthe, the Netherlands. *Acta Botanica Neerlandica* 37, 291–305.

Arnolds, E. (1990) Hygrocybeae. In: Bas, C., Kuyper, Th.W., Noordeloos, M.E. and Vellinga, E.C. (eds) *Flora Agaricina Neerlandica*, Vol. 2. A.A. Balkema, Rotterdam, Brookfield, pp. 70–115.

Arnolds, E. (1992a) Mapping and monitoring of macromycetes in relation to nature conservation. *McIlvainea* 10, 4–27.

Arnolds, E. (1992b) The analysis and classification of fungal communities with special reference to macrofungi. In: Winterhoff, W. (ed.) *Fungi in Vegetation Science. Handbook of Vegetation Science* 19(1). Kluwer, Dordrecht, pp. 7–47.

Arnolds, E. and Jansen, E. (1992) New evidence for changes in the macromycete flora of the Netherlands. *Nova Hedwigia* 55, 325–351.

Barkman, J.J. (1987) Methods and results of mycocoenological research in the Netherlands. In: Pacioni, G. (ed.) *Studies on Fungal Communities*. University of Aquila (Italy), pp. 7–38.

Dahlberg, A. and Stenlid, J. (1990) Population structure and dynamics in *Suillus bovinus* as indicated by spatial distribution of fungal clones. *New Phytologist* 115, 487–493.

Derbsch, H. and Schmitt, J.A. (1987) Atlas der Pilze des Saarlandes, Teil 2: Nachweize, Ökologie, Vorkommen und Beschreibungen. In: Minister für Umwelt des Saarlandes (ed.) *Aus Natur und Landschaft im Saarland*, Vol. 3. Saarbrücken.

Eveling, D.W., Wilson, R.N., Gillespie, E.S. and Bataille, A. (1990) Environmental effects on sporocarp counts over fourteen years in a forest area. *Mycological Research* 94, 998–1002.

Gams, W. (1992) The analysis of communities of saprophytic microfungi with special reference to soil fungi. In: Winterhoff, W. (ed.) *Fungi in Vegetation Science. Handbook of Vegetation Science* 19(1). Kluwer, Dordrecht, pp. 183–223.

Harley, J.L. and Smith, S.E. (1983) *Mycorrhizal Symbiosis*. Academic Press, London.

Hesler, L.R. and Smith, A.H. (1963) *North American Species of Hygrophorus*. University of Knoxville, TN.

Jansen, A.E. and Dighton, J. (1990) Effects of air pollutants on ectomycorrhizas. *Air Pollution Research Report* 30. EEC, Brussels.

Kreisel, H. and Ritter, G. (1985) Ökologie der Grosspilze. In: Michael, Henning and Kreisel (eds) *Handbuch für Pilzfreunde* 4, 3. Aufl. G. Fischer, Jena, pp. 9–47.

Krieglsteiner, G.J. (1977) *Die Makromyzeten der Tannen-Mischwälder des Inneren Schwäbisch-Frankischen Waldes (Ostwürttemberg) mit besonderer Berücksichtigung des Welzheimer Waldes*. Lempp Verlag, Schwäb. Gmünd.

Last, F.T., Pelham, J., Mason, P.A. and Ingleby, K. (1979) Influence of leaves on sporophore production by fungi forming sheathing mycorrhizas with *Betula* spp. *Nature* 280, 168–169.

Mason, P.A., Last, F.T., Pelham, J. and Ingleby, K. (1982) Ecology of some fungi associated with an ageing stand of birches (*Betula pendula* and *B. pubescens*). *Forest Ecology and Management* 4, 19–39.

Ohenoja, E. (1993) Effect of weather conditions on the larger fungi at different forest sites in northern Finland in 1976–1988. *Acta Universitatis Ouluensis*, series A, 243.

Rayner, A.D.M. and Todd, N.K. (1979) Population and community structure and dynamics of fungi in decaying wood. *Advances in Botanical Research* 7, 333–420.

Richardson, M.J. (1970) Studies on *Russula emetica* and other agarics in a Scots pine plantation. *Transactions of the British Mycological Society* 55, 217–229.

Termorshuizen, A. and Schaffers, A. (1991) The decline of carpophores of ectomycorrhizal fungi in stands of *Pinus sylvestris* L. in The Netherlands: possible causes. *Nova Hedwigia* 53, 267–289.

Winterhoff, W. (1984) Analyse der Pilze in Pflanzengesellschaften, insbesondere der Makromyzeten. In: Knapp, R (ed.) *Sampling Methods and Taxon Analysis in Vegetation Science. Handbook of Vegetation Science* 4. Junk, The Hague, pp. 227–248.

# Inventorying Microfungi on Tropical Plants

20

P.M. KIRK

International Mycological Institute, Bakeham Lane, Egham, Surrey TW20 9TY, UK.

This chapter addresses two aspects of inventorying; first, the magnitude of the problem when attempting to produce an inventory of microfungi on tropical plants and, second, the logistics of doing this.

First, the magnitude of the problem of sampling microfungi on tropical plants is considered with a review of the different types of fungi which may be found in this association. Although plants are an important element of almost all terrestrial ecosystems there are other important habitats in which fungi occur, often in large numbers. These include soil, water, both fresh and marine, and in or associated with animals. Each plant has a range of fungi associated with it, some specific, others less specific, either ubiquitous or restricted only to certain host genera or families. Those which are non-specific may still be restricted to the tropics whereas others may be of widespread occurrence. Mountains in the tropics may provide climatic conditions which allow the occurrence and survival of species which would otherwise be restricted to temperate regions. The part of the plant on which fungi occur is no less important than the type of plant or the species of plant on which they occur.

Starting at the top of the largest plants, trees, there are folicolous fungi, those associated with leaves, firstly when the leaves are still attached to the tree. These fungi may be parasites causing disease symptoms, which makes their collection, and often identification when they are host-specific, rather easy. Or they may be endophytes living in leaves, perhaps causing no symptoms and therefore only detectable by more elaborate techniques of culturing. They may be epiphytic, living on the surface of the leaf and apparently causing no damage other than that as a result of reducing the

© 1995 CAB INTERNATIONAL. *Microbial Diversity and Ecosystem Function*
(eds D. Allsopp, R.R. Colwell and D.L. Hawksworth)

photosynthetic potential of the leaf. Finally, they may be saprophytes, following a parasitic fungus and invading the necrotic tissue or gaining entry through damage to the plant as a result of, for example, insect damage. Consideration should also be given to what happens to the leaf when it falls from the tree. The parasites will still be there and they may start to form a different part of their life cycle. It should be noted that many fungi are pleomorphic; they exist in several different forms. Typically one form is characterized by meiotic spores (the sexual or perfect form for which the term teleomorph has been employed) and one or more forms are characterized by mitotic spores (the asexual or imperfect form, the anamorph). In the past there has been a tendency for different groups of mycologists to study these different parts of the life cycle. This has resulted in a kind of dual nomenclature for these fungi; they have two (or more) names or sets of names in the literature and in many cases these names or sets of names are not yet recognized as belonging to the same species, the holomorph, the whole fungus in all its forms. It is, therefore, very important to study both the living and decaying parts of a plant because such studies often provide the evidence for the linking of these names. The flowers are the next part of the plant to consider. Here are often found parasites of the anthers. The fruit, especially when ripe, may also have specific fungi associated with them because of the often high concentration of sugar.

The next major microhabitat on trees is the bark. So called corticolous fungi inhabit bark and this is typically the domain of the lichen, those strange fungi, because of their association with an alga, which have been, until recent times, studied independently from the non-lichen-forming species but which are, nonetheless, true fungi. Many lichens are sensitive to light intensity and, therefore, different species will occur in different parts of the tree dependent on the degree of exposure to light. In closed forests, the canopy is a rich area as are trees on exposed ridges and along rivers or other water courses. Some lichens are recognized as belonging to ancient groups and provide good indicator species for biodiversity richness in temperate and probably other forest types.

Fungi also occur in or on wood. They may be destructive parasites, such as those which cause root rots and invade the vascular tissues of living trees. Alternatively, and numerically more significant, there are those which are saprophytes, occurring on decaying wood.

Finally we arrive at the roots of the tree. Here we find the mycorrhizal fungi, critical in ecosystem maintenance. Most of these are macrofungi, belonging to the group including the mushrooms although some, the truffles and false truffles could be considered as microfungi in a loose sense. Surrounding the roots is an area termed the rhizosphere which is typically richer in fungi than soil distant from roots. Rhizosphere and soil fungi can only be studied satisfactorily by using culture techniques. Perhaps they are not technically 'on plants' but probably they are just as much so as the superficial folicolous fungi we started with.

Not all plants are trees; there are shrubs, herbaceous plants, grasses and bamboos, palms, climbers, epiphytes and others. In addition to the angiosperms there are the gymnosperms, the conifers and their allies, a most important plant group with their own specific fungi. These include, for a very particular reason, the rust fungi, a group of obligate plant parasites. The rusts are frequently characterized by forming more than one spore type; they are one of the groups of fungi exhibiting pleomorphism. In a significant number of the rusts this pleomorphism extends to the different spore types occurring on different hosts. Often this is on different families of angiosperms but in some groups one host belongs to the angiosperms and the other to the gymnosperms, the conifers, a rather bizarre life strategy considering the requirements of the parasite to deal with the many differences between angiosperms and gymnosperms.

Not all plants are flowering plants. There are the pteridophytes, the ferns and their allies. The bryophytes, the mosses and liverworts and their allies, are also an interesting plant group; the mosses, in particular, have recently been shown to support a quite unique fungal community. The following is an example of a problem involving ferns. In the tropics there is a rather common fungus which occurs on leaf litter of many types, for example, many species of *Moraceae*, *Dipterocarpaceae*, and *Leguminosae*. It occurs in South and Central America, tropical East Africa, India, South East Asia and Australasia; but only in the anamorphic form. The species is also known from the British Isles, where it occurs on the dead leaf fronds of *Dicksonia antarctica*, a tree fern from Australasia which can survive in the mild climate of the southwest of England and the northwest of Scotland, where the climate is significantly modified by the warming effects of the Gulf Stream. That it occurs under such conditions is surprising, and even more surprising is that the teleomorph, the perfect stage, is also present, in abundance, and this stage is known from here and nowhere else. A satisfactory explanation of these facts is still awaited.

Major decisions are required when planning inventories. They are major because of the significant cost and time overheads involved. All the collections of leaves, twigs, bark, wood and other samples representing hundreds of different species of fungi and, perhaps, an order of magnitude higher of individuals are available for the production of living cultures. These cultures are the raw material of biotechnology research and their importance cannot be denied. However, the resources required to obtain and process, to annotate and identify collections on the natural substratum are small compared with those required to prepare, nurture and preserve living cultures. There is no simple answer to this question of whether to culture or not to culture. One group of fungi of particular interest to the author, the *Zygomycetes*, is almost exclusively studied in culture. Here cultures are certainly required. However, other groups of fungi, such as the dematiaceous hyphomycetes, are mostly known from collections on the natural substratum. In many of the 5000 or so species, cultures are unknown.

There are over 360,000 collections in the dried reference collection at IMI representing some 32,000 species, probably almost half of all known species. The living collection, by comparison, contains about 20,000 isolates representing some 4500 species. From an applied, experimental aspect, the living collection is far more important than the dried reference collection; it is also more important from a business point of view. Cultures are certainly required for serious biochemical characterization work and the search for potentially important secondary metabolites. That there is such a disparity between the numbers, both of species and individuals, in the dried reference collection and the living collection is a direct result of the substantially greater costs involved in looking after living cultures.

The work discussed so far, making and maintaining collections, is relatively easy to accomplish compared with the second aspect of inventorying microfungi on tropical plants, which concerns the non-scientific logistics.

One major problem is that, whereas in the temperate countries there are some taxonomists, in tropical countries taxonomists are either absent or comparatively rare or widespread. A recent report on mycologists in African countries showed that 11 of 46 countries apparently had no mycologists and it is likely that as few as 10% of those with mycologists have taxonomic specialists (Hawksworth and Ritchie, 1993).

The logistics of inventorying microfungi on tropical plants fall into a number of categories.

First, the simple problem of finding time in the busy, if not overbusy, schedule of existing taxonomists. As funding for taxonomy has decreased it has been the author's experience that not only are there fewer taxonomists, but those that are left have to assume a greater administrative and/or support role as these essential posts become harder to justify. Even in these times of high unemployment, there are no armies of mycological taxonomists available and waiting for the call to produce inventories, either of microfungi on tropical plants or of any other group of fungi. Existing taxonomists can certainly be assigned to inventory production if funds are available to employ substitutes to carry out the majority of their existing duties.

Here we must also consider the problem of collecting expertise. Unlike the higher plants, which are generally easier to collect than fungi, certain fungal groups require an experienced eye. Therefore, an experienced field mycologist is required before all groups can be adequately covered. There are very few people even in the UK with adequate experience.

Second, one of the major problems is that of finding basic costs such as air fares and subsistence. The author's experience to date has been that obtaining funds for basic collecting or inventory-producing trips to the tropics has been very difficult, if not impossible. Opportunities to collect in the tropics have usually been as a result of teaching commitments. Some might argue that this is a good thing, i.e. going to teach taxonomy in countries where there are no taxonomists. However, the logistics of producing inventories are nothing compared with those associated with presenting successful training workshops in tropical countries. The usual scenario is several weeks of planning, spread over a year or more, during which time normal duties are ignored and a backlog begins to develop; several weeks away from the office when the backlog increases; followed by the dilemma, on return, of whether to deal with the material collected or tackle the backlog.

Third, detailed planning with in-country collaborators is very important if these activities are to succeed; the days are gone when we could just jet in and jet out. In the past the author has identified collaborators by directly contacting local mycologists, but this is probably not the best way to accomplish the task of producing inventories; higher-level contact is required. Equally important is the selection of sites and arranging the local transport, and, if required, local laboratory and other facilities. Here we can also include the acquisition of a plant taxonomist to identify the host plants, which presents difficulties for two reasons. One, a taxonomist who knows the local flora may not reside locally and two, the local flora may not be well known.

Fourth, the official documentation necessary to comply fully with all national and international laws and regulations is often difficult to obtain. That it is a necessary part of producing inventories where external resource persons are involved is not in question and all requirements must be complied with if misunderstandings or worse problems are not to befall any inventory production project.

Finally, there must be time allocated for working up the collections made to produce an identification of some type and, perhaps equally important, report writing. It may not be essential or desirable to carry out an identification which produces a species name. This is often very time consuming and, in many cases, the species present will be undescribed. A recent collecting trip to Kenya which concentrated on a group of fungi which are typically widespread in distribution showed that some 20% of the species collected were undescribed. Similarly, a recent extensive study of leaf litter microfungi in Malaysia has increased the number of species in a relatively common genus from 50 to 70, a 40% increase (Kuthubutheen and Nawawi, 1991).

The time taken to work up collections depends to a certain extent on the collecting techniques used. For the larger microfungi, those which can be seen with the unaided eye, the collector is usually making collections which can be identified as a fungus in the field. Each fungus collected can be enumerated and the collector will therefore have some idea as to how long it will take to work up the material collected. For those microfungi which cannot be seen with the unaided eye the collector is typically collecting suitable substrata. In this case only the subsequent examination of the collections with a microscope will reveal how many different species of fungi are present. With this latter technique, one day collecting in the field will generate sufficient material to require up to one month or more to work up the collections. The option of making living cultures from the material collected, and the time overheads involved, have already been referred to.

The question of where the working up should be done also needs to be addressed. Should the material remain in the country of origin or should it be removed, at least in part, to another country? Perhaps the question should be, would it be allowed to be removed to another country? This, of course, raises the question of property rights, which is discussed in Chapter 26. Will there come a time when certain countries will allow no natural products beyond their borders, even soil samples? Such a scenario is difficult to imagine with the increased and increasing amount of international travel.

It is hoped that these observations and illustrations have been of interest and have highlighted some of the non-taxonomic problems which taxonomists face when attempting to produce inventories.

References

Hawksworth, D.L. and Ritchie, J.M. (1993) *Biodiversity and Biosystematic Priorities: Microorganisms and Invertebrates.* CAB International, Wallingford, Oxon, 120 pp.

Kuthubutheen, A.J. and Nawawi, A. (1991) Key to *Dictyochaeta* and *Codinaea* species. *Mycological Research* 95, 1224–1229.

Viral Biodiversity   21

M.H.V. VAN REGENMORTEL

Institut de Biologie Moléculaire et Cellulaire, 15 rue Descartes, 67084 Strasbourg Cedex, France.

The diversity of viruses that are found in various geographic areas and function as pathogens or silent passengers in hosts, such as humans, animals, plants, protozoa, fungi and bacteria, is truly enormous. The Fifth Report of the International Committee on Taxonomy of Viruses (ICTV), published in 1991 (Francki *et al.*, 1991), recognizes 73 virus families and more than 4000 member viruses. An outline of the current classification of viruses is presented in Table 21.1.

After years of controversy regarding the status and nomenclature of taxa used in virus classification, the ICTV finally agreed in 1991 that the usual categories of species, genus and family should also be used in virus classification. As a result, a universal system encompassing orders, families, subfamilies, genera, subgenera and species will be described in the sixth report of ICTV to be published in 1995.

A further major difficulty encountered in the classification of microorganisms (Lenski, 1993) is the issue of whether viruses should be considered as a series of asexual clones or as freely intermixing populations (Bishop, 1985; Kingsbury, 1985, 1988). As both viewpoints could be defended in the case of different viruses, the question was eventually resolved when the ICTV accepted the following definition of virus species: 'A virus species is a polythetic class of viruses that constitutes a replicating lineage and occupies a particular ecological niche' (Van Regenmortel, 1989; Pringle, 1991). A major advantage of this definition, based on the concept of polythetic class, is that it can accommodate the inherent variability of viruses and that it does not depend on the existence of a single but illusory defining characteristic for each virus species. Members of a polythetic class

© 1995 CAB INTERNATIONAL. *Microbial Diversity and Ecosystem Function*
(eds D. Allsopp, R.R. Colwell and D.L. Hawksworth)

Table 21.1. Classification of viruses according to the Fifth Report of the International Committee on Taxonomy of Viruses (Francki et al., 1991).

Family	Subfamily	Genus	Type species	Host
Order 1: dsDNA, enveloped				
Poxviridae	Chordopoxvirinae	Orthopoxvirus 7 other genera	Vaccinia virus	Vertebrate
	Entomopoxvirinae	Entomopoxvirus A 2 other genera	Melolontha entomopoxvirus	Invertebrate
Herpesviridae	Alphaherpesvirinae 2 other subfamilies	Simplexvirus	Human herpesvirus 1	Vertebrate
Hepadnaviridae		Orthohepadnovirus Avihepadnavirus	Hepatitis B virus Duck hepatitis B virus	Vertebrate
Baculoviridae	Eubaculovirinae	Nuclear polyhedrosis virus Granulosis virus	Autographa californica nuclear polyhedrosis virus Trichoplusia ni granulosis virus	Invertebrate
	Nudibaculovirinae	Non-occluded	Oryctes rhinoceros virus	Invertebrate
Plasmaviridae		Plasmavirus	Acholeplasma phage L2	Mycoplasma
Lipothrixviridae		Lipothrixvirus	Thermoproteus phage TTV1	Bacteria
Polydnaviridae		Ichnovirus Bracovirus	Campoletis sonovensis virus Cotesia melanoscela virus	Invertebrate

Order 2: dsDNA, non-enveloped

Family	Genus	Example	Host
Iridoviridae	Iridovirus	Chilo iridescent virus	Vertebrate and invertebrate
	4 other genera		
Phycodnaviridae	Phycodnavirus	*Paramecium bursaria Chlorella virus* 1	Algae
Adenoviridae	Mastadenovirus	Human adenovirus 2	Vertebrate
	Aviadenovirus	Fowl adenovirus	
	Rhizidiovirus	*Rhyzidiomyces* virus	Fungus
Papovaviridae	Papillomavirus	Rabbit (Shope) papilloma virus	Vertebrate
	Polyomavirus	Polyoma virus (mouse)	
	Caulimovirus	Cauliflower mosaic virus	Plant
	Commelina yellow mottle virus	Commelina yellow mottle virus	Plant
Tectiviridae	Tectivirus	Phage PRD1	Bacteria
Corticoviridae	Corticovirus	*Alteromonas* phage PM2	Bacteria

Order 3: dsDNA, non-enveloped, tailed phages

Family	Genus	Example	Host
Myoviridae	T4 phage group	Coliphage T4	Bacteria
2 other families			

Table 21.1. continued.

Family	Genus	Type species	Host
Order 4: ssDNA, non-enveloped			
Parvoviridae	Parvovirus	Minute virus of mice	Vertebrate
	2 other genera		Invertebrate
	Geminivirus	Maize streak virus	Plant
	2 other genera		
Microviridae	Microvirus	Phage φX174	Bacteria
	2 other genera		
Inoviridae	Inovirus	Coliphage fd	Bacteria
	1 other genus		
Order 5: dsRNA, enveloped			
Cystoviridae	Cystovirus	*Pseudomonas* phage φ6	Bacteria
Order 6: dsRNA, non-enveloped			
Reoviridae	Orthoreovirus	Reovirus type 1	Invertebrate and vertebrate
	4 other genera		
	Plant reovirus 1	Wound tumour virus	Plant
	2 other genera		
Birnaviridae	Birnavirus	Infectious pancreatic necrosis virus	Vertebrate and invertebrate
Totiviridae	Totivirus	*Saccharomyces cerevisiae* virus L1	Fungus
	Giardiavirus	*Giardia lamblia* virus	Protozoa
Partitiviridae	Partitivirus	*Gaeumannomyces graminis* virus	Fungus
	Penicillium chrysogenum	*Penicillium chrysogenum* virus	

Order 7: sRNA, enveloped, positive-sense genome

Family	Genus	Example	Host
Togaviridae	Alphavirus	Sindbis virus	Vertebrate and invertebrate
	Rubivirus	Rubella virus	
	Arterivirus	Equine arteritis virus	
Flaviviridae	Flavivirus	Yellow fever virus	Vertebrate and invertebrate
	Pestivirus	Bovine viral diarrhoea virus	
	Hepatitis C virus	Hepatitis C virus	
Coronaviridae	Coronavirus	Avian infectious bronchitis virus	Vertebrate
	Torovirus	Berne virus	Vertebrate

Order 8: ssRNA, enveloped, negative-sense genome, single-stranded (Mononegavirales)

Family	Genus	Example	Host
Paramyxoviridae	Paramyxovirus	Newcastle disease virus	Vertebrate
	Morbillivirus	Measles virus	Vertebrate
	Pneumovirus	Human respiratory syncytial virus	
Filoviridae	Filovirus	Marburg virus	Vertebrate
Rhabdoviridae	Vesiculovirus	Vesicular stomatitis Indiana virus	Vertebrate, invertebrate and plant
	Lyssavirus	Rabies virus	
	Plant rhabdovirus	Lettuce necrotic dwarf virus	

Order 9: ssRNA, enveloped, negative-sense genome, multiple-stranded

Family	Genus	Example	Host
Orthomyxoviridae	Influenzavirus A	Influenza virus A	Vertebrate
	Influenzavirus C	Influenza virus C	
Bunyaviridae	Bunyavirus	Bunyamwera virus	Vertebrate and invertebrate
	3 other genera		
	Tospovirus	Tomato spotted wilt virus	Plant
Arenaviridae	Arenavirus	Lymphocytic choriomeningitis virus	Vertebrate

Table 21.1. continued.

Family	Genus	Type species	Host
Order 10: ssRNA step in replication			
Retroviridae	Lentivirus 5 other genera	Human immunodeficiency virus	Vertebrate
Order 11: ssRNA, non-enveloped, monopartite genomes, isometric particles			
Caliciviridae	Calicivirus Carmovirus	Vesicular exanthema swine virus Carnation mottle virus	Vertebrate Plant
Leviviridae	Levivirus Allolevivirus Luteovirus 4 other genera	Phage MS2 Phage Qβ Barley yellow dwarf virus	Bacteria Plant Plant
Picornaviridae	Enterovirus 3 other genera	Human poliovirus	Vertebrate and invertebrate
	Sobemovirus	Southern bean mosaic virus	Plant
Tetraviridae	Nudaurelia β virus Tombusvirus Tymovirus	Tomato bushy stunt virus Turnip yellow mosaic virus	Invertebrate Plant Plant
Order 12: ssRNA, non-enveloped, monopartite genomes, rod-shaped particles			
	Capillovirus Carlavirus 3 other genera	Apple stem grooving virus Carnation latent virus	Plant Plant Plant

Order 13: ssRNA, non-enveloped, bipartite genomes, isometric particles

Comovirus	Cowpea mosaic virus	Plant
5 other genera		Plant and invertebrate

Order 14: ssRNA, non-enveloped, bipartite genomes, rod-shaped particles

Furovirus	Soil-borne wheat mosaic virus	Plant
Tobravirus	Tobacco rattle virus	Plant

Order 15: ssRNA, non-enveloped, tripartite genomes, isometric particles

Bromovirus	Brome mosaic virus	Plant
2 other genera		Plant

Order 16: SSRNA, non-enveloped, isometric and bacilliform particles

Alfalfa mosaic virus	Alfalfa mosaic virus	Plant

Order 17: ssRNA, non-enveloped, rod-shaped particles

Hordeivirus	Barley stripe mosaic virus	Plant

Order 18: ssRNA, non-enveloped, tetrapartite genomes

Tenuivirus	Rice stripe virus	Plant

are necessarily defined by a number of different properties and no one property is essential or necessary for membership in a polythetic class. This is in sharp contrast to higher taxa such as families which are so-called universal classes and which do possess defining properties that are both essential and necessary.

The genomic plasticity of viruses, especially of RNA viruses whose genomes replicate in the absence of repair mechanisms, means that a virus species cannot be defined by a single genome sequence. RNA viruses are in fact usually considered to be so-called quasispecies populations. The quasispecies concept was developed by Eigen (1993) to describe the distribution of self-replicating RNAs believed to be the first genes on earth. RNA viruses evolve very rapidly due to very high mutation frequencies per site in the viral genome (between 10^{-3} and 10^{-5}). Since the RNA viral genome usually contains about 10 kb, a clone of an RNA virus always consists of a complex mixture of different genomes all of which compete during replication of the clone (Holland *et al.*, 1992). The term master sequence refers to the most fit genome sequence(s) within the quasispecies population replicating in a given environment. The term mutant spectrum refers to all competing virus variants that differ from the master sequence.

In a fairly constant environment, selection for fit master sequences may result in slow evolution. Unfortunately little is known of the selective forces that act on virus populations and shape the observed evolutionary changes.

The use of the term quasispecies for viruses should not be interpreted as implying that the term viral species refers to a single, invariant genome structure. A viral species is actually a complex self-perpetuating population of diverse, related entities that act as a whole (Eigen, 1993). The extensive genetic variability of viruses is responsible for their successful distribution and replication in all taxa of organisms. According to current estimates, as many as 30,000 different viruses or virus strains are currently under study in various reference centres and specialized laboratories.

References

Bishop, D.H.L. (1985) The genetic basis for describing viruses as species. *Intervirology* 24, 79–93.
Eigen, M. (1993) Viral quasispecies. *Scientific American* 269, 32–39.
Francki, R.I.B., Fauquet, C.M., Knudsen, D.L. and Brown, F. (1991) Classification and nomenclature of viruses. Fifth Report of the International Committee on Taxonomy of Viruses. *Archives of Virology Supplement* 2, 1–450.
Holland, J.J., de la Torre, J.C. and Steinhauer, D.A. (1992) RNA virus populations as quasispecies. In: Holland, J.J. (ed.) *Current Topics in Microbiology and Immunology. Genetic Diversity of RNA Viruses*. Springer-Verlag, Berlin, pp. 1–20.
Kingsbury, D.W. (1985) Species classification problems in virus taxonomy. *Intervirology* 24, 62–70.

Kingsbury, D.W. (1988) Biological concepts in virus classification. *Intervirology* 29, 242–253.
Lenski, R.E. (1993) Assessing the genetic structure of microbial populations. *Proceedings of the National Academy of Sciences of the USA* 90, 4334–4336.
Pringle, C.R. (1991) The 20th meeting of the Executive Committee of ICTV. Virus species, higher taxa, a universal database and other matters. *Archives of Virology* 119, 303–304.
Van Regenmortel, M.H.V. (1989) Applying the species concept to plant viruses. *Archives of Virology* 104, 1–17.

Exploration of Prokaryotic Diversity Employing Taxonomy

22

J. SWINGS

Laboratory for Microbiology, Ledeganckstraat, 35, 9000 Gent, Belgium.

When prokaryotic biodiversity is considered, it is important to be aware that there is only a small number of prokaryote 'species' described (less than 5000!), that, in nature, the non-culturable forms are dominant in numbers but not yet certain in function, and that, based on isolates cultured from the environment, less than 10% may be readily identified even by an experienced bacteriologist because of the lack of a suitable database.

In this chapter, prokaryotic genetic diversity will be covered from a polyphasic taxonomy point of view (Colwell, 1970), making use of classic, as well as molecular, techniques to study culturable organisms. The exploration of prokaryotic intrinsic, i.e. taxonomic, diversity is illustrated using three examples: one case that started as a classification study of the genus *Xanthomonas* and two that were begun to identify bacteria in environmental samples, involving PHB-degrading microflora and *Vibrio* isolates associated with fish hatcheries.

This chapter does not deal with a directed search for new organisms with given properties, non-culturables, or establishment of species-'richness' indices of given habitats.

Recent Developments in Prokaryotic Taxonomy

The estimation of the intrinsic genetic diversity of prokaryotes is a task of prokaryotic taxonomy. From the origin of bacteriology last century to the 1960s, the taxonomy of prokaryotes was dominated by a phenotypic approach, based mainly on morphological, biochemical, and physiological features for the description and differentiation of taxa. Taxonomic systems were based on subjective interpretations of 'important' phenotypic features. In many cases this led to artificial or special-purpose classifications. The pathovar system designed within the species *Xanthomonas campestris* is a typical example, illustrated further in this chapter. Since the 1960s, however, important advances have occurred in bacterial taxonomy including the following.

1. Introduction of computer-assisted numerical analysis of phenotypic data (Sokal and Sneath, 1963) has enabled bacterial taxonomists to interpret results in a more rational and unbiased way, by giving equal weight to each feature tested.
2. DNA–DNA homology has become a significant criterion for the definition of the prokaryotic species (Wayne *et al.*, 1987; Murray *et al.*, 1990).
3. DNA–rRNA hybridization and RNA sequencing (5S, 16S, 23S) techniques have introduced into prokaryotic taxonomy a phylogenetic dimension previously missing.
4. A polyphasic approach (Colwell, 1970), based on phenotypic, chemotaxonomic and genotypic data, has become routine. This approach is believed to lead to a stable, general-purpose taxonomy.
5. The development of fingerprinting techniques yielding large amounts of discriminating data from one single experiment. Most of these (commercial) systems are directed at practical diagnosis and are characterized by rapidity, automation, ease of standardization and the possibility to generate databases for identification.
6. The development of nucleic acid probes and PCR technologies for rapid identification.

The introduction of new techniques and concepts in taxonomy is proceeding in a very uneven way. Stackebrandt (1992), referring to the introduction of new phylogenetic insights obtained by 16S rRNA sequence data, stated that taxonomists are split into three groups: one group preferring the existence of traditional taxonomy for practical purposes, and a separate phylogenetic one, a second group, wishing to establish a system based on molecular genetics, visualizing the definition of taxa exclusively on signature oligonucleotides or signature nucleotides only. Until recently, there has been no serious attempt to delineate species or genera by rRNA sequences. Third are those taxonomists in search of an integrated system. These are taxonomists operating to develop polyphasic taxonomies.

Although prokaryote taxonomy should be basically phylogenetic, classifications should be useful for a variety of scientific purposes – first, for identification, and to produce databases that summarize as much relevant information about organisms as possible. According to Sneath (1989), scientific classifications are arrangements which aim to produce groups the

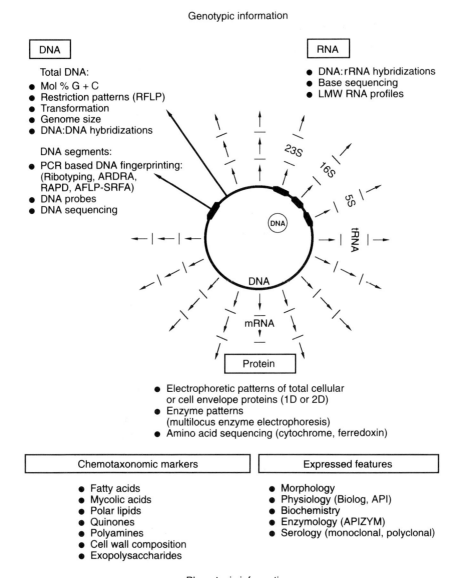

Fig. 22.1. Data for identification and classification.

members of which share a maximum number of common properties, and about which one may make the greatest number of predictive generalizations.

Information That Can Be Used in Prokaryotic Taxonomy

The information present at the cellular level which can be used in taxonomy is schematically represented in Fig. 22.1.

The Estimation of Genetic Diversity

In Fig. 22.2 different methods used in taxonomy are cited, together with their level of discrimination.

Within a species, characterization of individual strains may be accomplished by a number of techniques, including both classic (serological, phenotypic, enzymatic, phage- or bacteriocin-based) or more recently developed genotypic techniques (Restriction Fragment Length Polymorphism (RFLP); Low Frequency Restriction Fragment Analysis (LFRFA), Amplified Fragment Length Polymorphism (AFLP); Arbitrarily Primed PCR (AP-PCR); DNA Amplification Fingerprinting (DAF); Random Amplified Polymorphic DNA (RAPD); Amplified Ribosomal DNA-Restriction Analysis (ARDRA); probes; DNA sequences).

The genotypic techniques can be divided into three groups. The first group comprises RFLP, LFRFA, and ribotyping, which make use of restriction enzymes, but not of PCR. A second group makes use of restriction enzymes and PCR. The third group is based only on PCR. Either the whole genome is covered or only part of it. Obviously, genomic fingerprinting techniques represent a very rapidly evolving field.

In the past, strain differentiation was not always possible by phenotypic techniques, which had the disadvantage of being very labour-intensive. Nevertheless, a few phenotypic techniques were automated and commercialized, mainly for diagnostic purposes. Of these BIOLOG (BIOLOG Inc., Hayward, California; see Verniere *et al.*, 1993) and FAME (Fatty Acid Methyl Esters; Microbiol ID Inc., Newark, Delaware; see Yang *et al.*, 1993) are the most successful, because they are quick and generate useful databases. It should be added, however, that they both still suffer from 'childhood diseases', i.e. are in an early stage of development. Cellular protein patterns (SDS-PAGE) are cheaper than BIOLOG or FAME but are not fully automated. In our laboratory, we have in our databases the SDS-PAGE profiles of more than 4000 *Xanthomonas*, *Lactobacillus*, *Lactococcus*, *Bifidobacterium*, *Leuconostoc* and *Campylobacter* strains.

It is important to note that many of these techniques, e.g. BIOLOG, FAME, and SDS-PAGE, do not guarantee strain differentiation in every case.

A number of techniques do not appear in Fig. 22.2. The utility of some

Exploration of Prokaryotic Diversity

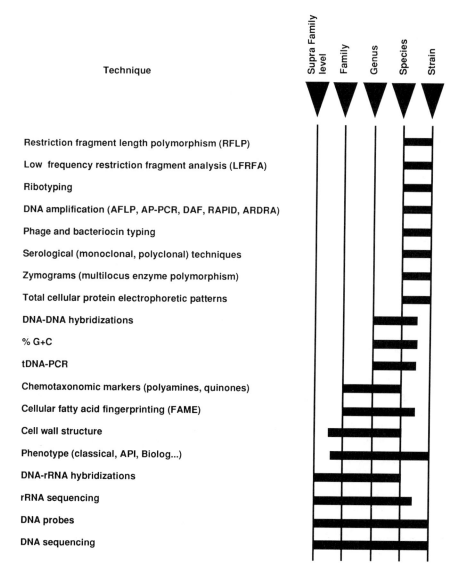

Fig. 22.2. Taxonomic resolution of currently used techniques.

techniques, e.g. 2D-protein patterns, has only rarely been explored (Goodfellow and O'Donnell, 1993). Figure 22.2 also indicates techniques that are not useful for strain differentiation, i.e. DNA–DNA hybridization, % G+C determination, DNA–rRNA hybridization, 5S and 16S rRNA sequencing.

As a taxonomist one should consider the two sides of a coin: identification is never better than the classification of a given group in the array or to put it in another way, the work of identification is, in a way, the quality control for a classification.

The following case histories are provided as illustrations of selected principles.

Case 1: The Genus *Xanthomonas*

This important genus comprising the plant-pathogenic bacteria, occurs worldwide, causing huge losses in important crops, e.g. rice, bean, citrus, manioc, etc., and includes seven species. The species *X. campestris* is further subdivided into more than 140 pathovars, each with a rather narrow host range (Swings and Civerolo, 1993). Over the last 10 years at least 25 manyears have been invested in analysis of the taxonomy of *Xanthomonas*. More than 1000 strains have been examined by phenotypic tests, FAME, polyamines, SDS-PAGE of soluble cell proteins, % G+C, DNA–DNA hybridizations, 16S rRNA sequencing and AFLP.

The most important results of this study can be summarized as follows:

1. DNA–DNA homologies reveal the existence of at least 21 DNA–DNA homology groups delineated at 60%. In the traditional taxonomy of this group, the species *X. campestris* comprises 16 DNA–DNA homology groups, achieved by lumping together a number of pathovars into one DNA homology group, in one case, and splitting pathovars into different DNA homology groups, in another case.
2. The heterogeneity of the species *X. campestris* and many of its pathovars has been confirmed by other techniques.
3. The pathovar system that has been devised for *X. campestris* is a typical example of a special-purpose taxonomy.
4. Results from the different methods applied to the taxonomy of this group are only partially congruent.
5. The non-pathogenic *Xanthomonas* represent a neglected group and are difficult to identify, evidenced by the databases established for the pathogenic *Xanthomonas*.

Case 2: Study of the Microflora Involved in the Breakdown of Polyhydroxyalkanoates (PHA)

PHAs are synthesized by several bacteria as an intracellular reserve. Because of their complete biodegradability and their thermoplastic properties, they were the subject of study during the past few years (Mergaert *et al.*, 1992).

For identification of the microbiota involved in the degradation of PHAs in different environments, e.g. soil, water, or compost, mainly SDS-PAGE, FAME and BIOLOG, combined with laboratory databases, have been used. The results are as follows.

1. In most environments a very complex PHA-degrading microflora exists, composed of Gram-positive and Gram-negative bacteria, together with streptomycetes and fungi.
2. A significant proportion of the organisms remain unidentified, particularly those isolated from freshwater and seawater environments.

Case 3: Characterization of *Vibrio* Isolates from Seabream Hatcheries

In recent aquacultural practice, a number of new problems have arisen, e.g. massive mortalities thought to be of bacterial origin and which constitute a serious threat to this otherwise particularly interesting and successful activity. In the seawater environment vibrios are important and are the first suspect cause of disease, since they include a number of known fish pathogens. It was the aim of our own study to isolate and characterize the *Vibrio* biota associated with seabream in two different hatcheries, at two different stages of growth, and in two different seasons. To study over 900 isolates, BIOLOG and FAME fingerprints were used as fingerprinting techniques. The results clearly showed that a single fingerprinting system could not be applied to all isolates and a polyphasic approach to identification was necessary.

Concluding Remarks

1. There is a need to fingerprint large numbers of strains, for example sets of 1000 strains or more, accurately and rapidly in a standardized way. Reliable identification methods are required for medical diagnosis, ecology, and industrial quality control; that is, all applications require rapid identification.
2. The importance of data capture, storage and numerical procedures to analyse data can hardly be overestimated, in view of the overwhelming

amount of data produced by the different fingerprinting techniques.

3. Correlation of results of the different fingerprinting techniques must be considered. For fingerprinting individual strains, excellent genotypic techniques are being introduced and their results compared, but correlation of results with those obtained by other methods needs further study, as does integration of the data into a classification.

4. 16S rRNA sequence data and other chemotaxonomic markers have added a phylogenetic basis to taxonomy.

5. Polyphasic classification has its weaknesses arising from a hierarchization of methods. When different methods are available, taxonomic practice tends to go from the application of cheap and rapid techniques for larger collections of organisms to more expensive and time-consuming techniques for small collections of strains; in the best of cases the latter are chosen for their representation, i.e. reference and type strains.

6. In many cases even identification becomes polyphasic, which is not always desirable.

7. There is a need for quality control of the products of classification, i.e. the establishment of the value of the schemes and the utility and reliability of databases generated for identification purposes during the process of classification.

8. There is a need to study further the taxonomy of bacteria and to define the bacterial species, particularly to examine the still unresolved question of whether a continuous spectrum of taxa occur in nature, or distinct taxa, or both (Sneath, 1985).

References

Colwell, R.R. (1970) Polyphasic taxonomy of bacteria. In: Uzuka, H. and Hasegawa, T. (eds) *Culture Collections of Microorganisms*. University of Tokyo Press, Tokyo, pp. 421–436.

Goodfellow, M. and O'Donnell, A.G. (1993) Roots of bacterial systematics. In: Goodfellow, M. and O'Donnell, A.G. (eds) *Handbook of New Bacterial Systematics*. Academic Press, London, pp. 3–54.

Mergaert, J., Anderson, C., Wouters, A., Swings, J. and Kersters, K. (1992) Biodegradation of polyhydroxyalkanoates. *FEMS Microbiology Reviews* 103, 317–322.

Murray, R.G.E., Brenner, D.J., Colwell, R.R., De Vos, P., Goodfellow, M., Grimont, P.A.D., Pfennig, N., Stackebrandt, E. and Zavarzin, G.A. (1990) Report of the ad hoc committee on approaches to taxonomy within the Proteobacteria. *International Journal of Systematic Bacteriology* 40, 213–215.

Sneath, P.H.A. (1985) Future of numerical taxonomy. In: Goodfellow, M., Jones, D. and Priest, F.G. (eds) *Computer-assisted Bacterial Systematics*. Academic Press, London, pp. 415–431.

Sneath, P.H.A. (1989) Analysis and interpretation of sequence data for bacterial systematics: the view of a numerical taxonomist. *Systematic and Applied Microbiology* 12, 15–31.

Sokal, R.R. and Sneath, P.H.A. (1963) *Principles of Numerical Taxonomy*. Freeman and Co., San Francisco.

Stackebrandt, E. (1992) Unifying phylogeny and phenotypic diversity. In: Ballows, A., Trüper, H.G., Dworkin, M., Harder, W. and Schleifer, K.-H. (eds) *The Prokaryotes*. Springer-Verlag, New York, pp. 19–47.

Swings, J. and Civerolo, E. (eds) (1993) *The Genus Xanthomonas*. Chapman & Hall, London.

Verniere, C., Pruvost, O., Civerolo, E.L., Gambin, O., Jacquemoud-Collet, J.P. and Luisetti, J. (1993) Evaluation of the Biolog substrate utilization system to identify and assess metabolic variation among strains of *Xanthomonas campestris* pv. Citri. *Applied and Environmental Microbiology* 59, 243–249.

Wayne, L.G., Brenner, D.J., Colwell, R.R., Grimont, P.P.D., Kandler, O., Krichevsky, M.I., Moore, L.H., Moore, W.E.C., Murray, R.G.E., Stackebrandt, E., Starr, M.P. and Trüper, H.G. (1987) Report of the ad hoc committee on reconciliation of approaches to bacterial systematics. *International Journal of Systematic Bacteriology* 37, 463–464.

Yang, P., Vauterin, L., Vancanneyt, M., Swings, J. and Kersters, K. (1993) Application of fatty acid methyl esters for the taxonomic analysis of the genus *Xanthomonas*. *Systematic Applied Microbiology* 16, 47–71.

International Biodiversity Initiatives and the Global Biodiversity Assessment (GBA)

23

V.H. HEYWOOD

UNEP Global Biodiversity Assessment, School of Plant Sciences, University of Reading, Whiteknights, PO Box 221, Reading RG6 6AS, UK.

In the lead-up to the United Nations Conference on the Environment and Development (UNCED), culminating in the signature at Rio de Janeiro in June 1992 of the Convention on Biological Diversity, many organizations and agencies initiated major programmes on various aspects of biodiversity. This chapter comments on the more significant of these initiatives and on another important project, the UNEP Global Biodiversity Assessment. To provide a context for this chapter, the leading international organizations that are involved in these initiatives are also identified.

The main international organizations that are involved in biodiversity are first the UN agencies (Box 23.1) such as:

1. UNESCO, mainly through its Man and the Biosphere Programme (MAB) with its world network of Biosphere Reserves;
2. UNEP (United Nations Environment Programme) which provides the Secretariat of the UN's Environment Programme created at the Stockholm

---

UNEP (United Nations Environment Programme)

UNESCO (United Nations Educational, Scientific and Cultural Organization)

FAO (Food and Agriculture Organization of the United Nations)

UNDP (United Nations Development Programme)

UNIDO (United Nations Development Organization)

Box 23.1. United Nations agencies.

© 1995 CAB INTERNATIONAL. *Microbial Diversity and Ecosystem Function*
(eds D. Allsopp, R.R. Colwell and D.L. Hawksworth)

Conference on the Human Environment, and covers a wide range of activities from providing the Secretariat for international treaties such as CITES and the Vienna Convention on the Ozone Layer (and more recently the interim secretariat for the Convention on Biological Diversity) negotiated under its auspices, to setting up information networks and systems of environment monitoring (Earthwatch including INFOTERRA and GEMS) to regional seas programmes and training; and

3. FAO (Food and Agriculture Organization of the United Nations) which covers a wide range of activities affecting biodiversity including its Forestry Programme and its Global System for the Conversion and Sustainable Use of Plant Genetic Resources (comprising the Commission of Plant Genetic Resources, the International Undertaking on Plant Genetic Resources and the International Fund).

The United Nations Development Programme (UNDP) and the United Nations Development Organization (UNIDO) are also involved but to a lesser degree. UNDP is one of the lead agencies of the Global Environment Facility (GEF) discussed later.

Other leading international organizations (Box 23.2) that do not belong to the UN family are the World Conservation Union (IUCN), a unique partnership of States, government agencies and a diverse range of non-governmental organizations (NGOs) with 720 members in 118 countries, providing them with concepts, strategies and technical support in the general areas of conservation and sustainable development; IUCN prepared the *World Conservation Strategy* and its successor *Caring for the Earth: A Strategy for Sustainable Living* and has regional offices in various parts of

---

IUCN (International Union for Conservation of Nature and Natural Resources; now known as IUCN – The World Conservation Union)

WWF (World Wide Fund for Nature)

WRI (World Resources Institute)

ICSU (International Council of Scientific Unions)

SCOPE (Scientific Committee on Problems of the Environment)

IUMS (International Union of Microbiological Sciences)

IUBS (International Union of Biological Sciences)

WCMC (World Conservation Monitoring Centre)

Box 23.2. Non-governmental international organizations.

the world. The World Wide Fund for Nature (WWF) is the world's largest private international conservation NGO with currently 28 affiliate and associate organizations around the world; its aim is to conserve nature and ecological processes by preserving diversity at the genetic, species and ecosystem level and by ensuring that natural resources are used sustainably. The World Resources Institute (WRI), based in Washington, is an independent research and policy institute created to help governments, environment and development organizations and others who are concerned with addressing the issue: How can societies meet human needs and nurture economic growth without destroying the natural resources and environmental integrity that make prosperity possible? It incorporates the Center for International Development. Its many publications include: *World Resources: A Guide to the Global Environment* and *Keeping Options Alive: the Scientific Basis of Conserving Biodiversity* and, in association with IUCN and UNEP, the *Global Biodiversity Strategy*. Conservation International (CI) is a private organization based in Washington which is dedicated to the preservation of natural ecosystems, especially the world's rainforests, and the species that rely on these habitats for their survival.

A number of intergovernmental bodies (Box 23.3), not UN agencies, are closely involved in some aspects of biodiversity such as the World Bank (more properly the International Bank for Reconstruction and Development) which is the world's largest multilateral aid donor and whose environmental policy has a major impact on biodiversity, and the Consultative Group on International Agricultural Research (CGIAR) funded by governments, multilateral lending agencies and other donors, which supports a network of international agricultural research centres including the International Plant Genetic Resources Institute (IPGRI, formerly IBPGR), the International Centre for Agroforestry (ICRAF) and the recently created Centre for International Forestry Research (CIFOR).

The World Bank (International Bank for Reconstruction and Development)

CGIAR (Consultative Group on International Agricultural Research)

IPGRI (International Plant Genetics Resources Institute)

ICRAF (International Centre for Research in Agroforestry)

CIFOR (Centre for International Forestry Research)

CAB INTERNATIONAL

etc.

Box 23.3. Intergovernmental bodies.

Another group of international organizations (Box 23.2) is the ICSU (International Council of Scientific Unions) family whose environmental programmes include the International Geosphere–Biosphere Programme, the World Climate Research Programme, and the Scientific Committee on Problems of the Environment (SCOPE), which was established in 1969 to advance knowledge of the influence of humans on the environment, as well as the effects of these environmental changes upon people, their health and welfare and to serve as a source of advice for the benefit of governments and NGOs. The SCOPE programme has an Ecosystems and Biodiversity cluster which includes projects on Ecosystem Function of Biodiversity and participates jointly with UNESCO and IUBS in the DIVERSITAS programme on biodiversity. Adhering ICSU members include IUBS and IUMS.

Some of the most significant international programme initiatives (Box 23.4) have been those associated in some way with the UNCED process, either in preparation for it, in support of it or as consequences of it. The *Global Biodiversity Strategy* deserves special mention: this WRI-IUCN UNEP publication was launched at the IV World Congress on National Parks and Protected Areas in Caracas in February 1992 and was the result of a remarkable collaboration involving 45 governmental and other partners, hundreds of collaborators and reviewers, and involved workshops and regional consultations in Nairobi (Kenya), San José (Costa Rica), Keystone (USA), Bogor (Indonesia), Brasilia (Brazil), London (UK), Bogotá (Colombia), Bangkok (Thailand), Perth (Australia) and Cambridge (USA). It represents, therefore, a broad consensus of the thinking and experience of 'scientists, community leaders, indigenous groups, industry, public administrators, and NGOs, providing the most comprehensive exploration of biodiversity issues and options that has taken place to date'.

The *Global Biodiversity Strategy* presents 85 proposals for action at local, national and international levels, to save, study and use the Earth's

---

GEF (Global Environment Facility – World Bank, UNDP, UNEP)

UNCED (United Nations Conference on the Environment and Development)

AGENDA 21

GLOBAL BIODIVERSITY STRATEGY
(WRI/IUCN/UNEP)

UNEP GLOBAL BIODIVERSITY ASSESSMENT (GBA)

DIVERSITAS (IUBS/SCOPE/UNESCO)

Box 23.4. International programmes and activities.

biotic wealth sustainably and equitably. It is or will shortly be available in English, Spanish, Portuguese, Chinese, Czech, Japanese, Indonesian and French. Hindi, Russian and Swedish versions are in preparation. It should be read by all those who are concerned with biodiversity, its conservation and sustainable use.

The Convention on Biological Diversity, negotiated under the auspices of UNEP, was signed by nearly 160 countries at Rio de Janeiro and the European Community and has been ratified by 107 countries subsequently. The Convention, in the words of Article 1, has as its objectives 'the conservation of biological diversity, the sustainable use of its components and the fair and equitable sharing of the benefits arising out of the utilization of genetic resources, including by appropriate access to genetic resources and by appropriate transfer of relevant technologies, taking into account all rights over those resources and to technologies, and by appropriate funding'. The Global Environment Facility (GEF) has been agreed as the interim funding mechanism for the Convention.

The Convention was the result of extensive scientific and technical discussions and difficult and protracted political negotiations. It is very much an Outline Convention and the various articles and clauses will require detailed interpretation. Many of the basic topics of biodiversity covered by the Convention are still the subject of debate, even controversy, and in July 1992 two *ad hoc* working groups of the GEF Scientific and Advisory Panel decided that it would be desirable to carry out as early as possible a global assessment of biodiversity. A project called a Global Biodiversity Assessment (GBA) was prepared by UNEP and approved as a GEF project. After appropriate discussions the project was approved for implementation in May 1993 and the first meeting of the Steering Group appointed by UNEP took place in Trondheim at the end of May 1993.

The object of the GBA is 'to provide an independent, critical, peer-reviewed scientific analysis of the issues, theories and views regarding the origins, dynamics, assessment, measurement, monitoring, economic valuation, conservation and sustainable use of biodiversity globally'. It will follow a procedure similar to that used for the Intergovernmental Panel on Climate Change to produce an Assessment of Biodiversity that is the standard reference work on the main themes and issues of biodiversity and to provide support for the Convention on Biological Diversity when it comes into force. It is aimed at international, regional and national environmental organizations, both governmental and non-governmental, as well as at policy makers and a wide range of scientists working in the field of biodiversity.

The GBA will be divided into 12 sections (Box 23.5) covering the characterization and measurement of biodiversity, its origins and dynamics, magnitude and distribution, inventorying and monitoring, ecosystem function, the multiple uses of biodiversity and human influences on it,

> **Outline of Sections**
>
> 1. Introduction
> 2. Characterization of biodiversity
> 3. The origins, dynamics and future of biodiversity
> 4. The magnitude and distribution of biodiversity
> 5. Inventorying and monitoring
> 6. Biodiversity and ecosystem function: basic principles
> 7. Biodiversity and ecosystem function: ecosystem analyses
> 8. Economic values of biodiversity
> 8bis Multiple values of biodiversity
> 9. Human influences on biodiversity
> 10. Conservation, restoration, sustainable use, and maintenance methods
> 11. Biotechnology
> 12. Data information management and communication
>
> Policy Makers' Summary
> Executive Summary

Box 23.5. UNEP Global Biodiversity Assessment.

conservation, restoration and sustainable use, biotechnology and data, and information and communication (Box 23.5).

Like the *Global Biodiversity Strategy*, the GBA will be very broadly based and leading experts from the fields of biology, conservation biology, systematics, environmental economics, sociobiology, anthropology, environmental monitoring and other related disciplines have agreed to participate as Coordinators of the different sections and others as lead authors. Honorary Advisers include Dr E.O. Wilson and Dr M.S. Swaminathan. The Chairman of the GBA is Dr Robert T. Watson, who was Chairman of the Scientific Assessment Panel of the Montreal Protocol, and the Executive Editor Professor is Vernon H. Heywood. The GBA will be extensively peer-reviewed internationally by about 1000 specialists. It is planned to complete the Assessment and have it ready for publication by autumn 1995.

These, then, are the main international initiatives, apart from the DIVERSITAS programme mentioned in the Introduction to this volume. We must do our best to ensure that they succeed and lead to effective action to conserve, maintain and use sustainably the biodiversity that still exists.

THE RESOURCE BASE IN MICROBIOLOGY VII

Living Reference Collections 24

H. SUGAWARA[1] AND JUN-CAI MA[2]

[1] WFCC World Data Center on Microrganisms, The Institute of Physical and Chemical Research, RIKEN. 2-1 Hirosawa Wako, Saitama 350-01, Japan and [2] Institute of Microbiology/Chinese Academy of Sciences, Beijing, People's Republic of China.

Introduction

Animals, plants and microorganisms are living resources which interrelate with each other. Whereas some microorganisms do harm to plants and animals, other microorganisms do them good. It is known that microorganisms are important links in food chains. They have been a major subject of research in agriculture, fermentation, and medicine for many years. They have become tools for scientists to elucidate mechanisms of life in the age of molecular biology. Today they are used in forms of YAC, BAC and PAC in human genome projects. Microorganisms are as valuable living resources as plants and animals. Therefore, they have been collected, evaluated, and preserved in the culture collections of research institutes and industries.

Many culture collections are long-established. There are now 479 working culture collections in the world according to the WDCM World Data Center on Microorganisms. Some of these were recently set up by national projects to meet a need from the life sciences and biotechnology. However, it is not feasible to establish a new huge centre which collects all living resources from every corner of the world. Even in a small country like Japan, more than 20 collections collaborate to function as an integrated network of living reference collections. The numbers of strains held and distributed by them are summarized in Table 24.1 (Okada, 1992). The largest collection in Japan is a foundation named IFO (Institute of Fermentation, Osaka), which has the longest history in Japan. JCM (Japan Collection of Microorganisms) in RIKEN was set up in 1980 with support from scientific communities which wanted a national reference collection

Table 24.1. Status report of JFCC.

Culture collections	Number of strains preserved						Number of strains distributed		
	Filamentous fungi	Yeasts	Actino-mycetes	Bacteria	Viruses	Algae and protoza (animal cells)	In Japan	Abroad	Total
AHU	1326	835	115	329	0	0	217	12	229
ATU	720	295	117	403	0	0	22	0	22
FERM	16	2	0	2	0	0	21	0	21
GIFU	0	0	0	12,800	0	0	465	83	548
HUT	590	420	268	99	0	0	59	0	59
IAM	1325	435	80	1067	0	0(586)	905	129	1034
IFM	2151	709	512	0	0	0	414	107	521
IFO	7221	2750	1452	2175	70	0(347)	7742	1148	8890
IID	0	1	5	1415	0	0	101	1	102
IMRG	0	0	0	63,622	0	0	91	8	99
JCM	672	1460	1556	3511	0	0(653)	2108	239	2347
MAFF	4251	417	181	3605	455	0(21)	850	32	882
NIAH	0	0	6	1430	447	510(5)	178	36	214
NIES	0	0	0	0	0	0	529	39	568
NIG	0	0	0	16,570	22	0	1196	13	1209
NRIC	413	1589	9	2374	7	0	561	0	561
OUT	100	3700	0	700	0	0	62	10	72
REMD	0	0	0	1050	0	0	67	11	78
RIB	889	156	0	129	0	0	44	1	45
RIFY	25	500	2	400	0	0	37	0	37
TIMM	1557	3562	49	0	0	0	161	0	161

NIG (National Institute of Genetics) is a special collection of mutants of *Escherichia coli*. Those collections in Table 24.1 have different backgrounds and specialities. They publish an integrated catalogue in the name of the Japan Federation for Culture Collections (JFCC) to emphasize that they comprise an invisible network of living reference collections. It is not a giant centre but a network of living resources centres that contributes to the maintenance of microbial biodiversity.

This chapter describes the current status of culture collections by use of databases of the World Data Center on Microorganisms (WDCM), which could be a model of an infrastructure of networks. Japanese efforts in the fields of plants and animals are also described in contrast with the efforts of WDCM.

Status of Culture Collections

WFCC World Data Center on Microorganisms (WDCM)

WDCM originated from the findings of a survey by Dr S. Martin of the National Research Council of Canada in 1967 of institutions that maintained microbial cultures. WDCM was materialized by the late Professor V.B.D. Skerman of the University of Queensland in Australia sponsored and supported by UNEP, UNESCO, FAO, WHO, UNU and the university. WDCM was endorsed by WFCC (World Federation for Culture Collections), IUCN, IUBS, IUMS, CODATA (Committee on Data for Science

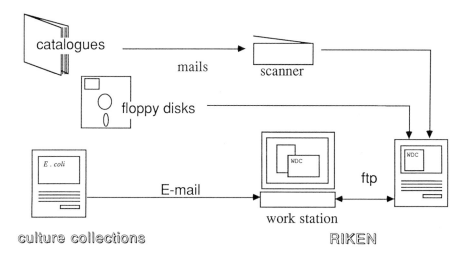

Fig. 24.1. Flow of capturing data in WDCM.

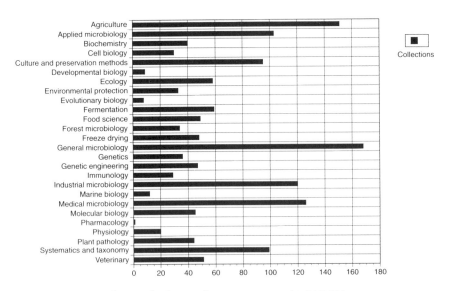

Fig. 24.2. Main subjects of culture collections registered in WDCM.

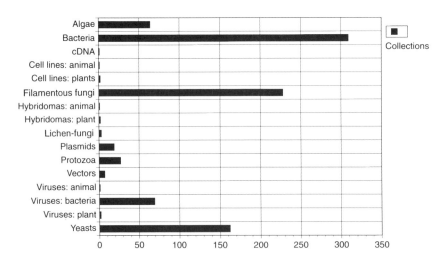

Fig. 24.3. Cultures held by culture collections registered in WDCM.

and Technology) and ICRO (International Cell Research Organization). Professor Skerman and his colleagues published the World Directory of Collections of Cultures in 1972, 1982, and 1986. In 1986, WDCM was transferred from Australia to RIKEN (The Institute of Physical and

Chemical Research) based on the guidance by WFCC which confirmed that WDCM is a component of WFCC (Komagata, 1987).

WDCM develops databases of culture collections (CCINFO), names of strains held (STRAIN) and algal strains (ALGAE) (Miyachi *et al.*, 1989), all of which are accessible on-line in Japan and from abroad. CCINFO includes institutional information from 479 culture collections, STRAIN gives names of 97,142 microbial strains in total and ALGAE is an on-line catalogue of 11,162 strains from 33 algal collections in the world (Sugawara *et al.*, 1993).

The databases on CCINFO and STRAIN are updated from time to

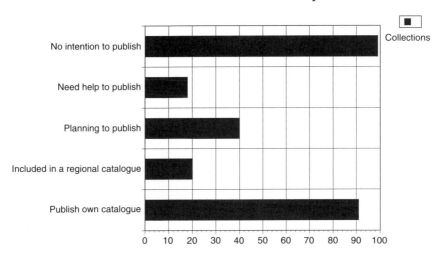

Fig. 24.4. Publication of catalogues by culture collections registered in WDCM.

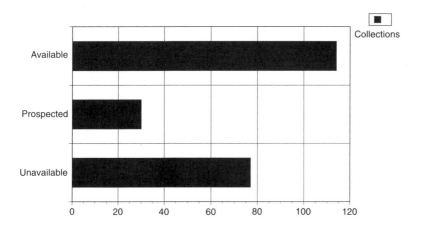

Fig. 24.5. Availability of computers in culture collections registered in WDCM.

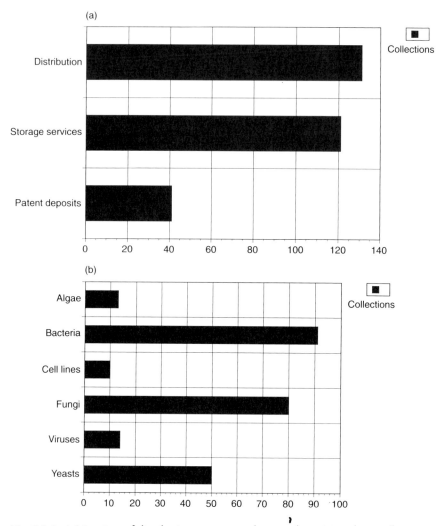

Fig. 24.6. (a) Services of distribution, storage and patent deposit in culture collections registered in WDCM. (b) Identification services in culture collections registered in WDCM. (c) Training in culture collections registered in WDCM. (d) Consultation in culture collections registered in WDCM.

time when new data are available from culture collections. WDCM prepares an input format (CCINFO format) for institutional information and asks culture collections to provide data. For the database STRAIN, WDCM captures data either by hard copies of catalogues, floppy disks or E-mail (Fig. 24.1). Each medium has advantages and disadvantages. Files by E-mail are easiest to be integrated into a master file of a centre but 'noise'

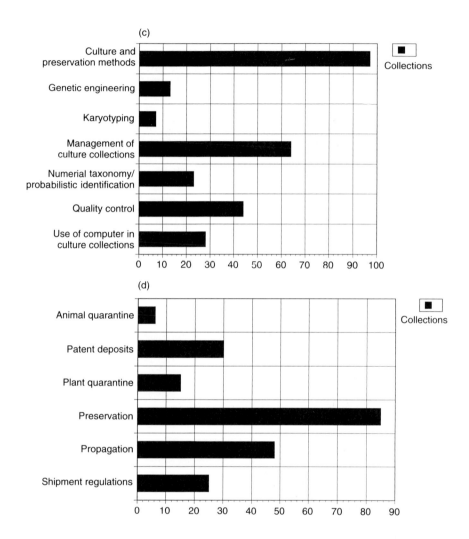

is a problem on some computer networks. The problem of floppy disks is the variety of formats and database management systems which vary between culture collections. A simple ASCII text file is a common format for data exchange, though WDCM is not able to take advantage of structured data in the case of flat files. Languages of AWK and SED help sometimes. Catalogues are easy for users to read, but they are not suitable for electronic data exchange. A machine misreads 1–5% of the text in a catalogue when WDCM converts the data contents into disk form, and data contents may be too old when WDCM gets a catalogue.

Status of Culture Collections

The database CCINFO provides a good overview of culture collections in the world. In Figs 24.2–24.6 main subjects, categories of cultures held, publication of catalogues, availability of computers in their place, services of distribution/storage/patent deposit, identification services, training, and consultation are summarized as histograms based on the data from CCINFO.

Culture collections gave many answers to the question of their 'Major subjects' shown in Fig. 24.2. The term 'biodiversity' does not appear there, because the term is not listed in the set of choices in the CCINFO format; but it can clearly be seen that culture collections provide major services in biodiversity studies.

Culture collections that were repositories of bacteria, filamentous fungi and yeasts have enlarged the range of their holdings. From Fig. 24.3, it is clear that culture collections now are actually gene banks which hold cell lines, plasmids, vectors and cDNA, although WDCM does not collect information of some cell banks and DNA banks. Seed banks are also not included in WDCM. Nevertheless, culture collections registered in WDCM already compose a large living genetic resources reservoir.

Some collections publish catalogues to publicize their activities and holdings. However, three quarters of the collections (Fig. 24.4) registered in WDCM do not publish catalogues because they do not have enough resources for publication or they only want to exchange strains in a framework of collaborative research. It is a role of WDCM and WFCC to encourage and/or help the publication of catalogues of collections with limited resources.

As shown in Fig. 24.1, WDCM converts hard copies into machine-readable form by use of a scanner. The necessity of this procedure is explained in Fig. 24.5. There are still many collections which do not use computers, even though, in some cases, their mother institutes are computerized. There are many collections where it is not easy for staff to access the international packet switching system and/or computer networks like INTERNET. Therefore, WDCM will publish a new edition of the World Directory of Collections of Cultures in 1993 sponsored by UNEP and in 1994 distribute floppy disks for PC users which contain data from CCINFO providing international funding becomes available. In Fig. 24.7(a)–(c) screen formats of the main menu, the edit menu and the input to the question of 'Main Subjects' are given as examples of the PC version of CCINFO. A user follows the instruction of menus and input data mostly by selecting terms in subwindows as shown in Fig. 24.7(c) where choices are displayed in the narrower window when a user types '?' in the main window.

The fundamental roles of culture collections, especially service collections, are the distribution/storage/patent deposit of living material and identification, training and consultation, which are summarized in Fig. 24.6(a)–(d). In Fig. 24.6(a), it can be seen that only about 130 collections answered that their main function is 'distribution'. This indicates that about 27% of the culture collections registered in WDCM are service collections and the rest are in-house or research-oriented collections (Kirsop, 1988). These may be in industries, universities and research laboratories and will distribute to closed communities or collaborative research groups. Nevertheless, the value of the rest is not less than the service collections in terms of the maintenance of microbial biodiversity.

Identification, training and consultation will increase in importance in the services offered by culture collections when 'biodiversity' becomes a major issue in the scientific and public community as a need for identification and classification of microbial strains increases. To meet this need, RIKEN is developing an information base in which data from different sources are integrated and useful knowledge for identification and classification is produced. Users assemble data from their laboratories and published papers into a database and apply numerical taxonomy to any subsets of the database. They are able to simulate a taxonomic structure by deletion/addition of OTUs on a dendrogram referring to raw data.

Utilization of WDCM

WDCM maintains a directory of culture collections and acts as a coordinator of information flow among culture collections and their users. Thus WDCM is a key component of WFCC and strengthens the network of reservoirs of living genetic resources. With access to an international computer network, the user is able to locate a collection as shown in Fig. 24.8(a)–(c). In Fig. 24.8(a) software called Gopher information client, which is in the public domain, is shown. Software is available on-line from the University of Minnesota, USA. When the user selects one of the databases and types any keyword, the result is a list of records. Items from the list can be selected (Fig. 24.8b) to obtain more detailed information (Fig. 24.8c). If the user is not able to access INTERNET on which the software 'Gopher' is usable, it is possible to dial up the WDCM system via the international packet switching system and search the databases by use of an information retrieval system implemented by WDCM. Useful addresses of WDCM are:

>gopher fragrans.riken.go.jp 70
>telnet 134.160.52.3
>http: //www.riken.go.jp
>dial up 4401 438 4135.

WDCM is also reachable via MOSAIC and WWW (Sugawara et al., 1994).

(a)

```
             Culture Collections in the World
                       Main Menu

                     ┌─────────────────┐
                     │  Introduction   │
                     │                 │
                     │      Edit       │
                     │                 │
                     │    Agreement    │
                     │                 │
                     │      View       │
                     │                 │
                     │      Print      │
                     │                 │
                     │      Copy       │
                     │                 │
                     │      Quit       │
                     └─────────────────┘

    The agreement to give the information above to be used by WDC

F1= HELP  |  F5= COLOR/MONO  |
```

(b)

```
           A List of Culture Collections in the World
                       Edit Function

                 ┌──────────────────────────┐
                 │    Basic Information     │
                 │  Status of your collection│
                 │  Institution Information │
                 │   Personnel Information  │
                 │       Main Subjects      │
                 │       Subcollections     │
                 │       Cultures held      │
                 │         Catalogue        │
                 │         Services         │
                 │   Computer Information   │
                 │           Quit           │
                 └──────────────────────────┘

      The Information on name, address of your collection

F1=HELP  |
```

(c)

```
                            Main Subjects
                                       ┌─The Number of records:  26─┐
 1.  ?enetics                          │        DATA DICTIONARY      │
 2.  Applied microbiology              ├─────────────────────────────┤
 3.  Biochemistry                      │Agriculture                  │
 4.  Industrial microbiology           │Applied microbiology         │
 5.  Culture and preservation methods  │Biochemistry                 │
 6.  Fermentation                      │Cell biology                 │
 7.  Freeze drying                     │Culture and preservation metho│
 8.  General microbiology              │Developmental biology        │
 9.  Systematics and taxonomy          │Ecology                      │
10.                                    │Environmental protection     │
11.                                    │Evolutionary biology         │
12.                                    │Fermention                   │
13.      ═══════ ATTENTION ═══════     │Food science                 │
14.     │ Press <Uparrow> or <Dnarrow> │Forest microbiology          │
15.     │ to locate the record         │Freeze drying                │
16.     │ Press <RETURN> to select     │General microbiology         │
17.     │ Press <S> to search          │Genetic engineering          │
18.     │ Press <Q> to quit            │Genetics                     │
19.      ═══════════════════════════   │Immunology                   │
20.                                    │Industrial microbiology      │

F1=HELP  |
```

Plants and Animals in Japan

In the 1980s, activities in the life sciences and biotechnology were the stimulus for the introduction of gene banks. Japan Collections of Microorganisms (JCM) and cell banks are newly established and the plant germplasm bank has been enlarged. The National Institute of Agrobiological Resources (NIAR), which is under the Ministry of Agriculture, Forestry and Fisheries (MAFF), used to be a national centre seed bank. In 1986, the Genetic Resources Centre of MAFF was set up in NIAR, which became a centre bank of 15 institutes of MAFF; that is, a network of living resources reservoirs was developed based on existing institutions without starting a new huge centre. This Genebank Project covers crops, forest trees, animals, fisheries, microorganisms, and DNA.

The information system of MAFF is constructed by use of a relational database management system and computer networks. The system was developed mainly for the seed bank, and has three subsystems for passport data, stock control and characteristics evaluated. The passport data subsystem is named after travel passports which are needed to cross borders. Detailed characteristics are linked to the passport data if available. This two-level data structure reflects the hierarchy of germplasm data and will be applicable to information systems for biodiversity. Data items common worldwide could be structured as passport items. Strain characteristics, which are maintained in the places where they are evaluated, could be linked to the passport items. Client-server models will be useful for this kind of data structure.

In the case of plants in botanical gardens, a database has been developed since 1989, mostly by universities. So far, 16,000 records have been compiled from five botanical gardens attached to universities and one public botanical garden. The number of data items is 29, most of which correspond to the passport data of MAFF in principle. Detailed data could be described in one of the data items. The database is in Japanese, installed on PCs, and is not yet public. It is hoped to exchange data with botanical gardens abroad in the future; however, there is likely to be a problem in translation of data.

Data in more than one language must be discussed seriously in biodiversity projects, as translation of terms is not always direct and living resources present different phenotypes depending on environmental conditions. In the latter case, collaborative research based on exchange of strains should be carried out.

Fig 24.7. (opposite) (a) Main menu of PC version of CCINFO. (b) Edit menu of PC version of CCINFO. (c) Input format for the data item of 'Main subjects'.

(a)

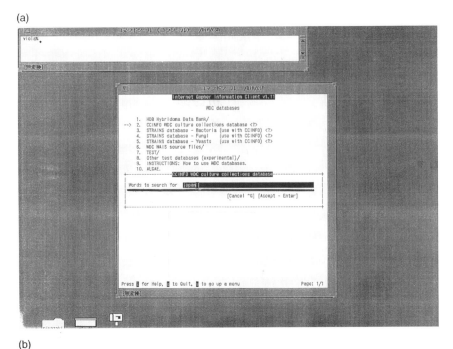

(b)

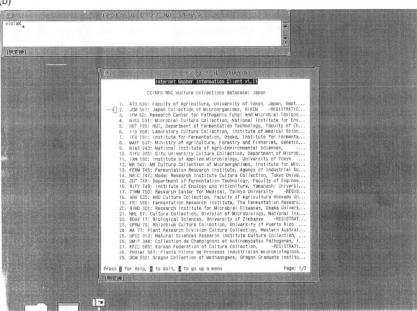

Fig. 24.8. (a) A menu of WDCM databases in WDC Gopher server. (b) A list of culture collections hit by a keyword 'Japan'. (c) Detailed information of JCM which was in (b). (The poor quality of these figures is a result of reproducing the actual visual displays.)

(c)

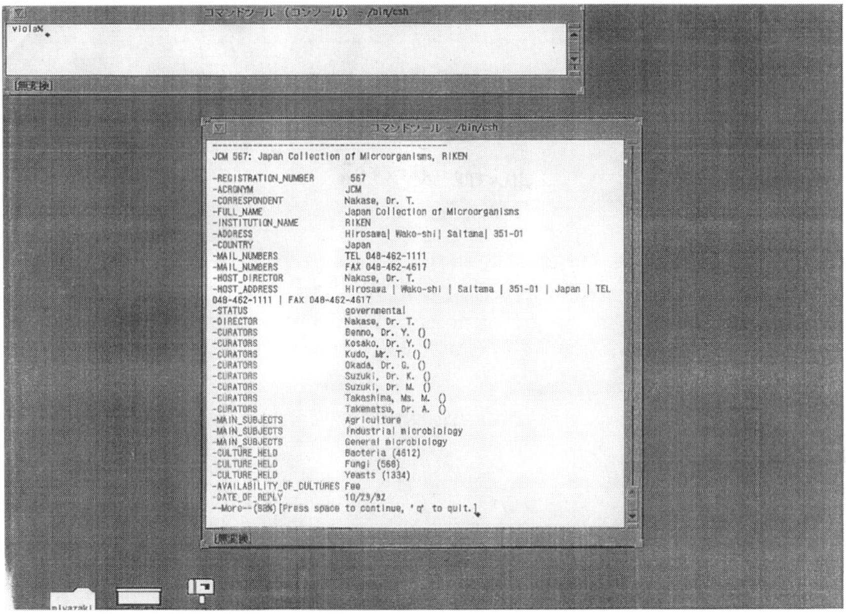

Fig. 24.8. continued. See opposite for legend.

RIKEN has been involved in a survey on the usage of laboratory animals (JALAS Working Group, 1992), which are key living resources to biology and medicine. The number of mice used increased from 4,099,426 in 1988, 4,575,999 in 1989 to 5,192,126 in 1990. The number of rats is stable around 2 million. Guinea-pigs, chickens and rabbits are the next three most used animals. However, more microbiologically and genetically controlled strains are used year by year. It is important to use 'authentic' strains for experiments in biology and medicine and an information system is needed to locate them. The Japan Society of Laboratory Animals (JSLA) is developing an on-line database on commercially available strains. In the database, the status of microbiological control of strains is, of course, included.

In the real world, microorganisms interact with plants and animals. Therefore a database which is a model of the real world has to implement correlation among those organisms. For example, a user of the database of JSLA will want to refer to databases on microbial strains. Cross-reference among databases will be a key function of an information system for 'biodiversity'.

Conclusion

We are in an ironic situation in that we study evolution while we destroy living resources by our own activities. It is not too late yet. Civilization brought us a concept of 'biodiversity', based on which we may be able to conserve living resources. Culture collections will be as important as botanical gardens and zoos, though they do not attract much public attention. It is necessary for culture collections to increase their visibility to the outside world and evolve further. In July 1993, the Japan Federation for Culture Collection changed its name to the Japan Society for Culture Collections to achieve a good harmony of services and research and to make a strong representations to scientific communities. It is also very important that the culture collections which have their own specialities become more active than before and be nodes of a comprehensive 'natural history' reservoir. It is obvious that data/information centres, which are hubs of information flow, are indispensable to this global 'invisible' museum.

References

JALAS Working Group for Laboratory Animal Data Bank and Life Science Research Information Section (1992) Number of animals used in experiments in 1990 – Results of a survey. *Experimental Animals* 41, 101–106.

Kirsop, B.E. (1988) Resources centres. In: Kirsop, B.E. and Kurtzman, C.P. (eds) *Living Resources for Biotechnology: Yeasts.* Cambridge University Press, London, pp. 1–35.

Komagata, K. (1987) Relocation of the World Data Center, *MIRCEN Journal* 3, 337–342.

Miyachi, S., Nakayama, O., Tokohama, Y., Hara, Y., Ohmori, M., Komagata, K., Sugawara H. and Ugawa, Y. (1989) *World Catalogue of Algae.* Japan Scientific Societies Press, Tokyo.

Okada, S. (1992) Status report of JFCC. *Bulletin of the Japan Federation for Culture Collections* 8, 203.

Sugawara, H., Ma, J., Miyazaki, S., Shimura, J. and Takishima, Y. (1993) *World Directory of Collections of Cultures of Microorganisms.* WFCC World Data Center on Microorganisms/RIKEN, Wako, 1148 pp.

Sugawara, H., Shimura, J. and Miyazaki, S. (1994) WFCC World Data Center on Microorganisms (WDCM) in the age of INTERNET. *Microbiology and Culture Collections* 10, 4–7.

Dried Reference Collections as a Microbiological Resource

D.W. MINTER

International Mycological Institute, Bakeham Lane, Egham, Surrey TW20 9TY, UK.

Introduction

Collections of various sorts have been used by scientists for hundreds of years. In the natural sciences, these collections are the archives on which knowledge of the natural world is based. They provide a record of change and form an important part of our cultural heritage. At the end of the 20th century, as habitats disappear and species become extinct, more and more specimens in natural science collections are assuming a new significance as a non-renewable resource of great importance to humanity (Groombridge, 1992; Duckworth *et al.*, 1993).

In microbiology as in other branches of the natural sciences, the earliest collections were of dried material. Some of these still exist after more than two centuries. By comparison, living reference collections, in the form of culture collections, have developed as a major microbiological resource only in the 20th century. Their rapid growth in certain areas of microbiology has meant that the significance of the older dried collections has sometimes been eclipsed. Thus, in reviews of natural science reference collections within the last decade (for example Morgan, 1987), living collections of microorganisms have received far more attention than their dried counterparts. This emphasis may be understandable from the viewpoints of bacteriology, protozoology and virology: in these disciplines, because of limitations imposed by the nature of the organisms under study, dried reference collections have been little used. Where they exist, such dried collections are dispersed, poorly known, often even more poorly advertised, and tend to contain few items, mainly specimens providing examples of symptoms of plant diseases.

In terms of estimated numbers of species, however, bacteriology, protozoology and virology are much smaller than mycology and algology (Hawksworth and Ritchie, 1993), and in these two branches of microbiology dried reference collections constitute a long-established and important resource. This chapter examines that contribution from the viewpoint of mycology. It should be remembered that, as traditionally defined, mycology embraces organisms which are now generally accepted as belonging to a range of microbiological kingdoms. In this chapter, words such as 'fungi' and 'mycological' are used in the broad and traditional sense which includes those organisms now redisposed in other kingdoms.

The Heritage of Fungal Dried Reference Collections

Because historically fungi were regarded as members of the plant kingdom, almost all fungal dried reference collections began as parts of larger botanical collections. Many are still maintained jointly with these botanical collections. This generally makes sense: the most common organisms associated with fungi are higher plants, and many other fungi grow as lichens in close mutualistic association with algae. Botanical and mycological literature has therefore always been closely linked: much mycological research is published in botanical journals, so that libraries of botanical collections are themselves a major mycological resource.

Mycology has benefited from the tradition – established at least in the second half of the 18th century – of scientists making expeditions and depositing their discoveries in these dried reference collections. Most of the great European and North American general botanical collections also contain large numbers of mycological specimens resulting from this tradition and collected from many parts of the world. Furthermore, the flowering plant collections themselves are a huge and largely ignored resource for the mycologist: on leaves, buds, flowers, fruits, twigs, and other organs of many of these specimens, microfungi are often present, collected unwittingly. Insect collections, incidentally, are another largely forgotten resource for the species-rich, but specialized and under-recorded, group of obligate entomogenous ascomycetes, the *Laboulbeniales*..

There can also, however, be disadvantages to the association between mycological and botanical dried reference collections. These mycological collections tend to be isolated from scientists working in other areas of microbiology, where dried reference collections are not normally kept, leading to suboptimal awareness about their utility. Furthermore, in general collections dominated by the plant kingdom, the many other ecological associations fungi are known to have are often inadequately represented, for example fungi from soil, industrial, marine and freshwater environments, and fungi pathogenic to animals. Other disadvantages may be administrative: more than one fungal collection has been maintained as a poor relation of its botanical sibling.

Sources of Information About Fungal Dried Reference Collections

In assessing fungal dried reference collections as a resource, it is important to establish their size and nature. There are two main sources from which information about fungal dried reference collections can be obtained: the second edition of the *International Mycological Directory* (Hall and Hawksworth, 1990), and the eighth edition of *Index Herbariorum, Part 1: The Herbaria of the World* (Holmgren et al., 1990) (see also addendum on p. 413).

The *International Mycological Directory* is a specialist volume which aims to provide a directory to 'sources of information, reference materials and pertinent organizations'. The latest edition is based on 122 responses received to a questionnaire sent to 280 mycological organizations and institutions worldwide. Entries were restricted to those organizations having a 'predominantly mycological orientation', but in countries where no mycological organization existed, a university, government or other contact was provided wherever possible.

Index Herbariorum, by comparison, is a substantial but generalist directory which attempts to list all dried reference collections of all organisms covered by the International Code of Botanical Nomenclature. The latest edition tries to provide, wherever possible, for each entry, information on the locality of the collection, the staff responsible for its curation, together with their interests, information on historically significant specialist collections within the general collection, and a summary of the general nature of the collection, its history, its size.

Problems in Using These Sources of Information

The second edition of the *International Mycological Directory* omitted to mention some institutions known to have important mycological interests 'because no response was received from them, despite reminders being sent'. It was hoped that publication of the new edition of this directory would stimulate interest, and would generate more contributions for future editions. While this policy of including only those collections from which a response had been received was laudable, in that it ensured all information published was up-to-date, there was the unfortunate side effect that some important mycological collections were not included, because no response had been received.

Because *Index Herbariorum* is a 'broad-spectrum' compilation, it is often difficult to identify dried reference collections with a mycological element: it is not uncommon, for example, to encounter entries describing a collection as 'containing lichens', but specifically 'excluding fungi'! In abstracting figures from this work, therefore, a dried reference collection was assumed to have a mycological element if it had been listed as containing fungi (including lichens, myxomycetes and other 'vague' mycological terms), or if it was reported as having a member of staff with broadly defined mycological interests.

The Number and Distribution of Fungal Dried Reference Collections

The *International Mycological Directory* identified almost 50 mycological dried reference collections, and for each collection listed, it attempted to provide an estimate of the size of the collection, an indication of the range of organisms covered, and an assessment of what indexing, paper-based or mechanical, had been carried out for the collection.

More than 500 dried reference collections worldwide were found in *Index Herbariorum* using the criteria previously described. Almost all of those derived from the *International Mycological Directory* were also present in *Index Herbariorum*, but much less information could be abstracted from the latter's entry. It was, for example, usually impossible to judge from the *Index Herbariorum* entry what indexing was available for the fungal holdings.

While the criteria used could not produce an exact assessment, it was possible to obtain perhaps the best estimate to date of numbers of collections and numbers of dried specimens they contain. Some collections curated by staff interested in mycology may nevertheless have little or no mycological content. Many more collections, particularly large general botanical ones, may contain significant holdings of the fungi, which are not mentioned in *Index Herbariorum*, and may have no curatorial staff identifiable as being interested in mycology. It was clear from scanning the entries in *Index Herbariorum* that, above the 500 recognized here, there must be several hundred more collections with mycological specimens which the criteria used here were not able to identify.

A simple breakdown by continent of the 500 identified collections revealed that 13 were located in Africa (most in South Africa or north of the Sahara), 58 in Asia (most in the People's Republic of China, India and Japan), 22 in Australasia and Oceania (including Hawaii), 235 in Europe, 142 in Canada and the USA (excluding Hawaii), and 53 in the rest of the Americas. The picture these figures provide tallies with our knowledge of the distribution of expertise in systematic mycology in particular (Hawksworth and Ritchie, 1993).

The Total Number of Items in Fungal Dried Reference Collections

For almost all collections identified solely through *Index Herbariorum*, information on numbers of items held in individual collections was simply not available. A typical entry in *Index Herbariorum* provides an estimate of the total number of specimens held in a given collection, with no breakdown by kingdom, so that it is usually impossible to know how large the fungal collections are, though one might often suspect that number to be very large indeed. Even where mycology was clearly identified as an interest, these numbers were rarely available. It was clear from scanning *Index Herbariorum* that for many entries such numbers were not known even by the staff responsible for curation.

The exceptions in contrast were the 50 collections identified from the *International Mycological Directory*. These provide some interesting statistics. The total number of items held by these collections was over three million. The number of items held in individual collections ranges from almost one million in the largest to about one hundred in the smallest listed. The three largest collections were as follows: USDA (Beltsville, Maryland), 975,000; Farlow (Boston, Massachussetts), 600,000; IMI (Egham, Surrey), 360,000. On their own, these account for almost two million items.

It is worth pointing out that no statistics were available for the following collections: Naturhistoriska Riksmuseet (Stockholm); Laboratoire de Cryptogamie (Paris); Royal Botanic Gardens (Kew); Rijksherbarium (Leiden); The Natural History Museum (London); Cryptogamic Herbarium (St Petersburg). All of these collections are known to have huge mycological holdings. On the basis of the information available, it is possible to make a 'guestimate' that perhaps more than eight million mycological specimens are held in dried reference collections worldwide.

Fungal Dried Reference Collections as a Resource

The value of fungal dried reference collections lies in the information contained in them. This has two principal components: the information contained in the stored specimen itself, and information present on the packet label or in notes and illustrations accompanying the specimen.

Most of the specimens held in fungal dried reference collections take the form of dried material secured in labelled paper packets. The dried material itself is usually part of a plant or a sample of soil or rock bearing the fungus, or an invertebrate or, less frequently, a part of a vertebrate animal or some other material bearing the fungus, or a dried culture in a special box. The labelled paper packets themselves are often fixed several to a sheet, and the sheets sit in folders. Larger specimens may be specially boxed. Microscope slide preparations, colour transparencies, photographs, drawings, field notes, and correspondence often accompany these specimens. They are usually located inside the same paper packets, but substantial slide or photographic collections may also be located in separate cabinets. Floppy disks and other forms of electronic data storage are also now starting to form a part of such collections.

The information on the packet label, together with notes and illustrations accompanying the specimen is usually handwritten, less often typed or printed. Drawings and notes may be on separate sheets of paper, or may be made directly on the packets themselves. The legibility of the handwriting, and the amount of information and its quality vary greatly from packet to packet. There is no standard format for presenting packet information. The language and alphabet used on the packet vary. Older handwritten packets can be problematic particularly when the Latin alphabet has not been used.

As a minimum, each packet usually bears the name of the fungus. In addition, the name of an associated organism, a short description of the substratum, the locality and date of collection, and the names of the collector and identifier are usually present. Brief ecological notes, the date of identification, estimations of abundance, assessments of the maturity of the material enclosed, and other categories of information may also be present. The packet labels furthermore often contain information indicating the special nomenclatural status of types and other authentic specimens, and exsiccati (examples from single collections formally distributed to other collections) and other specimens, together with their packet notes and illustrations provide a valuable insight into the reasoning of earlier researchers.

Using the Resource

The single most important use of fungal dried reference collections as a resource is to enable names to be provided for material requiring identification. The correct naming of organisms is of crucial importance in many aspects of scientific work with natural resources. These include: ensuring the correct mycorrhizal fungus is selected and inoculated in new plantations of forest trees; understanding the function of decomposer fungi in maintaining soil fertility; developing fungi for waste utilization; using fungal pathogens to control insect pests; using antagonistic fungi as natural enemies of fungal plant pathogens; employing specialist fungi as fermenters or in the manufacture of bread, cheeses and alcoholic drinks; controlling international movement of fungal pathogens through quarantine; identifying and obtaining full information on fungal plant and animal pathogens (Hawksworth and Ritchie, 1993).

Biosystematics: Taxonomy, Nomenclature and Identification

Dried reference collections are an essential resource for all aspects of biosystematic work in mycology. The three principal aspects of biosystematics may be defined as follows: taxonomy – the production of classifications of living things which accurately reflect their evolutionary history and the degree of genetic difference between them (this is usually achieved through the examination and comparison of many different specimens); nomenclature – the process by which the units the taxonomist decides to distinguish are provided with names; identification – the process by which particular specimens are referred to particular units (e.g. species) recognized by taxonomists.

Compared with other microorganisms, most fungi produce complex fruiting structures which constitute a large and valuable source of scientific information. Some of this information can only be obtained from structures freshly collected or growing in pure culture, but other highly significant amounts of information can survive the process of preservation for many years in the form of dried and preserved specimens. Stored specimens can retain anatomical and structural characters, as well as details of colour, size, shape, reaction to different cytological stains, and development. As a result, fruit bodies and other fungal structures are well suited for storage in dried reference collections, and the characters so preserved are among the main features used in mycology in taxonomic work.

Having decided to distinguish individual units within a classification scheme, the taxonomist provides each with a name through the second part of biosystematics: the process of nomenclature, which, in mycology, is governed by the International Code for Botanical Nomenclature (ICBN). The machinery of the ICBN works through the system of 'types', a 'type' being defined as the material with which a name is inseparably linked. In practice, almost all 'types' of fungi are specimens in dried reference collections.

The process of identification requires the biologist first to make comparisons between the particular specimen and material already named, before assigning it a name. While much of this may be achieved through use of published identification keys for better-known groups of organisms, direct comparison with material in dried reference collections remains essential for many of the more poorly documented or critical groups. As with taxonomy, therefore, the process of identification relies heavily on material in dried reference collections. The three main aspects of biosystematics (taxonomy, nomenclature and identification) are thus the main purposes for which dried reference collections are used.

Molecular and Cultural Studies

The polymerase chain reaction (PCR) technique developed in recent years allows one to amplify a specific DNA fragment from a heterogeneous mixture of sequences. The amount of DNA required for this technique is minute. It has been used to copy genes from single human hairs and a single sperm. It has also been used to amplify unique sequences from highly degraded DNAs derived from 140-year-old museum skins, 980-year-old grains, and 7000-year-old human brain tissue. In mycology the technique was recently used successfully to amplify and sequence DNA from specimens of basidiomycetes collected up to 50 years previously and in different fungal dried reference collections (Bruns *et al.*, 1990).

The results demonstrated that the technique was fast, convenient, and worked well even with very small quantities of material. Bruns and his colleagues reported that the DNAs extracted from preserved specimens were all degraded to various degrees, but, with one possible exception, amplification was possible. Interestingly, they noted that collections from all sources yielded enough DNA for amplification, which suggested that most or all modern methods of drying and maintaining fungal specimens in dried reference collections adequately preserve DNA for such studies.

In a full review of similar studies, Haines and Cooper (1993) also noted that it was possible to amplify DNA from specimens of fungi collected up to 80 years previously and now in dried reference collections. They noted, however, that older techniques for drying and preserving specimens from the ravages of insects could have a deleterious effect on DNA content, and they observed that all of their test results indicated that long chains of DNA start to deteriorate between 2 and 6.5 years after collection, and that the process of breakdown into segments of less than 600 base pairs is almost complete in most specimens by 33 years of museum storage.

Clearly there is now a need for further studies to determine which storage factors contribute to the breakdown of DNA in such specimens, so that future collections can be stored with minimal breakdown of their nucleic acids. Nonetheless, this technique opens up tremendous possibilities for using molecular data from dried reference collections for biosystematic work and biodiversity studies at the gene level.

Monitoring of Change

Dried reference collections have a particular value when monitoring environmental change over the last two centuries. The mere existence in a collection of a particular specimen from a particular locality may, for example, help to establish that changes have occurred in geographic distribution of the species it represents since the time when the specimen was collected. Detailed analysis of the specimen, with knowledge of the species' ecological requirements may help to establish whether that change was a result of altered climatic conditions, ecosystem destruction, or higher levels of pollution. Without such collections, there would now be no possibility of establishing baseline levels of radioactivity against which to measure changes associated with the Chernobyl Atomic Power Station accident in 1986 (Wasser and Grodzynskaya, 1993).

Biodiversity Studies

As already noted, the labels on specimen packets in mycological dried reference collections contain information identifying the fungus, its host or other associated organism, the substratum, the locality and date of collection, and the name of the collector. These are the main elements of information needed for many studies in biodiversity, and clearly these dried reference collections comprise a huge data resource of great potential value in such studies. For that resource to be useful, however, the information on the packet label has to be easily accessible. At present, this is almost universally not the case, as accessibility implies availability of the information in machine-readable form.

Only very few dried reference collections have yet started systematically to keyboard this information into computerized databases. At the International Mycological Institute, for example, information on all new specimens added to the dried reference collection has been stored on computer since 1989, amounting to over 30,000 records. There remains, however, the backlog of 330,000 precomputerized records which await keyboarding. The problems involved in establishing computerized databases and keyboarding such backlog records have been very ably summarized by Farr (1994).

Comparison of Dried and Living Reference Collections

While the living collection enables fungi to be kept in culture for use as living organisms, its scope is limited by the biology of the organisms it seeks to preserve. Some fungi are difficult to cultivate *in vitro*, for example many lichen-fungi and obligate parasites (like the rusts). To retain living material of such fungi, there are few alternatives. They may be cultivated in special 'plant disease gardens' analogous to botanical gardens (an option generally regarded as impractical because of the high financial overheads and quarantine implications), or their spores may be preserved in suspended animation (a technology which does not yet have the capability for widespread application, and cannot in any case conveniently preserve structures larger than the spores themselves). Dried reference collections are the only realistic option for retaining the large amounts of voucher material required for the serious scientific study of these fungi.

Many other fungi can be isolated and grown in pure culture, but either do not fruit under these conditions, or produce structures which differ strikingly from those occurring in the wild. Yet more fruit when first isolated, but lose that capability after several subculturing cycles. While such fungi may be retained in culture collections, they cannot yet be identified purely by cultural characteristics. This is a serious problem for culture collections, because it is very difficult to verify that each tube still contains the fungus originally isolated. Many fungi, perhaps even most of them, are therefore not suited for storage purely in living collections.

There is also a strategic value in maintaining dried reference collections. Compared with living reference collections, overall running costs are low, and the maintenance costs for individual items are far less. The high cost of running a living reference collection, together with the difficulty in maintaining many fungal species in a living condition, has meant that only a very small proportion of the probable total number of fungal species is in the world's culture collections (Nisbet and Fox, 1991). In the largest mycological living reference collections, the number of items held runs to tens of thousands, compared with the hundreds of thousands of items held in dried reference collections: in terms of storage alone, the dried reference collections are an order of magnitude greater than their living counterparts. The low running costs and comparative ease of maintenance also help dried collections to be more resilient than living collections to upheaval caused, for example, by political instability.

Conclusions

In mycology, dried reference collections have constituted, and will continue to constitute, an essential resource for many aspects of research on fungi. Their contribution complements rather than overlaps that of living collections. Using dried reference collections as a resource is hampered by difficulties in knowing where they are. A third edition of the *International Mycological Directory* is needed to provide much better access to this information. Dried reference collections are suitable as a focus for biodiversity studies, and keyboarding information from specimen labels is one of the most cost-effective, though not universally appreciated, ways of establishing major databases on biodiversity. There is, however, a need to establish data standards so that information keyboarded in different collections is compatible. The absence of adequate dried reference collections in the tropics and subtropics, especially Africa, hinders progress in mycology.

Addendum

Since presentation of this paper, a third edition of the *International Mycological Directory* has been published (Hall and Minter, 1994). This volume contains many more entries than the previous edition, and each entry is far more detailed than before. It also includes entries for institutions known to have a mycological interest, even when no response was received to requests for information. Indexing is much improved, and significantly larger numbers of dried reference collections are identified. Estimates of collection sizes received for the third edition are more accurate, reflecting moves towards computerizing of collection data and, perhaps, a greater awareness of the importance of the collections.

Estimates of collection sizes are available for the first time for the following major collections: University of Uppsala (Uppsala), 800,000; the Royal Botanic Gardens (Kew), 700,000; New York Botanic Garden, 475,000; Rijksherbarium (Leiden), 450,000; Botanischer Garten und Botanisches Museum (Berlin), 450,000; Natural History Museum (London), 400,000; the Canadian National Mycological Herbarium (Ottawa), 280,000; Komarov Botanical Institute (St Petersburg), 250,000; Royal Ontario Museum (Toronto), 250,000. In addition, the new estimate for the USDA (Beltsville, Maryland) collection is 1,210,000 (235,000 greater than in the second edition), and many other collections are listed as having between 100,000 and 200,000 items. It is thus clear that the 'guestimate' of eight million referred to in this chapter was very conservative.

References

Bruns, T.D., Fogel, R. and Taylor, J.W. (1990) Amplification and sequencing of DNA from fungal herbarium specimens. *Mycologia* 82, 175–184.

Duckworth, W.D., Genoways, H.H. and Rose, C.L. (1993) *Preserving Natural Science Collections: Chronicle of Our Environmental Heritage.* National Institute for the Conservation of Cultural Property, Washington, DC, USA, 140 pp.

Farr, D.F. (1994) Information resources for pest identification: databases. In: Hawksworth, D.L. (ed.) *The Identification and Characterization of Pest Organisms.* CAB International, Wallingford, UK, pp. 139–152.

Groombridge, B. (ed.) (1992) *Global Biodiversity. Status of the Earth's Living Resources. A report compiled by the World Conservation Monitoring Centre.* Chapman & Hall, London, UK, 585 pp.

Haines, J.H. and Cooper, C.R. (1993) DNA and mycological herbaria. In: Reynolds, D.R. and Taylor, J.W. (eds), *The Fungal Holomorph: Mitotic, Meiotic and Pleomorphic Speciation in Fungal Systematics.* CAB International, Wallingford, UK, pp. 305–315.

Hall, G.S. and Hawksworth, D.L. (1990) *International Mycological Directory*, 2nd edn. CAB International, Wallingford, UK, 163 pp.

Hall, G.S. and Minter, D.W. (1994) *International Mycological Directory*, 3rd edn. CAB International, Wallingford, UK, 163 pp.

Hawksworth, D.L. and Ritchie, J.M. (1993) *Biodiversity and Biosystematic Priorities: Microorganisms and Invertebrates.* CAB International, Wallingford, UK, 120 pp.

Holmgren, P.K., Holmgren, N.H. and Barnett, L.C. (eds) (1990) *Index Herbariorum, Part 1: The Herbaria of the World*, 8th edn. New York Botanic Garden, Bronx, New York, USA, 693 pp.

Morgan, P.J. (1987) ['1986'] *A National Plan for Systematic Collections?* National Museum of Wales, Cardiff, UK, 192 pp.

Nisbet, L.J. and Fox, F.M. (1991) The importance of microbial biodiversity to biotechnology. In: Hawksworth, D.L. (ed.) *The Biodiversity of Microorganisms and Invertebrates: Its Role in Sustainable Agriculture.* CAB International, Wallingford, UK, pp. 229–244.

Wasser, S.P. and Grodzynskaya, A.A. (1993) Content of radionuclides in macromycetes of the Ukraine in 1990–1991. In: Pegler, D.N., Boddy, L., Ing, B. and Kirk, P.M. (eds) *Fungi of Europe: Investigation, Recording and Conservation.* Royal Botanic Gardens, Kew, UK, pp. 189–210.

Microorganisms, Indigenous Intellectual Property Rights and the Convention on Biological Diversity

26

J. KELLEY

International Mycological Institute, Bakeham Lane, Egham, Surrey, TW20 9TY, UK.

Introduction

During negotiations preceding the United Nations Conference on Environment and Development which was held in Rio in June 1992 developing countries used their positions as major 'owners' of genetic resources to promote biotechnology transfer (Porter, 1992).

It was estimated in 1982 that 70% of the 3000 organisms known to have anticancer properties had been found in tropical forests (National Academy of Sciences, 1982). Roughly 74% of the 121 plant-derived compounds in the global pharmacopoeia were discovered through research based on ethnobotanical information from indigenous peoples (King, 1991). Statistics of this nature and the feeling that recent developments in biotechnology have made the potential of tropical forest resources 'limitless' stimulated negotiations and helped bring about aspects of the final structure of the Convention on Biological Diversity which was signed at the Rio Conference. Developing countries saw the treaty as a means of financing development whereas developed countries hoped that it would help to preserve threatened environments and biodiversity.

The use of 'biodiversity' as a tool for assisting in preservation of the environment and for raising income in developing countries has its supporters and detractors. Putting the case for the Convention as a means of fund-raising Sedjo (1992) points out that all conservation activities have a price attached: 'If preservation were costless, than all genetic resources

© 1995 CAB INTERNATIONAL. *Microbial Diversity and Ecosystem Function*
(eds D. Allsopp, R.R. Colwell and D.L. Hawksworth)

would be preserved.' The industrialized world has argued in the past that wild genetic resources are global resources and that the development of better lines of food plants, pharmaceuticals, etc. generates global benefits. However, a landowner (be it an individual or a nation) who holds the habitats for a unique genetic resource currently has no claim to its benefits whereas if it contained oil or bauxite or timber they would have a saleable resource. On the other hand, any company that obtains a commercial product from this genetic material can protect it via patent or plant rights, thus providing substantial income. Sedjo argues that the lack of private or national property rights to genetic material means that almost all efforts directed toward preservation and protection of them are altruistic, i.e. most proposals for preserving 'biodiversity' involve third-party action by governments, charities or the 'international community' attempting to persuade or coerce developing countries into protecting habitats.

In some areas, this has worked, in others it has had limited success. The argument follows that by harnessing the environment as a source of genetic material, rather than of renewable or non-renewable materials and by giving countries a financial interest in its maintenance and preservation then the international community as a whole will benefit.

The other supporting argument is that of 'natural justice', the feeling that the history of developed countries exploiting the natural resources of developing nations is now extending to the exploitation of genetic resources and the intellectual property (i.e. local knowledge) of their peoples. These concerns were expressed in the Declaration of Belem. In June 1988 anthropologists, biologists, chemists, sociologists, and representatives of indigenous populations met in Belem, Brazil, at the 'First International Congress of Ethnobiology'. The resulting declaration called for development aid to be steered toward inventory, conservation and management programmes together with recognition of indigenous knowledge, the dissemination and exchange of information on conservation and sustained utilization of resources.

Cunningham (1991) draws industrial parallels in that, if a private company accumulates 'unique and useful' knowledge through research, it patents that knowledge and receives a percentage of the profits from its use. Traditional knowledge, like industrial knowledge, has also been accumulated by research, but has been made public with no patent rights attached.

In general, then, it was felt that the people who live in areas rich in biodiversity could be its leading guardians if more of the value of their knowledge of their environment was returned to them. Recognition of intellectual property rights could play a role in this (Tickell, 1993).

There are, however, parties who are concerned about the content and implementation of the Convention and the concept of indigenous property rights. The editorial in the 18 June 1992 issue of *Nature* (immediately post Rio) suggested that developing countries had been erroneously led to believe that 'compensation for custodianship will meet the capital costs of development. That is a gigantic and cruel mistake.' Gollin (1993) puts this into perspective by suggesting that revenues from intellectual property rights will not exceed several hundred million dollars per annum whereas conservation costs may be tens of billions of dollars.

Schwartz (1992) looking at the convention from the pharmaceutical company perspective expressed concern that these efforts to obtain a part of the 'pharmaceutical profit pie' will add new elements to the already rapidly escalating costs of finding and developing new therapeutic compounds. Eventually this may inhibit the search for new compounds, and the time involved may limit the already short (in market terms) profit-making life of the product.

Intellectual Property Rights (IPR)

Gollin breaks down IPRs as they may apply in the implementation of the Convention on Biological Diversity into traditional and untried concepts.

Traditional rights include patents and trade secrets whereas the newer concepts which could be applied to the search for genetic material would be discoverer's and geographical rights.

Justification for intellectual property rights is that they provide incentives to innovators, they promote the public disclosure of new information, they reward 'invention' and they facilitate technology transfer. The feeling is that they form the only logical framework around which the Convention can function.

Convention on Biodiversity

Sands (1993) pointed out the ambitious nature of this convention in that previous international agreements such as the Berne Convention and the Convention on the International Trade in Endangered Species (CITES) aimed to protect only particular species or particular ecological niches. The Convention on Biological Diversity, however, is seeking to protect whole ecosystems.

It is fairly clear that higher plants and to some extent animals were the main considerations taken into account during the drafting of this Convention. Yet in many ways its significance to the microbiology world is greater than to higher plants and animals.

Looking at the articles with a microbiologist's eye the following commentary can be made.

Article 1: Objectives

This article outlines the three main objectives: conservation of biodiversity, sustainable development of genetic resources and 'fair and equitable' sharing of the resulting benefits. Very laudable objectives.

Article 2: Use of Terms

Problems begin to arise at this early stage. Most of the basic terms defined are fairly straightforward and non-controversial. However, the definitions of 'Country of origin of genetic resources' and 'Country providing genetic resources' both imply that there will be a single country involved. This is unlikely when we consider plants and animals, but even less likely when microorganisms are involved. The Convention does not cover likely problems – for example, when an active molecule is found after screening culture collection material from country X and a more active strain is then found after further selective isolations from country Y. Gollin (1993) suggests that it will be necessary to focus 'rewards' and technology transfer on individual people or organizations with knowledge and access to resources rather than on countries; but this still does not solve the question posed above.

Article 6: General Measures for Conservation and Sustainable Use

This article calls for signatories to develop 'national strategies, plans or programmes for the conservation and sustainable use of biological diversity'.

Article 7: Identification and Monitoring

The identification, sampling and monitoring of species and habitats is called for in this section together with the 'identification of activities which have or are likely to have significant adverse impacts on the conservation and sustainable use of biological diversity'. This appears to be a very attractive set of proposals, particularly as the establishment of databases to hold the information is also required. However, Gollin suggests that the key question will be whether such information will be in the public domain. Databases and publications can be covered by copyright and trade secret rights. The Convention does not outlaw privately funded inventories and allows private ownership of resulting copyrights and trade secrets.

Article 9: Ex Situ Conservation

Article 9 encourages *ex situ* conservation in gene banks, culture collections, etc., preferably in the country of origin in order to complement *in situ* conservation. Material in collections may have already been screened for useful activities. It has been suggested (Gollin) that this fact may make wild sources of new genetic material even more valuable and hence indirectly may actually undermine conservation measures. This is unlikely in the case of microorganisms as collecting to extinction is not a major problem; loss of habitat and substrate is more important.

Article 11: Incentive Measures

This article requires signatories to 'adopt economically and socially sound measures to act as incentives for the conservation and sustainable use of components of biological diversity'. Intellectual property rights can provide such incentives. Developing countries are often reluctant to allow the patenting of natural products but are now being encouraged to strengthen their laws in these areas and in the areas of national conservation and natural resources law.

Article 12: Research and Training

Article 12 calls on contracting parties to establish education and training programmes for the identification, conservation and sustainable use of biological diversity, taking particular account of the needs of developing countries. They should also promote and encourage research and promote and cooperate in the use of scientific advances. Providing sufficient funding is made available, this can only be applauded.

Article 15: Access to Genetic Resources

This article is one of the key sections from the microbiologist's point of view. It recognizes the sovereign rights of States over their genetic resources, but commits countries to allow access for environmentally 'sound' uses. This is to be the subject of national legislation. Access should be on 'mutually agreed terms' and will require a nation's 'prior informed consent'.

Gollin points out that this article relates only to genetic resources and that it could therefore be argued that countries could prevent access to other biological resources such as chemical extracts, but feels such an interpretation would be at odds with the spirit of the Convention.

Article 15 (3) emphasizes that the Convention applies primarily to genetic resources from countries of origin or from countries that acquired genetic resources under the terms of the Convention; in other words the Convention is not retrospective and does not apply to collecting from non-signatory countries.

How local legislation will develop is yet to be seen. The INBIO agreement between Costa Rica and Merck is the most advanced example.

Article 15 (6) suggests that one aspect of access agreements will be participation in scientific research on genetic resources, and that 'fair and equitable sharing of any proceeds from the commercial use of genetic resources should be on mutually agreed terms'. It is suggested that such measures should include an intellectual property framework.

Article 16: Access to and Transfer of Technology

This is a key article in the area of property rights and technology transfer. It complements the access to genetic resources requirements detailed in article 15; it specifically links technology transfer to biodiversity conservation policy.

The United States felt that this connection was not necessary for biodiversity conservation and this was one of the main reasons for the USA refusing to sign the Convention in 1992. Gollin feels that the wording of this article is so 'convoluted and ambiguous that the obligations of a signatory nation are not immediately clear'. However, he suggests that the interpretation of the language should in fact be that international agreements should be promoted and encouraged, but are not required.

This article is aimed at governments, not private individuals. Private agreements can be made without the Convention being invoked. However, Duesing (1992) suggests there is merit in requiring countries to promote voluntary biodiversity prospecting agreements to 'prime the pump'.

One of the initial fears was that this article required compulsory licensing of products and technology. However, whereas the convention allows domestic legislation to require compulsory licensing it also allows, for example, the US laws which exclude most types of compulsory licensing.

Article 16 (1) calls for a two-way flow of technology, with high technology and taxonomic expertise balanced by local traditional knowledge. Inevitably the flow will be mainly from high technology- to low technology-based economies.

Article 16 (2) appears to reinforce the above observation by providing for 'most favourable, concessional, or preferential terms for transfer to developing countries'. However, this is qualified by the statement that transfer should be equally fair unless 'mutually agreed' for most technologies which are covered by patents and other intellectual property rights 'on terms which recognize and are consistent with the adequate and effective protection of intellectual property rights', which means in effect that existing local law presides (Gollin, 1993).

Section 16 (3) suggests that the country providing genetic resources should receive access to technology which makes use of those resources. It relates only to genetic resources, not other related technologies. Initial interpretations suggested that developed country governments would be required to take a compulsory licence from their own biotechnology companies and then provide this technology to a developing nation. But Gollin again points out that it is on 'mutually agreed terms' so that nothing is compulsory. However, Sands (1993) suggests that existing property rights may have to be limited. He interprets this section as requiring the source country to have access to results of research and development irrespective of patents.

Section 16 (4) calls for encouragement for private firms to make technology transfers – again on mutually agreed terms, again nothing compulsory.

Section 16 (5) looks at the influence of patents and intellectual property rights on the Convention and suggests that such rights 'are supportive and do not run counter to its objectives'.

Gollin's interpretation of this Article suggests that rather than mandating unfavourable technology transfer it can be read as promoting non-compulsory inducements such as tax incentives, trade agreements, grants and awards to private companies agreeing to transfer biotechnology to countries providing genetic resources.

Article 17: Exchange of Information

Article 17 encourages the exchange of public domain information. Gollin points out that care should be taken to ensure government agencies do not inadvertently provide data on protected information.

Article 18: Technical and Scientific Cooperation

This article requires the promotion of cooperation, particularly through training and the exchange of experts and by the establishment of a clearing house which would collect fees from recipients of samples and distribute them to the suppliers.

Section 18 (4) specifically recognizes indigenous technologies. It is suggested that the best way to protect such information is through intellectual property rights.

Article 19: Handling of Biotechnology and Distribution of its Benefits

This article reinforces article 16 with particular reference to biotechnology but is again peppered with 'as appropriate' and 'where feasible'. Gollin again interprets this article as an agreement for non-compulsory 'two-way technology transfer'.

Section 19 (3) recognizes that genetically modified organisms could have adverse effects on biodiversity and suggests that biotechnology could be inherently unsafe. The USA was particularly unhappy with this inference.

Article 22: Relationship with Other International Conventions

Interpretations of this article are contradictory. It states that 'the Convention shall not affect the rights and obligations ... deriving from any existing international agreement'. There is, however, a qualifying statement which says 'except where the exercise of these rights and obligations would cause serious damage or threat to biological diversity'. Lawyers seem to interpret this from both sides, with Sands highlighting 'the primacy of the Convention over other international law' whereas Gollin stresses 'the Convention is subordinate to existing international agreements'. Case law will decide which view prevails.

There are a number of problems which can be seen with the implementation of the Convention, some of which are specific to or enhanced in the case of microorganisms.

Ownership of Intellectual Property

Local knowledge of microorganisms does not often play a part in guiding screening companies to new molecules – this is usually from random sampling and isolations. Problems arise with 'cross-boundary' isolations. Schwartz (1992) cites the case of cyclosporin, which was first identified in cultures from Canadian soil samples. However, the significant 'hit' came from Norwegian soils. Who in this case would receive the income? Could it be shared and if so how would this be apportioned? This can be further complicated by culture collection material. Suppose a useful molecule is found in material from a deposited culture screened on the recommendation of the collection staff. A better producer may then be obtained by a selective isolation programme based on this initial hit. Who is entitled to a percentage of royalties or other payment here? The original isolator, the depositor (who may not be the same person), the collection, the country from which the final strain was isolated?

Many molecules, although initially derived as natural products, are later produced synthetically; should payments then continue and if so for how long? Curare was first used medically during World War II but more effective synthetic compounds are now employed. Would the Indians whose intellectual property was first employed still be drawing royalties?

Any potential antibiotic originally obtained from a soil isolate will then undergo extensive research where hundreds of similar compounds will be synthesized before a useful one is obtained. Who could claim payment here and at what percentage?

What Will Happen to 'Open' Science?

The requirement to apply in advance and the need to sign collecting agreements may inhibit the collecting of material for systematic purposes. Kirsop (1993) points out that taxonomic research requires large numbers of strains to be studied. It is important that the free circulation of type specimens and herbarium material continues. Is this going to become more difficult? Spontaneous collecting will certainly be inhibited, but it will not be easy to police the export of small amounts of material intended for microbial isolation studies, which raises the spectre of sample smuggling.

How Do Workers Transfer Technology, Training and Royalties?

The treaty requires governments to establish legislation and routes for cooperation and technology transfer. However, is it likely that payments will filter through from government level to benefit directly the institutes or indigenous peoples involved? Many companies and institutes are already establishing policy which requires payment to and liaison with institutes as first preference. This will probably prove to be the most reliable way of ensuring benefits reach the correct recipients.

How Will Gene Banks and Collections Maintain Open Access to *Ex Situ* Organisms?

The concept of indigenous intellectual property rights adds greatly to the responsibilities of culture collections. Extra administration will be required to ensure that all new deposits from signatory countries are logged as such and their sale and distribution monitored. Kirsop suggests the establishment of a master database to monitor material flow and designated receiving collections which would be identified by national governments functioning in much the same way as the Budapest Treaty currently does to hold and distribute cultures from signatory countries.

Taken to its ultimate conclusion royalties should be paid on every culture isolated from a signatory country sold by a collection. However, most collections are not-for-profit organizations which only operate through grants, subsidies or by being part of a larger government organization.

At the very least collections will need to establish good documentation. It is also suggested that editors of publications should be required to ask for a full isolation history and a collection number for all microorganisms reported on. Although this is good practice now, it may become compulsory in the future.

Adherence to IPR and Convention Requirements

Many pharmaceutical companies and gene banks already have a policy in place or in preparation. A number of ideas have emerged to ensure compliance.

Pharmaceutical Companies

Often companies will only purchase material for screening from national or international organizations which themselves have a biodiversity policy in place. They will reimburse the collecting country and/or institute for their efforts and expertise. They will pay royalties on profits from any commercially marketed drug. However, the question of technology transfer is more difficult. The Merck–INBIO agreement is probably the only example where training and some aspects of technology transfer are being fully implemented at national level at the moment.

Gene Banks and Culture Collections

Most of these organizations will be adding value to the material by naming, inclusion on a database, development of maintenance methods for live material and possibly by carrying out some initial screens. They will share royalties on any commercially marketed drugs with the collector and/or depositor. Many of these collections are already heavily involved with training in taxonomy, collection development and maintenance in developing countries. This may be enhanced, increased and more directed in the future.

Collecting and Extracting Companies

Such companies may only purchase from national institutes or recognized collectors. They will pay royalties on useful products. At least one company is establishing local extracting laboratories to enhance technology transfer and may in the future carry out initial screening in local laboratories.

These are examples of guidelines which have been established by individual companies before legislation requires them to do so. Once local legislation is in place and some case law is established things may have to change.

It has been suggested that one potential avenue would be to extend the market exclusivity given to drugs derived from natural products so that extra funds may thus be made available to developing countries for a longer period (Aylward, 1993).

The Future

Concern has been expressed that, unless the level of pharmaceutical prospecting increases dramatically in the next few years, the use of the pharmaceutical potential of biodiversity as an economic argument for biodiversity conservation will be unfounded (Aylward, 1993).

Others are even more concerned. If conservation is too closely linked to income generation and this 'gold mine' fails to materialize then the reason for maintaining threatened environments and genetic material would disappear (Carr, 1993).

New biotechnology techniques could remove the need for screening natural products in the not too distant future, so we are looking at medium-term gains. There must be other reasons for conserving our threatened habitats and sources of finance must be in place before this happens.

In the meantime Indigenous Intellectual Property Rights must be addressed by everyone working in microbiology; and policy and legislation must be established by governments as a matter of urgency.

References

Aylward, B. (1993) The economic potential of genetic and biochemical resources. In: *Intellectual Property Rights, Indigenous Cultures and Biodiversity Conservation.* Summary of a meeting held by Green College Centre for Environmental and Policy Understanding, 14 May 1993.

Carr, G. (1993) Implication of advances in biotechnology. In: *Intellectual Property Rights, Indigenous Cultures and Biodiversity Conservation.* Summary of a meeting held by Green College Centre for Environmental and Policy Understanding, 14 May 1993.

Cunningham, A.B. (1991) Indigenous knowledge and biodiversity. *Cultural Survival Quarterly*, Summer 1991.

Duesing, J.H. (1992) The convention of biological diversity: its impact on biotechnology research. *AGRO Food Industry Hi-Tech* 3(4), 19.

Gollin, M.A. (1993) An intellectual property rights framework for biodiversity prospecting. In: *Biodiversity Prospecting.* World Resources Institute, Washington, DC.

King, S.R. (1991) The source of our cures. *Cultural Survival Quarterly*, Summer 1991.

Kirsop, B. (1993) The Biodiversity Convention: some implications for microbiology and microbial resource centres. WFCC Document.

Kirsop, B. and Hawksworth, D.L. (eds) (1994) *The Biodiversity of Microorganisms and the Role of Microbial Resource Centres.* World Federation for Culture Collections, UK, 104 pp.

National Academy of Sciences (1982) *Ecological Aspects of Development in the Humid Tropics.* Washington, DC.

Porter, G. (1992) *The False Dilemma: The Biodiversity Convention and Intellectual*

Property Rights. Environmental and Energy Studies Institute, Washington, DC.

Sands, P. (1993) The Biodiversity Convention: Legal issues and mechanisms for equitable trade in biological resources. In: *Intellectual Property Rights, Indigenous Cultures and Biodiversity Conservation.* Summary of a meeting held by Green College Centre for Environmental and Policy Understanding, 14 May 1993.

Schwartz (1992) Redistributing discovery's profits - the Schwartz commentary. *SCRIP* 1733, 16.

Sedjo, R.A. (1992) Resources for the future. *Journal of Law and Economics* 35, 199–213.

Tickell, C. (1993) The policy context. In: *Intellectual Property Rights, Indigenous Cultures and Biodiversity Conservation.* Summary of a meeting held by Green College Centre for Environmental and Policy Understanding, 14 May 1993.

Note in Proof

Since this workshop took place a document, *The Biodiversity of Microorganisms and the Role of Microbial Resource Centres* (Kirsop and Hawksworth, 1994) has been produced by the World Federation for Culture Collections in collaboration with the United Nations Environment Programme. It updates and expands on many of the points made in this chapter and proposes solutions to some of the queries.

Extent and Development of the Human Resource 27

E.J. DaSilva

Responsible for Life Sciences, Division of Basic Sciences, UNESCO, 1, rue Miollis, 75015 Paris, France.

An important feature of AGENDA 21 is the building up of capacity to manage biodiversity. The capacity of human resources and specialized institutions to deploy biodiversity on the path to development was considered to be an important element especially in the developing countries. Much of the emphasis has focused on diverse flora and fauna with some attention on the world of microorganisms which constitutes an essential element of the planet's biological fabric. Different components of the microbial world – algae, bacteria, mycoplasmas, fungi (including lichens and yeasts), protozoa and viruses – are vital to the function and maintenance of the Earth's ecosystems and biosphere. Different components of the microbial world contribute, through the interlocking biogeochemical cycles, the recycling of nutrients in food chains in aquatic and terrestrial niches, and the bioconversion of waste residues reflecting varied skeins of physiological activities in the maintenance and management of the environment. The extent and development of the human resource, therefore, is directly linked to the constitution and extent of microbial diversity.

The groundwork for economic development is daily being constructed through research advances coming from a number of biotechnological fields such as cell biology, microbiology, recombinant DNA, virology, etc. Evidence of such catalytic research is to be found in the proliferation of research contributions to journals and books that cover fields ranging from gene therapy and pharmaceutical biotechnology to the release of genetically engineered microorganisms and issues of intellectual property rights.

Linked to such development are priorities for biosystematic research in support of biodiversity, with emphasis on microorganisms and invertebrates, in developing countries (Hawksworth and Ritchie, 1993).

Characterized by its contribution to environmental conservation and economic growth, microbial diversity is being encountered as the cornerstone of new policy initiatives and environmental regulations. Examples are the advocation of ethanol as a fuel by the new US administration and the closing, by governmental authorities in Hong Kong, of a bleaching and dyeing factory that had a long history of dumping untreated effluent into the Ho Chung river.

Microbial diversity is closely linked with human existence. Since times immemorial, it has been a source of food, medicine and industrial material. It has been used to some extent in ensuring food security by investing in Nature's bank of rhizobacteria. Modest investments yield high dividends. It is widely estimated that microbial diversity accounts for the fixing of 175 million tons of nitrogen annually in comparison to that of 50 million tons provided by artificial fertilizers. Again, in the field of human health, antibiotics, derived from diverse microbial species, belong to classical biotechnology and represent 11% of the world therapeutical market. Their selling volume was about $12 billion in 1990. In this field, chemical synthesis cannot compete with fermentation.

Today, the development of 'genetically engineered organisms' by *in vitro* DNA procedures is playing an important role in agriculture, medicine and industry. Nevertheless, despite the research devoted to microorganisms that are of direct relevance to humanity, substantial gaps remain.

In discussing bacterial descent from earlier centuries Postgate (1990) discussed the significance of the phenomenon of cryptobiosis. Furthermore, Postgate (1991) argued that just as bacteria are artefacts of the culture media, so also the bacterial world can be viewed as an artefact of the rest of the biosphere.

Bull *et al.* (1992), in drawing attention to the significance of microbial diversity as a major resource for biotechnological products and processes, made the paradoxical observation that, despite the commercial revenue realized from the exploitation of microorganisms, knowledge concerning the extent of microbial diversity is rather scarce. Scientific evidence of such existence is to be found in *Epulopiscium* – the world's largest bacterium – and *Armillaria* – the champion thallus that has been reported in the popular press. Another report concerns the flourishing of bacteria, *Pyrodictium brockii*, a hyperthermophile, at marine and terrestrial geothermal sites; and more recently the emergence of a new strain 0139 of *Vibrio cholerae* in Thailand and the Bay of Bengal. The myxobacteria have been investigated as a source of new antibiotics (Reichenbach *et al.*, 1988). Hori (1992) has recently discussed the evolutionary outline of living organisms as deduced from 5S ribosomal RNA sequences and relationships between the eubacteria, the metabacteria, and the archaebacteria.

En passant it should be noted that the ethics of wiping out the world's most deadly virus, i.e. the agent of smallpox, and by implication rendering it extinct, has been debated (Hawkes, 1993).

The microbiologist's equivalent of botanical gardens, seed banks, zoos and aquaria are culture collections. These collections constitute reservoirs of microbial diversity proliferating in artificial habitats that mimic *in situ* conditions. Moreover, a culture collection, likened to a museum that serves both molecular biology and microbiology, is a history of science (Lamana, 1976).

Most developing countries lack proper endogenous capacities and facilities to manage, conserve and preserve microbial diversity. In addition, issues such as ownership and protection of intellectual property need to be discussed and explained with a view to drawing up a rationale that fosters the creation, development and transfer of the novel biotechnologies. Furthermore, patent law in most developing countries is either outdated or archaic, a situation that negates the attraction of incentives to invest in development of biotechnologies. Furthermore, biosystematics and taxonomic studies are either based on classical reference texts or check-lists established by experience.

In several developing countries, efforts in building up capacities to explore, document, preserve and conserve, and ultimately use microbial diversity for development, are conducted through a variety of training programmes at university level. These efforts, employing standard microbiological curricula, deal with routine isolation and identification of microbial species. In the more reputed universities of long standing, and in which staff have been exposed to modern approaches in taxonomy, systematics and culture collections, treatment of strain data, the application of immunological principles and numerical tables in taxonomy are likely to be encountered. Nevertheless, an important aspect of such activities is that the existing collections of microbial diversity have been built up from species used in communicating and teaching an understanding of the microbial world; or from specialized research being conducted by postgraduate students; or again, from contract research with local industry and from interaction with those sectors dealing with health, food, the treatment of domestic wastes, energy production, and the diagnosis and treatment of diseases. Recently new ways of injecting microbes into mainstream education have been described with a view to inculcating a better appreciation of the availability of microbial wealth (Zook, 1990).

In the extent and development of the human resource, information on the collection, storage and dissemination of microbial resources is vital. Information by itself constitutes an important means in the safeguarding and sharing of the range of microbial genetic resources.

The rapid deployment in the areas of agriculture, industry and medicine of microbial genetic resources worldwide underscores the urgency for the exchange of information. This is coupled to the need for easy access to computerized data in gene banks, bibliographic services and acquisition

Table 27.1. The world network of microbial resources centres (MIRCEN).

Speciality	Location	Place and country
Rhizobium	Department of Soil Science and Botany, University of Nairobi	Nairobi, Kenya
Rhizobium	Instituto de Pesquisas Agronomicas	Porto Alegre, Brazil
Fermentation, food and waste recycling	Thailand Institute of Scientific and Technological Research	Bangkok, Thailand
Biotechnology	Ain Shams University	Cairo, Egypt
Biotechnology	Central American Research Institute for Industry	Guatemala City, Guatemala
Rhizobium	NifTAL Project, College of Tropical Agriculture, University of Hawaii	Hawaii, USA
Biotechnology	Karolinska Institutet	Stockholm, Sweden
World data centre	Life Science Research Information Section, RIKEN	Saitama, Japan
Rhizobium	Centre National de Recherches Agronomiques	Bambey, Senegal
Biotechnology	Planta Piloto de Procesos Industriales Microbiologicos (PROIMI)	Tucuman, Argentina
Rhizobium	Cell Culture and Nitrogen-fixation Laboratory	Maryland, USA
Fermentation technology	Institute of Biotechnology, University of Osaka	Osaka, Japan
Biotechnology	International Institute of Biotechnology, Canterbury	Kent, UK
Mycology	CAB INTERNATIONAL, International Mycological Institute	Surrey, UK
Biotechnology and agriculture	University of Waterloo	Waterloo, Ontario, Canada
Marine biotechnology	Department of Microbiology, University of Maryland	Maryland, USA
Biotechnology	Centre de Transfert	Toulouse, France
Biotechnology	University of Queensland	Brisbane, Australia
Microbial technology	Institute of Microbiology, Academia Sinica	Beijing, China
Microbial biotechnology	Caribbean Industrial Research Institute	Tunapuna, Trinidad and Tobago

Table 27.1. *continued.*

Culture collections and patents	German Collection of Microorganisms and Cell Culture	Braunschweig, Germany
Culture collections	Department of Microbiology, University of Horticulture and Food Industry	Budapest, Hungary
Biotechnological Information Exchange System (BITES)	UNESCO International Centre for Chemical Studies	Ljubljana, Slovenia
Bioconversion technology	Department of Applied Biology, The Chinese University of Hong Kong	Shatin, Hong Kong
Culture collections	American Type Culture Collection	Maryland, USA
Culture collections	National Collection of Type Cultures	London, UK
Culture collections	Iranian Research Organization for Science and Technology	Tehran, Iran

of microbial species for academic and industrial purposes.

The sharing of information is critical since such sharing leads to increased preservation and rational use of microbial diversity, to avoidance of wasteful duplication of labour and financial resources, and to the linkage of scientists in developed and developing countries in managing and conserving the planet's microbial genetic heritage (Table 27.1).

Information is of no value whatsoever if it does not reach the constituency of researchers working with microbial diversity and curators of culture collections handling, conserving and preserving microbial diversity for beneficial use. Furthermore, the inventorization, retrieval and sharing of information is crucial for the development of the human resource in designing efficient training and research programmes. The availability of such information spurs curriculum development in the microbiological sciences and helps keep the teaching and student communities abreast of new innovations in techniques, standards and protocols involved in the categorization and use of microbial diversity.

Human resource development calls for the use of reliable data in deploying microbial diversity for technological advancement. Bower (1989) has provided a comprehensive review on the main collections and databases that are associated with microorganisms, animal cells, hybridomas, plant cells, mouse strains, rare and crop plants, seeds, DNA probes and DNA sequence data. Interactive and commercial biotechnology resources have been reviewed by Johnson and Rader (1990).

In the area of bibliographic services, the NifTAL MIRCEN, for example, provides a computerized service entitled 'The Legume/ *Rhizobium* Symbiosis: A Continuing Bibliography'. Provided to MIRCENs specializing in biological nitrogen fixation, the purpose of this bibliography of current publications on the legume/*Rhizobium* symbiosis is to provide developing country researchers with access to the document service available to them through the NifTAL project. Furthermore, this MIRCEN also issues a series of single-sheet protocols entitled 'Illustrated Concepts in Agricultural Biotechnology' covering topics such as 'A Simple Transfer Chamber for Aseptic Work with Microorganisms' and 'Inoculating Tree Legume Seeds and Seedlings with Rhizobia'. Similar technical information sheets concerning work in culture collections have been released by the World Federation of Culture Collections (WFCC).

Data banks of microbial genetic resources have an important role to play in the development of human resources. The production of audiovisual materials (films, slides, etc.) and technical flow-sheet schemes portraying the collection, handling, inventorization, dissemination, and retrieval of the information and date of microbial resources is a necessity for many developing countries that have 'first and last mile' problems in accessing data banks of categorized microbial diversity.

Many useful techniques and technologies that contribute to an awareness of microbial technology can be available to the developing world. Their transfer is, however, inhibited by a lack of awareness of both their existence and their availability. Models and schemes for researchers and curators to move away from the limitations of day-to-day maintenance and protection activities towards more information-based preservation computerized programmes would greatly enhance the appreciation and value of the range and extent of the microbial resource hitherto known, and perhaps to be enriched by discovery of the unknown.

The most important element in building up capable and competent human resources is the vehicle of training, which serves as one of the pillars of sustainable development. The conservation and use of microbial genetic resources and the building up of new knowledge on these resources is dependent upon a cadre of trained personnel. Training, indisputably, constitutes a sustainable approach in conserving, managing, tapping, and using the potential of microbial resources for environmental management, socioeconomic advancement and improvement in the quality of life.

Microbial genetic resources are an asset and a tool to catalyse economic development. Research, which is complementary to training, brings an added feature – VAT, i.e. value-added training, which brings exposure of trained personnel at the basic level to new research skills, to new research techniques, to new teaching methods, and to a new understanding of the health and wealth of microbial resources for deployment in agriculture, industry, medicine and technology.

The transfer of research knowledge and skills helps to underpin research activities in developing countries. Capacity-building is best built up through the use of existing institutions. A number of strong research institutions linked to *ex situ* conservation of microbial diversity already exist in developed and developing countries as is evident from different editions of the catalogues issued by the World Data Centre MIRCEN.

The availability of trained personnel has always been one of the key requirements for successful technology transfer. Developed country scientists are required not only to aid the transfer process, but to use their expertise to perform a key part of the task itself. This is the rationale of the UNESCO/BAC and UNESCO/MIRCEN Professorship schemes which help to ensure the transfer of the knowledge and skills collected during years of experience. The difficulties of transferring skills for use in microbial diversity programmes are compounded by the lack of sufficient staff. The pressing demand for staff with basic technical skills in some developing countries will, perhaps, be solved over time, as appreciation and economic value of the microbial gene pool grows through research and training programmes.

Scientific research, using the microbial labour force as the tools of advancement for human progress, plays an educational role concerning identification, loss and recovery of microbial diversity. Training and research are highly complementary activities that link and increase the economic value of genetic resources. Training programmes must have an integrated approach. Too often research staff have skills in one discipline but lack those required for an overall programme.

Thus, biotechnologists in handling, managing and using microbial diversity should also be knowledgeable in other areas that interlink or overlap with their own fields of specialization. Training and teaching methods should be integrated in such a way that the curriculum is broadly based and introduces students to different areas of biotechnology, giving them a broad view of opportunities for specialization. The importance of such an approach to training is to be found in culture collections that provide services to industry, government, and academic institutions (Allsopp and Simione, 1988).

Integrating individual fields into a biotechnologies-diverse curriculum has several advantages because there is a need for specialists who also have skills in genetic engineering, microbiology, enzymology, biochemistry, chemistry, biophysics and cell biology. As an example, the Guatemala Microbial Resources Centre (MIRCEN), in conjunction with the International Organization of Biotechnology and Bioengineering, offers an interdisciplinary course that deals with the principles of biochemistry, microbiology, fermentation technology, and biodegradation. This course also provides a broad understanding of the various technical languages used in the different disciplines.

Thus, developing countries, focusing on capacity-building, should be able to exploit the range of industrial opportunities presented by biotechnology by using the combined resources of the biology, biochemistry, microbiology, and chemical engineering departments. By tailoring the training to the country's need, attention could be directed to those areas of applied research related to local resources, conditions and needs of the home countries.

Biotechnology for many of the non-industrialized countries is expected to be the answer to all the problems of their country. Certainly not the 'cure-all', biotechnology, nevertheless, can be harnessed to improve the quality of life and sustain economic growth. For this potential to be realized an academic infrastructure is an absolute necessity.

Universities and research centres are a prerequisite, where basic and further education in the field of biotechnology are given prominence, and which are staffed with academic personnel who are qualified for teaching and research purposes. Many developing countries can easily fulfil these conditions keeping in mind the needs and requirements of these countries in terms of national development, national markets and national economic growth.

It is important for developing countries to train and develop their own scientists and engineers, to resolve the problems with which they are acquainted. In such a manner, and with political will, developing countries in the immediate future could become active, equal and stable partners in addressing issues of the environment and development that are confronting humankind.

The widespread support for culture collections dates back to the early catalytic assistance provided by UNESCO in 1946. The initial impetus for the exchange of valuable culture holdings and information on the occurrence and utility of worldwide culture collection research was provided by UNESCO through its support to the pioneering *International Bulletin of Bacteriological Nomenclature and Taxonomy*, an activity that today is being pursued by electronic means.

The important tasks and services of culture collections in the collection, maintenance, preservation, management, use and supply of microbial genetic resources have been covered in collective volumes in a series of publications resulting from a number of international conferences organized under the auspices of the World Federation for Culture Collections. Continuous updates on such information may be found in journals or news bulletins published by the World Federation for Culture Collections, the UNESCO global network of 27 microbial resources centres (Table 27.1), the World Intellectual Property Organization (WIPO), the Information Centre of European Culture Collections, the Microbial Strain Data Network, the Microbial Information Network Europe and the European Laboratory Without Walls Programme on phages and plasmids. The Commission of the European Communities has made a strong contribution through the release of the *European Biotechnology Information Service Newsletter*.

One of the milestones at the 16th Session of UNESCO's General Conference in 1970 was a resolution from the Governments of Denmark,

Finland, Norway and Iceland calling for the establishment of specialized microbial research centres in developing countries. Four years later, following the United Nations Conference on the Human Environment at Stockholm, 1972, experts from UNEP, UNESCO and the international microbiological scientific community represented by ICRO met at UNEP Headquarters at Nairobi, to jointly formulate a worldwide programme aimed at preservation of microbial gene pools and at making them accessible to developing countries through a project entitled 'Development of an Integrated Programme in the Use and Preservation of Microbial Strains for Deployment in Environmental Management' (July 1975–October 1984). An important feature of the UNESCO MIRCEN network is to build and strengthen relevant academic infrastructure, to train human resources, to help formulate the research agenda of institutions, and to attract the full participation of all shareholders in the management of the treasury of microbial diversity.

Against this background, several commendable initiatives (Table 27.2) need to be further elaborated with suitable background research programmes built around the existing nucleus of international cooperation *vis-à-vis* microbial diversity.

More recently, within the framework of joint UNESCO/UNDP collaboration, the concept of a Biotechnological Information Exchange System (BITES) was discussed and accepted in 1988 on the occasion of a European MIRCEN Biotechnology Symposium at Ljubljana, Yugoslavia. The Symposium had a twofold strategy.

1. The establishment of a European Network of Microbial Resources Centres built around existing microbial culture collections with the prime aim of attracting and consolidating existing regional and international *ad hoc* cooperation in the development and expansion of culture research, maintenance, use and eventual application for evolution of bio-based industries. The focal point of this network is based at the National Collection of Industrial and Agricultural Microorganisms, Budapest, Hungary, which is also recognized under the Budapest Treaty as an International Depository Authority for patents.

2. The establishment of a Biotechnological Information Exchange System (BITES) that complements the work of the European MIRCENs and which would be composed of network partners linked to current relevant information systems thus providing a means of filling information gaps and developing higher levels of biotechnological information exchange services, with particular emphasis on priority research application areas such as: agricultural processing, production of speciality chemicals, aquaculture, waste/wastewater management, and new enabling technologies. The focal point of BITES is the UNESCO International Centre for Chemical Studies, Ljubljana, Slovenia.

Table 27.2. Some examples of information sources dealing with microbial resources.[a]

Entity	Purpose
MINE – Microbial Information Network Europe	The MINE project, supported by the Commission of European Communities in its BRIDGE programme, aims at the centralization of data on strains present in the European culture collections in one database, thus ensuring easy and quick accessibility of these data. One centralized and integrated database has been created, holding data on bacteria, filamentous fungi and yeasts. This database is available on-line at the Deutsche Institut für Medizinische Dokumentation und Information (DIMDI), Cologne, Germany.
MSDN – Microbial Strain Data Network	The Microbial Strain Data Network, based in the UK, is an international information and communications network for microbiologists and biotechnologists. It provides electronic mail, bulletin boards, computer conferences, databases, training, software distribution, user support. It is sponsored by IUMS, CODATA, COBIOTECH, WFCC. Financial support has been received from time to time from UNEP, EEC, NSF, EPA, USDA, Environment Canada, the National Institute of Dental Research and UNESCO.
ICECC – Information Centre for European Culture Collections	The major tasks of the Information Centre, based in Germany are: to provide a permanent central contact point for European scientists and any institutions seeking advice and information on biological material and on culture collection related matters; to publicize the resources within the culture collections in terms of materials and scientific expertise, by preparing printed and visual material for distribution; and to encourage scientists to deposit microbial strains and other biological specimens of biotechnological and/or of general scientific significance in European culture collections as an important future resource.
BITES – Biotechnological Information Exchange System	Built with the UNESCO software CDS-ISIS, BITES is a database on biotechnological processes and applications covering data concerning biotechnological raw materials, processes and products, with highly specialized branches on (1) antibiotics for animal use, (2) biotechnological processes and products based on starch, (3) bioremediation of polluted waters, (4) microencapsulation in biotechnology and (5) ethanol fermentation. BITES is located in Ljubljana, Slovenia.

Table 27.2. *continued.*

Regional Genetics Resources Centre and Gene Bank	Established in 1985 by the SADCC group of countries and located in Lusaka, Zambia, the Regional Gene Bank interacts with the Nordic Gene Bank and FAO/IBPGR.
AIPO – African Intellectual Property Organization	Established by the Libreville Agreement of 1962, the African Intellectual Property Organization is concerned with the availability of patents for microbiological processes and products of such processes, data technology and information on intellectual property rights.

*a*See also Bower (1989) and Johnson and Rader (1990).

UNESCO has also taken the lead in consulting with the International Union of Microbiological Societies (IUMS) and its international commission on food microbiology and hygiene on streamlining postgraduate teaching in advanced food microbiology and recommending a core curriculum.

In other initiatives, UNESCO's Division of Science Teaching and Environmental Education has developed materials that encourage an international exchange of ideas and information in science education, including agriculture and biology teaching, genetically based technologies, the uses of the sea and its organisms, biology and human welfare, and systems thinking in biology education.

In collaboration with UNESCO, the Commission for Biological Education of the International Union of Biological Sciences has produced *Teaching Biotechnology in Schools*, which is complemented by a UNESCO publication, *Microbiological Techniques in Schools*. The former book provides detailed classroom lessons that can be incorporated into many biology courses, complemented by case studies for classroom analysis, including those related to ethical and policy issues.

Even a quick glance at the number of organizations cited in this contribution suggests that substantial efforts are being made in establishing cooperation in the fields of acquisition, study, distribution, and proper usage of microbial resources available through culture collection and allied institutions. These efforts are being made on national, regional, and international levels.

Also it is perhaps evident that the geography of microbes still needs to be charted in terms of natural occurrence and migration in and from natural or artificial habitats. The vast holdings of diverse microbial germplasm and diversity held in culture collections the world over enjoy to some degree the spirit of international cooperation and scientific goodwill. Thanks to the Herculean efforts of the pioneering curators, and sustained by the contributions of successive generations of microbiologists worldwide, culture collections today are breaking new ground in furthering the horizons of bioinformatics, i.e. the handling of biological information by computers and its storage in databases.

Furthermore, culture collections will be in the vanguard of conserving the rich diversity of germplasm, and of bringing countless species away from the brink of extinction. Much has been done, but much remains to be done, despite the vast array of organizations, educational and financial problems.

In conclusion, however, let us return to the basic problem. In a sense the diversity of microbial germplasm and life on Earth, either known or still unknown to humankind, constitutes a common heritage.

In sustaining this heritage of microbial diversity, there is no gainsaying the fact that the extent and development of human resources will always be a challenge.

References

Allsopp, D. and Simione, F. (1988) Culture collection. In: Hawksworth, D.L. and Kirsop, B.E. (eds) *Living Resources for Biotechnology – Filamentous Fungi*. Cambridge University Press, Cambridge, pp. 162–172.

Bower, D.J. (1989) Genetic resources worldwide. *TIBTECH* 7, 111–116.

Bull, A.T., Goodfellow, M. and Slater, J.H. (1992) Biodiversity as a source of innovation in biotechnology. *American Review of Microbiology* 46, 219–252.

Hawkes, N. (1993) Scientists debate ethics of wiping out world's most deadly virus. *The Times* 10 August, p. 2.

Hawksworth, D.L. and Ritchie, J.M. (1993) *Biodiversity and Biosystematic Priorities: Microorganisms and Invertebrates*. CAB International, Wallingford, UK, 120 pp.

Hori, H. (1992) Evolutionary outline of living organisms as deduced from 5S ribosomal RNA sequences. In: Solbrig, O.T. van Emden, H.M. and van Vordt, P.G.W.J. (eds) *Biodiversity and Global Change*. IUBS Monograph 8, IUBS, Paris, pp. 95–104.

Johnson, L.M. and Rader, R.A. (1990) An overview of biotechnology information resources. *TIBTECH* 8, 318–323.

Lamana, C. (1976) Role of culture collections in the era of molecular biology. In: Colwell, R.R. (ed.) *Role of Culture Collections in the Era of Molecular Biology*. American Society of Microbiology, Washington, DC, p. 3.

Postgate, J. (1990) The microbes that would not die. *New Scientist* 21 July, 127(1726) 46–49.

Postgate, J. (1991) The malleable microbe. *New Scientist* 16 February, 129(1756) 38–44.

Reichenbach, H., Gerth, K., Irschick, H., Kunze, B. and Hofle, G. (1988) Myxobacteria: a source of new antibiotics. *TIBTECH* 6, 115–121.

Zook, D. (1990) *Microcosmos: New Ways of Injecting Microbes into Mainstream Education. ASM News* 56, 424–426.

Biodiversity Information Transfer: Some Existing Initiatives and How to Link Them

28

B. KIRSOP[1] AND V. CANHOS[2]

[1]*Biostrategy Associates, Stainfield House, Stainfield, Bourne, Lincs PE10 0RS, UK and* [2]*Base de Dados Tropical, Fundacao Tropical de Pesquisas e Tecnologia 'Andre Tosello', Rua Latino Coelho, 1301 – Parque Taquaral, 13087-010 Campinas, Brazil.*

Introduction

At the first workshop held by IUBS/IUMS in Amsterdam, September 1991, there was considerable discussion about how all the initiatives to inventory and monitor data on biodiversity could be identified and how a mechanism to link this vast quantity of information, dispersable worldwide, could be set up.

A proposal from the World Federation for Culture Collections (WFCC) to help organize a workshop to bring together experts in networking was accepted and, with the support of UNEP and other funding organizations (see below), the workshop took place at the Tropical Data Base (BDT), Campinas, Brazil, in July 1992, immediately following the United Nations Conference on Environment and Development (UNCED).

This chapter describes some of the bioinformatics initiatives taking place and the recommendations of the workshop and its subsequent activities.

Current Initiatives

Following the UNCED Conference and the signing of the Convention on Biological Diversity, many initiatives have been set up to inventory and record biodiversity information and many more are currently under discussion. These are supplementing programmes that were in existence before the Convention, creating a global resource of information that is being amassed on different computers, using different software, available in different forms (CD-ROM, online, printed directories/lists). Data collection is being organized according to biological discipline (botany, zoology, microbiology) or organism (viruses, legumes, birds, insects), or as part of international programmes (such as DIVERSITAS, UNEP country studies, Antarctic biological research programmes, BIOMASS and BIOTAS). It is also being accumulated in universities and research institutes as part of their on-going research.

A further information resource is developing as a result of 'search and discovery' programmes with industry (INBio, in Costa Rica; Biotics Ltd, UK; Missouri Botanical Gardens, USA; Royal Botanic Gardens, Kew; National Cancer Institute, USA). In these cases, information may be made available only to the contractors, may be restricted for limited periods to give contracting companies lead time, or may be restricted in part (e.g. the origin of extracts from samples).

National programmes are also underway as part of the commitment of countries to biodiversity conservation. In the USA, a proposal has been made to establish a National Biological Survey; in the UK the research councils are supporting the development of a directory of information on global change (the GENIE Project); the Environmental Research Information Network, ERIN, in Australia is well advanced in its activities to harmonize and link information on the environment.

At the UNEP/Norway meeting in Trondheim (May, 1993), proposals were made to consider the establishment of a global biodiversity monitoring network, to be part of a global environment warning system to detect undesirable perturbations beyond normal variations. Another proposal was made that a biodiversity information network should make data available to industrial users in return for recompense to countries of origin that provide the information, should it lead to commercial developments. These proposals were discussed once more at the Global Biodiversity Forum meeting in Geneva in September 1993.

Following a recent workshop in Philadelphia, May 1993, there has been a proposal to develop an inventory of all taxa to the species level for five years in a single highly biodiverse protected site. The objective of such an intensive study (known as All Taxa Biodiversity Inventory, ATBI) is to provide benchmarks against which to measure methods used by other studies in the future.

In microbiology, a number of information initiatives exist, particularly in the culture collection world. The WFCC's World Data Center for Collections of Microorganisms (WDCM) forms a valuable and well-established resource on the microbial resources of the world, together with additional related data, all now available at an Internet gopher (see Chapter

Biodiversity Information Transfer   441

24). Similarly available, and linked to the WDC, is the Microbial Germplasm Database that is maintaining information on US research microbial collections associated with agriculture. In Europe, the Microbial Information Network Europe (MINE) has been working to make data on strains available in the major European service collections available in a common format. The Microbial Strain Data Network (MSDN) has been set up to link catalogue and other microbiological information.

It is clear from this list and the proceedings of the Campinas workshop, neither of which is in the least comprehensive, that there are vast and growing amounts of information already available on biodiversity, which will be expanded exponentially. What is to be done to link this gigabyte resource and make it accessible and comprehensible to the equally growing band of users in the scientific, administrative and commercial worlds?

The Biodiversity Information Network Workshop

To take a first step towards answering this question, a biodiversity information network workshop was held in July 1992 at the Tropical Data Base (BDT), Campinas, Brazil under the sponsorship of IUBS, IUMS and WFCC. It was financially supported by UNEP, Insituto Brasileiro do Meio Ambiente e dos Recursos Naturais Renovaveis, Programa de Formacao de Recursos Humanos para Areas Estrategicas, Conselho Nacional de Desenvolvimento Cientifico e Tecnologico Financiadora de Estudos e Projetos and the British Council.

Its objective was to consider ways in which existing and future biodiversity information could be linked into a comprehensive resource. It was concerned with networking and not with the parallel issues of data collection, formatting and updating, which, it was recognized, were additional major topics for debate.

The workshop brought together 40 experts in information networking from a number of international organizations. Before, during and for two weeks after the workshop, the proceedings of the workshop were made available online. Over 200 people accessed these online proceedings and some 40 additional contributions were received and discussed during the workshop.

The large number of contributors, present and online, led to the publication of a valuable 'state of the art' document, the workshop proceedings, which will be of value in continuing discussions.

The published Recommendations were as follows:

1. The initiative will be known as the Biodiversity Information Network 21. The network will facilitate access to all levels of information (from molecular to biosphere) and will combine the knowledge within each discipline, furthering the understanding of biodiversity of living systems. Such an effort will identify and seek to fill the gaps, leading to new research and more informed policy decisions.

2. The goal will be to exchange information by electronic means whenever possible, but to include other ways of communication as needed by the network participants. To achieve a global electronic access, support should be provided to regions where facilities do not exist.

3. It will be a distributed network that will link many different sources of information across the world and will operate on a not-for-profit basis. Such a design is scientifically, economically and politically practical, allowing effort and resources to be shared.

4. The network will be open to a wide range of user groups including, but not limited to, scientists, teachers, natural resource managers, policy makers, regulatory and legislative agencies and public interest groups. The needs of the user community will be actively sought to enable their requirements to be met more effectively.

5. The network will actively encourage the free exchange of information on a worldwide basis and will also encourage the standardization of information and methodology.

6. A Secretariat will be established as a focal point and clearing house to facilitate and coordinate the flow of information among those with an interest in biodiversity.

7. Cooperating groups will be established with the purpose of encouraging participation and regional development. Collaboration with existing centres will be encouraged in order to prevent duplication of efforts already underway and to promote efficient use of funds. Support for developing countries to ensure global participation will be an important element.

8. Initially, an interim Steering Committee will be set up to coordinate immediate activities and seek funding. It will be supported by a number of Working Groups. These working groups will advise in areas such as: Technical Issues, Outreach, Training and Editorial/Moderating Functions.

9. An initial activity will be to design and develop a *Directory of Biodiversity Information Resources*, drawing on existing directories. It will be made widely available by all possible means.

10. The involvement and support of other organizations and initiatives working within biodiversity will be solicited.

Activities Since the Workshop

Following the workshop and the establishment of an interim steering committee and specialist committees, the following activities have taken place:

1. Editing of the Workshop Proceedings.
2. Publishing of the Workshop Proceedings by UNEP (2000 copies).
3. Distribution of the Proceedings via volunteer distributors (workshop participants) in different geographical regions; additional distribution of 500 copies by UNESCO.
4. Report of the Workshop to UNEP/Nairobi meeting (January 1993), Green Centre, Oxford conference (May 1993), Trondheim Norway/UNEP meeting (May 1993) and at other appropriate events.
5. Setting up of a biodiversity List Server on the Internet network (biodiv-1, accessed regularly by nearly 300 subscribers). Biodiv-1 is managed by the Tropical Data Base, Campinas, Brazil.
6. Establishment of an interim BIN21 Secretariat at the Tropical Data Base, pending a call for proposals for a permanent Secretariat.
7. Expansion of the Technical Committee to form a Technical Working Group (TechWG) composed of people active in the world of linking information resources via the Internet gopher systems and other mechanisms. The TechWG to advise on the best mechanisms for (i) linking resources and (ii) establishing a Directory of information resources as a permanent resource.
8. Assessment of financial needs for (i) a meeting of the TechWG and (ii) a permanent Secretariat.
9. Preparation and submission of proposals for financial support.

Technical Working Group

Since the workshop it has become clear that there have been and are major developments in the collection and distribution of biodiversity information via the Internet network. Groups around the world have been highly active in taking advantage of recent software developments to organize and publicize their information, using software management systems known as gophers.

The gopher management centres have made it possible to make large information resources publicly available on the Internet network in forms that are readily accessible and compatible with other resources. Thus, information management systems in Australia, Brazil, Japan, the USA and other regions have been established that go a long way towards providing a mechanism that will substantially assist linking different biodiversity information.

By using the gopher systems, it is possible to use a globally recognized system to move from resource to resource, without charge, on the Internet network. Databases and information resources at such centres as ERIN (Australia), TAXACOM (Harvard), Microbial Germplasm Network (Oregon), the World Data Center on Collections of Microorganisms (Japan) (see

Chapter 24) and the Tropical Data Base (Brazil) have become linked electronically and are available to Internet users in a similar form. Other centres are under development and will certainly become available in due course. Collectively, they will form an enormous information pool that is available to ALL users, now that the initially academic network is accessible by anyone through various intermediaries.

The Interim Steering Committee has now taken steps to enlarge the Technical Committee by the addition of people involved in the development of these gopher centres. Most people contacted have agreed to work together in this capacity, and a few responses are still awaited. The next step is to bring these people together to discuss how the gopher systems can work together in support of a global biodiversity information network. Proposals for financial support for a meeting of this kind have been made to appropriate organizations and the outcome of these is awaited. 'Second generation' gopher managers in selected countries showing potential to act in this way will be invited to attend if finances allow. In the meantime discussions are continuing online via the biodiversity list server, managed on the Internet by BDT.

Interim Secretariat

BDT has agreed to act as Interim Secretariat of BIN21, serving as a focal point for network activities. In this capacity it is not only managing the biodiv-1, but has offered to host the planned meeting of the gopher managers.

The Future

Sponsorship of the network has been sought from appropriate organizations. The World Conservation Union (IUCN) has agreed to sponsorship and looks forward to working with and using the network in the future. Formal agreement from other organizations is awaited.

It is anticipated that the gopher managers' meeting scheduled for the autumn of 1993 will shed light on the technical possibilities facing the biodiversity information community, make recommendations for future developments of BIN21 and assess the financial needs.

Once the opinions of the leading activists in the field are known, the BIN21 Steering Committee will invite proposals for a permanent Steering Committee and a permanent Secretariat.

Meanwhile, other bioinformatics initiatives will be continuing and it is hoped that a collective move will gradually emerge to assist new resources to be linked to the mechanism, the backbone of which will have been put in place.

It is clear that, since the original BIN21 workshop was held, many new roads to dispersed information resources have been built. The need in the future will be for the identification and support of topographical gaps and the creation of signposts and maps to show the routes that already exist. Training in the use of the available mechanisms will be required. Promotion and awareness will have to be addressed.

It is clear that interconnectivity is advancing rapidly and resources are increasingly available to users in the remotest regions of the world. Satellite links are becoming more and more possible and will be assessed as cheaper means to link to the rest of the world in areas where telephone systems are poor and ill maintained. Radio will be used to overcome telephone inadequacies, and printed information will continue to play an important role in information distribution.

It is the opinion of many experts in the communications field that understanding of the means to link resources is already well advanced. Further support will be required to spread the network worldwide and to provide training, technical support, and financial resources for workers unable at present to drive to the major highways. It is hoped that these requirements will form an integral part of existing and new initiatives involving the accumulation of biodiversity data. The collection of information serves little purpose if the resulting databases, books, inventories and directories remain unknown and so unused.

BIN21 will before long provide the necessary links; as these are pronounced and made widely known, it is to be hoped that those involved in biodiversity projects will find ways, through contact with projects already linked to BIN21, to forge their own links to the network. Only then will we ensure that diversity conservation decisions are based on comprehensive knowledge, rather than knowledge based only on information from organizations with the technological skills and training.

Indigenous Rhizobia Populations in East and Southern Africa: A Network Approach

N.K.N. KARANJA[1], P.L. WOOMER[2] AND S. WANGARURO[1]

[1] *Department of Soil Science, University of Nairobi, PO Box 30197, Nairobi, Kenya, and* [2] *Tropical Soil Biology and Fertility Programme, c/o UNESCO ROSTA UN Complex, PO Box 30592, Gigiri, Kenya.*

Introduction

An estimation of the total viable rhizobia in soils and other test substrates can be obtained through application of the most-probable-number (MPN) technique. This technique is used to estimate microbial population sizes when direct quantitative assessment of individual cells is not possible. First, a suitable legume host is cultured in rhizobia-free media, then a test substrate is serially diluted and applied to the root systems of the legume hosts. After 21 days, the pattern of presence and absence of root nodules is recorded and these data are used to derive a population estimate of the original test substrate. The MPN technique is based on the mathematical approaches of Halvorson and Ziegler (1933). Cochran (1950) later identified estimation of error and mean separation procedures. Computer programs are now available that combine these operations, resulting in greater flexibility of experimental designs and scientific procedures (Woomer *et al.*, 1990). Tests of experimental technique that allow researchers either to accept or to reject experimental results are also available (Halvorson and Moeglein, 1940; Woodward, 1957; deMan, 1975; Woomer *et al.*, 1988b).

The need for the application of rhizobial inoculants and the magnitude of the response to applied rhizobia are determined by the species diversity and population sizes of indigenous soil rhizobia and the availability of mineral nitrogen within the soil system (Thies *et al.*, 1991). The legume/

rhizobia symbiosis is characterized by specificity between the host and microsymbiont; these are identified as cross-inoculation groups (FAO, 1984). The selection of a range of legumes characteristic of individual cross-inoculation groups as hosts for MPN assays allows for the species composition and population sizes to be characterized (Woomer *et al.*, 1988b). The lack of rhizobia within a soil allows for the comparison of inoculated and uninoculated treatments as a direct measure of biological nitrogen fixation. To summarize, the overall objectives of the *Rhizobium* Ecology Network of East and Southern Africa (RENEASA) are to characterize the biodiversity and population sizes of indigenous rhizobial populations in soils of East and Southern Africa, to develop predictive understanding of when and where an economic response will be obtained from the application of rhizobial inoculants and to facilitate communication and data sharing among *Rhizobium* ecologists throughout East and Southern Africa.

Materials and Methods

A standard experimental protocol was formalized, based on work done at the University of Nairobi MIRCEN, and an experimental package assembled and distributed to 12 cooperating investigators in eight countries in East and Southern Africa. These are listed in Table 29.1.

Table 29.1. *Rhizobium* Ecology Network of East and Southern Africa (RENEASA) investigators, their institutes, countries and representative agroecological zones.

Investigator	Institute	Country	AEZ[a]	Sites
Edmundo Barrios	ICRAF	Kenya	2,3	2
A. Hakizimana	ISAR	Rwanda	1	2
I.S. Haque	ILCA	Ethiopia	1,3,4	4
Patrick Jjemba	CIAT	Uganda	3	1
Nancy Karanja	MIRCEN	Kenya	1,2	7
Patrick A. Ndakidemi	ARI	Tanzania	1	2
Charles Nkwiine	Makerere	Uganda	1,3	7
Mr Michael Nyika	SPRL	Zimbabwe	2	4
Godfrey Msumali	SUA	Tanzania	4	2
Carlota Quilambo	EMU	Mozambique	3	1
Howard Tembo	Makulu	Zambia	3,4	4

[a]AEZ, agroecological zone (1, moist highland; 2, semiarid highland; 3, moist lowland; 4, semiarid lowland).

Host Species

In MPN plant-infection procedures, nodulation of the legume host serves as a selective medium in determining the presence or absence of the rhizobia at a given dilution level. Therefore, the choice of a specific legume host greatly influences the experimental results. More information on the specificity of legume hosts and the selection of these hosts as an indicator of rhizobial species may be obtained from FAO (1984) and Woomer et al. (1988a). The list of host species and their respective rhizobial microsymbionts is presented in Table 29.2.

Plant Preparation and Inoculation

Host legumes were cultured in 'growth pouches' fabricated locally in Nairobi. Seeds were surface sterilized by immersion in a 1.5% solution of sodium hypochlorite for 4 minutes followed by five rinses with sterile distilled water. 'Hard-seeded' legumes were surface sterilized by immersion in concentrated sulphuric acid for 4–25 minutes followed by eight rinses of sterile water. The seeds were transferred to a sterile germination vessel. Upon emergence of the radicle, two germinated seedlings were planted in the trough of the growth pouch containing 75–100 ml of sterile N nutrient (Woomer et al., 1988a) solution of rhizobia-free water, taking care that the radicle was in partial contact with the growth pouch paper wick. Host plants were cultured for approximately 1 week prior to inoculation in a clean glasshouse or growth room. A 1 ml portion of a six-step, fivefold dilution series was inoculated into four replicate legumes representative of the various legume–rhizobia cross-inoculation groups (Table 29.2). The plants

Table 29.2. The host legume species, associated rhizobia, and respective cross-inoculation groups used in the *Rhizobium* Ecology Network.

Host species	Rhizobial microsymbiont	Cross-inoculation group
Vigna unguiculata	Bradyrhizobium sp.	Cowpea miscellany
Glycine max	Bradyrhizobium japonicum	Soybean rhizobia
Phaseolus vulgaris	Rhizobium phaseoli[a]	Bean rhizobia
Gliricidea sepium	Rhizobium loti	Leucaena/Gliricidia group
Trifolium semipilosum	Rhizobium trifolii	E. African clover group
Pisum sativum	Rhizobium leguminosarum	Pea/Vicia/Lathyrus group

[a]*Rhizobium phaseoli* has been reclassified as *R. leguminosarum* bv. *phaseoli*; *R. trifolii* as *R. leguminosarum* bv *trifolii*; and *R. leguminosarum* as *R. leguminosarum* bv. *vicea*.

were allowed to grow an additional 21 days after which nodulation was scored as either positive or negative, depending on the presence of root nodules. Population estimates were assigned to the results using MPNES software (Woomer *et al.*, 1990).

Results and Discussion

A total of 154 legume infection counts of rhizobia were conducted using soils collected from 30 sites including natural pastures, savannas, forests, fallows and croplands. The ranges of selected environmental data are presented in Table 29.3. The soil orders present at the sites were classified (United States Department of Agriculture, USDA) as alfisols (5), andisols (2), sandy entisols (1), inceptisols (2), mollisols (2), oxisols (4), ultisols (7) and vertisols (3).

Population sizes of indigenous *Bradyrhizobium* and *Rhizobium* spp. grouped by general agroecological zone are presented in Fig. 29.1 and Table 29.4. When grouped by zone, the population sizes of *Bradyrhizobium* sp. were 1.25–2.40 $\log_{10}$ cells (g soil)$^{-1}$. This rhizobial species is associated with a diverse range of symbiotic leguminous plants indigenous and naturalized to the region including *Acacia* spp., *Cajanus cajan*, *Vigna* spp. and *Arachis hypogaea*, hence its wide distribution and high density in all soils. The population size of *R. leguminosarum* bv. *phaseoli* was largest in highland soils, but the microorganism was present in many lowland soils. The common bean (*Phaseolus vulgaris*), although not indigenous to Africa, has been widely cultivated for several centuries and presumably along with the rhizobial species which co-evolved with that legume in Latin America (Duke, 1981). These have had the opportunity to colonize many soils in Africa. Population sizes are greatest where *Phaseolus vulgaris* is most widely

Table 29.3. Ranges of selected environmental parameters measured at RENEASA study sites.

Parameter	n	Minimum	Maximum
Mean annual precipitation (mm)	25	209	1400
Mean annual temperature (°C)	25	15.0	29.1
Total soil organic carbon (%)	30	0.14	6.87
Total soil nitrogen (%)	28	0.04	0.47
Extractable phosphorus (ppm)	26	0.93	178
Soil pH (1:2 H$_2$O)	30	4.4	3.5
Clay content (%)	30	15	59

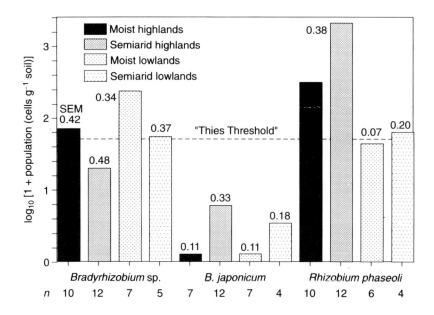

Fig. 29.1. Distribution and population sizes of selected rhizobial species in East and Southern Africa.

cultivated, specifically in the drier highlands. Response by beans to rhizobia inoculation is usually very low or does not occur at all. *B. japonicum* has co-evolved with selected *Glycine* species in southern China where its principal hosts, *Glycine max* and *G. soya*, originated (Duke, 1981). Unlike *P. vulgaris*, *Glycine max* (soyabean) is not frequently cultivated in East Africa and, as a result, rhizobial populations are non-existent or very small. On the other hand, soyabean production is rapidly increasing in southern Zambia, where it is often inoculated with rhizobia. At one site population sizes of *B. japonicum* were 2350 cells per g soil (data not shown).

The naturalized population size of *R. leguminosarum* bv. *phaseoli* shown in Fig. 29.1 suggests that no response to applied rhizobia may be observed in many highland sites. By comparing the results of response to inoculation trials and rhizobial population sizes, Thies *et al.* (1991) identified the critical rhizobial population sizes in soils which preclude observed responses to inoculation as >50 cells (g soil)$^{-1}$ for an individual host. Generally other rhizobia species which include *R. leguminosarum* bv. *trifolii*, *R. loti*, *R. leguminosarum* bv. *viceae* and *Bradyrhizobium japonicum* were poorly distributed in all the soil samples from the region. The host legumes for these rhizobia species were introduced from the temperate region and their cultivation remains limited to a few farms. The example of *B. japonicum* is

Table 29.4. Population (cells (g soil)$^{-1}$) of *Rhizobium* and *Bradyrhizobium* in various ecological zones[a]

	Highlands		Lowlands	
	Moist	Dry	Moist	Dry
Bradyrhizobium sp.	1631	1,456	751	131
B. japonicum	0.71	230	0.71	3.3
R. leg. bv. *viceae*	2187	NR	137	12.2
R. leg. bv. *trifolii*	0	0	3	7
R. leg bv. *phaseoli*	5965	10,361	46	80
R. loti	2	0	7	2.3
R. meliloti	2	NR	0	NR

[a]Moisture regimes classified after S. Jagtap, IITA Climatologist (personal communication).

shown in Fig. 29.1 where the population in the four ecological zones was below the 'Thies threshold', which may explain the high response to rhizobia inoculation by the soyabean. (Thies *et al.* (1991) showed that indigenous populations of rhizobia greater than 50 significantly reduced observed responses to the use of rhizobial inoculants on a range of legume hosts.) With the introduction of new legume plants for improving pastures, establishment of agroforestry farming systems, and high-yielding grain legume cultivars, these results do show that the new crops may lack their complementary rhizobia species in the soils of East and Southern Africa; hence a need to emphasize the use of inoculants.

No clear covariation was observed between rhizobial populations and climatic variables. Indigenous organisms are by their very nature acclimatized to the principal stresses of their environment and it is most likely that greater covariance would be observed between the cover of legume hosts rather than with climatic variables (Woomer *et al.*, 1988b). One exception to this is with soil cation and CEC measurements. The population sizes of many *Rhizobium* spp. demonstrated strong covariation with CEC and nutrient cation content of the soil (Table 29.5). This relationship was not universal, however, and the population sizes of *Bradyrhizobium* spp. could not be accounted for by climatic or soil data.

In conclusion, rhizobial population composition and size vary widely in East and Southern Africa. Rhizobia associated with indigenous legumes are the most widespread but not necessarily the most numerous owing to intensive cultivation of exotic food legume species. Population sizes of individual rhizobial species were poorly explained in terms of climatic and soils data, and are more likely regulated by the abundance of host legumes,

Table 29.5. Correlation coefficients (r^2) of abiotic factors and population sizes of indigenous Rhizobium and Bradyrhizobium spp.

Rhizobial species	Abiotic factors[a]				
	CEC	Ca	Mg	K	Sum
R. loti	0.58*	0.43	0.61**	0.81**	0.51*
R. leg. bv. viceae	0.99**	0.74**	0.84**	0.85**	0.79**
R. leg. bv. trifolii	0.38	0.77**	0.80***	0.70**	0.79***
R. leg. bv. phaseoli	−0.07	−0.14	−0.15	−0.29	0.27
Bradyrhizobium sp.	−0.22	−0.17	−0.15	−0.19	−0.15

[a]Probability level of relationship; * = $P < 0.05$, ** = $P < 0.01$, *** = $P < 0.001$.

namely, the colonization of *B. japonicum* in Southern Africa from intensified cropping and inoculation of soyabeans.

Acknowledgements

This paper would not have been possible without the enthusiastic participation of the RENEASA scientific investigators and the funding of that network by UNESCO and the Rockefeller Foundation.

References

Cochran, W.G. (1950) Estimation of bacterial densities by means of the 'most probable number'. *Biometrics* 6, 105–116.

deMan, J.C. (1975) The probability of most probable numbers. *European Journal of Applied Microbiology* 1, 67–78.

Duke, J. (1981) *Handbook of Legumes of World Economic Importance*. Plenum Press, New York.

FAO (Food and Agriculture Organization of the United Nations) (1984) *Legume Inoculants and Their Use*. FAO, Rome.

Halvorson, H.O. and Moeglein, A. (1940) Application of statistics to problems in bacteriology. V. The probability of occurrence of various experimental results. *Growth* 4, 157–168.

Halvorson, H.O. and Ziegler, N.R. (1933) Applications of statistics to problems in bacteriology. I. A means of determining bacterial population by the dilution method. *Journal of Bacteriology* 25, 101–121.

Thies, J.E., Singleton, P.W. and Bohlool, B.B. (1991) Influence of the size of indigenous rhizobial populations on the establishment and symbiotic performance of introduced rhizobia on field-grown legumes. *Applied and Environmental Microbiology* 57, 19–28.

Woodward, R.L. (1957) How probable is the most probable number? *Journal of Applied Bacteriology* 1, 505–516.

Woomer, P., Singleton, P.W. and Bohlool, B.B. (1988a) Ecological indicators of native rhizobia in tropical soils. *Applied and Environmental Microbiology* 54, 1112–1116.

Woomer, P., Singleton, P.W. and Bohlool, B.B. (1988b) Reliability of the most-probable-number technique for enumerating rhizobia in tropical soils. *Applied and Environmental Microbiology* 54, 1494–1497.

Woomer, P.L., Bennett, J. and Yost, R. (1990) Overcoming inflexibilities in most-probable-number procedures. *Agronomy Journal* 82, 349–353.

Progress in the Synthesis and Delivery of Information on the Diversity of Known Bacteria

J.G. Holt[1], M.I. Krichevsky[2] and T. Bryant[3]

[1]*Department of Microbiology and Center for Microbial Ecology, Michigan State University, East Lansing, Michigan 48824, USA;* [2]*Bionomics International, 12221 Parklawn Drive, Rockville, Maryland 20852, USA; and* [3]*Medical Statistics and Computing, University of Southampton, South Academic Block, Southampton General Hospital, Southampton SO9 4XY, UK.*

The purpose of this chapter is to explain how bacteriologists are facing the challenges of the current rush to describe the living world. Our problems are potentially overwhelming from the point of view of sheer numbers of taxa that exist and the need to identify bacteria that may be uniquely beneficial, cause grave problems for the human race, or simply be considered as part of the world's genomic diversity. The difficulty is one of keeping track of all the described organisms and disseminating the information to scientists and technicians worldwide.

Much of the communication of biological diversity information reports the findings in terms of the incidence and distribution of species. Thus, the utility of such communication depends on the availability and quality of compendia of taxonomic information.

We in bacteriology have adopted a successful system of storage and retrieval of taxonomic information, a model that is easily adapted to other biological groups. The system we wish to describe is represented by the operation of the Bergey's Manual Trust, a very successful, self-perpetuating organization dedicated to disseminating taxonomic information to the microbiological community.

Historical Perspective and the Development of the Bergey's Manual Trust

From 1916 to 1918 a number of seminal papers were published by R.E. Buchanan (1916–1918) which, for the first time, proposed a classification of all the known bacteria. Soon after, a committee was formed by the Society of American Bacteriologists (SAB), chaired by C.E.A. Winslow, which was charged with the task of formulating an overall classification of the bacteria which could form the basis of usable keys for identification. One of the committee members, David Bergey, began to compile a manual for identification. He was joined by four others and in 1923 the first edition of *Bergey's Manual of Determinative Bacteriology* was published (Bergey et al., 1923). This was soon followed by new editions every few years. The book was moderately successful in the United States, but was not widely used outside that country. Royalty income from sales through the fourth edition in 1934 amounted to about $20,000 and was kept in the SAB treasury. Bergey and his co-editor, Robert Breed, asked the SAB to provide them with some modest funds from the royalty income to pay for expenses for the next edition. The SAB leadership refused the request and a long bitter fight ensued to release the money. Finally in 1935 the SAB relented and turned the total proceeds over to Bergey, who promptly put the money into a trust fund. Thus, on 2 January 1936, the Bergey's Manual Trust was formed under the laws of the Commonwealth of Pennsylvania with the legal mandate to use the royalty income to defray the expenses of future editions.

After Bergey's death in 1937 the editorship was assumed by Breed, who also served as Chairman of the Board. The mantle of Editor and Chairman went to Buchanan after Breed's death in 1956, and the editorial offices were moved to Ames, Iowa. Up to this point royalty income was never sufficient to support the operation and Buchanan, who had retired from Iowa State University in 1948, came back to the Department of Bacteriology as a Research Professor and ran the office with the generous support of grants from the National Library of Medicine. Royalty income did not exceed expenses until the publication of the eighth edition in 1974, one year after the death of Buchanan. Since that time royalty income has been sufficient to maintain the editorial office. The headquarters of the Trust were moved to a permanent location at Michigan State University, East Lansing, MI in 1990.

The publication of the eighth edition of the *Manual* was a departure from past editions in that an effort was made to make the book more international in composition. The Board of Trustees, which was increased from three to nine in the 1950s, took on non-North American members and the authorship was drawn from an international pool of experts. This effort has fixed the *Manual* as the international authority in bacterial systematics.

Lastly, the Trust has recently published (1984–1989) a four-volume encyclopaedic treatment of the bacteria titled *Bergey's Manual of Systematic Bacteriology*. This systematic treatment of a large group of organisms in one compendium is a unique effort in biology and one which should be repeated in other fields.

One other event took place in bacterial systematics that has made their taxonomy unique in biology. That was the adoption of a new starting date for bacterial nomenclature. When bacterial nomenclature was formalized in the 1930s, all names of bacteria dated from the publication of Linnaeus' 1753 book on plants. The 1976 revision of the Bacteriological Code (Lapage *et al.*, 1975) designated a new starting date of 1 January 1980. This meant that all the names proposed prior to that date had no standing in nomenclature, except those that were well recognized and with adequate type material. These names were put into Approved Lists of Bacterial Names (Skerman *et al.*, 1980), and were the only names recognized as validly published. The rules for valid publication were tightened up and essentially required that validity could only be achieved by publication or notice of effective publication in a single journal, the *International Journal of Systematic Bacteriology*. The effect of this change has been manyfold: the number of names applied to bacterial taxa has been reduced by a factor of 10; there is good control over the publication of new names; and it is now easier to keep track of nomenclatural changes. The process was a cathartic one and should be considered by other fields of biology. It has certainly made the compilation of taxonomic information of bacteria more straightforward.

The point of this historical discussion is to demonstrate a model for the development of systematic surveys of microorganisms which are financially successful and have a highly perpetual nature. It is a model which should be adopted by other groups to handle the burgeoning amounts of data generated by studies of biodiversity. With the bacteria there is a system in which the classification is contained in one self-supporting reference and which has a clean nomenclature that is not saddled with large, cumbersome lists of synonyms.

We would now like to describe an effort that the Trust and its associates, Williams & Wilkins and Bionomics International, have initiated to handle taxonomic data on bacteria through the use of information technology. The goal of this project is to convert the printed material in the *Manual* into a computerized database, which can then be easily expanded and converted into accessible forms, including print and electronic media.

The Printed Manuals

The variety and diversity of bacteria means that their classification is an enormous task which relies on international cooperation. The four volumes of *Bergey's Manual of Systematic Bacteriology* were compiled using 290 contributors from around the world and took 12 years to produce. The complexity in organizing this cooperation is self-evident. A practical issue facing recurrent production of updated *Manuals* is the cycle time necessary to produce new versions. The reason is that it takes a long time and much labour to produce such a compendium of information by the traditional manuscript production and the attendant necessity for multilayered editorial and redaction steps.

The format for the *Bergey's Manuals* has evolved from considerable proportions of 'natural language' descriptions towards more tabular material and lists of items which conceptually represent 'collapsed tables'. Within the tables, the representation of frequency of occurrence of features within taxa evolved from a five-placed ordered multistate convention (i.e., '+' = 90% or more strains positive, 'd' = 11–99% strains positive, '−' = 90% or more strains negative, 'D' = different reactions in different taxa and 'v' = strain instability (NOT equivalent to 'd')) to a three-state convention (i.e., positive, variable and negative). In the ninth edition of *Bergey's Manual of Determinative Bacteriology* (Holt et al., 1994), tables of the percentage frequency of occurrence of phenotypic characteristics in species were introduced where such data were available.

Bacterial taxonomic information can be organized in two main ways. The first, found in the early *Manuals*, is taxon by single taxon; all the information on each taxon is gathered and organized as a set in one record. (Local introduction of tables within taxa or groups of taxa does not change this logic.) The alternative is lists in which the information for all taxa on each attribute is given, attribute by attribute (e.g., a list of all organisms that are pathogenic). This latter organization is impractical for the printed work.

Thus, the taxonomic descriptions on the printed page are most useful for obtaining a 'gestalt' of a taxon. They are cumbersome to impossible for making comparisons among large groups of taxa or searching for all examples of taxa with a given set of attributes in common.

Conceptual Framework for Database

Compendia of species descriptions such as the *Bergey's Manuals* may be viewed as two-dimensional tables with the dimensions being taxa (rows) by descriptive attributes (columns). 'Attributes' is meant in the broadest sense to include all information known, including history, habitat, uses, etc. When viewed as a series of horizontal rows, the items take on a taxon by taxon

Table 30.1. An arbitrary representation of the information set recorded on the bacteria. N taxa are listed in N rows. The attributes are listed column by column in an open-ended array.

Taxon no.	Taxon	History	Uses	Features
1				$--++$
2				$+-++$
3				$--+-$
•				$+---$
N				$+-+-$

character. Considering the vertical columns, the organization becomes one of lists of attribute by attribute for each taxon. That is, the total information on a single taxon is obtained by reading across the row. The occurrence of an attribute for all taxa is read by reading down a column (Table 30.1).

The information cells in the tables at the row/column intersections inform the reader as to whether the species exhibits that attribute and to what degree and/or frequency. For some attributes the cell may contain lists, usually of alternative information. This is most common in historical attributes such as sources of isolation, biotechnological uses. Isolates of a species may be found in various locations and habitats, each to be listed.

The data are coded as characters (alphanumeric), numbers, or quaternary states (yes, no, variable, unknown). Because of missing data and variability within taxa, true binary coding cannot be used. The formats for the particular columns are chosen based on what is convenient or natural to the kind of data in the columns. Examples are: literature references, taxa, source and geographic place of isolation are entered as alphanumeric; titres, minimal inhibitory concentrations, GC content and feature frequencies entered as numeric; and most phenotypic attributes entered as quaternary.

The format of the species descriptions has evolved towards formal tabular and list presentations. Natural language text is used for introductory and explanatory material and to enhance readability. The natural language material would be written in the traditional manner by contributors, or where appropriate, generated (at least in part) by computer from the tables.

Use of modern electronic information technology has two benefits. The production of future manuals can be accomplished without the use of the traditional manuscripts, thereby permitting much more frequent and timely versions to be produced. The information set need only be supplemented or changed by virtue of the advent of new information. Second, the potential exists for disseminating the information in purely electronic form as a direct consequence of using databases in managing the information.

The advent of electronic 'publishing' does not obviate the necessity for producing the more traditional printed book format, however. There are many circumstances where consultation of a book is more practical and desirable than invoking a computer system. Returning to the row and column analogy, reading the description of a single species (row) is simpler from the printed page. However, the process of comparison of attributes for aiding identification is clearly better done with the aid of computer searching.

Design Considerations for Electronic Bergey's Manuals

The scale of the problem in creating a system to handle the data that are currently found in the printed *Manuals* needs to be understood. One cannot use a general-purpose database package. The whole data set does not lend itself to storage and manipulation as a rectangular matrix, although subsections of it do. The data form a sparse matrix with many holes therein due to the use of certain attributes for describing one group of bacteria and different attributes for another group. The total number of different attributes used to describe the bacteria exceed the limitations of many existing commercial database management packages. This fact, along with others such as insufficient searching capabilities, makes them impractical for this task.

The first requirement for communication between two persons is a mutually understood vocabulary. This may be an agreed-upon spoken language or a set of specifically constructed symbolic conventions. Scientific communication is most precise and unambiguous when using symbolic constructs as in physics, chemistry, mathematics, geology, etc. With the exception of the various taxonomic codes of nomenclature, biology has not evolved such universally accepted constructs. As described earlier, bacteriologists have developed a most useful process for managing taxonomic nomenclature that covers the whole discipline.

However, the same is not true of the vocabulary used to describe the attributes of the species. The same word may mean very different things in different taxa, e.g. 'spore' may refer to a survival or a reproductive structure or both. Conversely, the same concept may be described by different words such as 'oxidative' or 'aerobic' metabolism. Natural language is the most difficult form of data coding to manage in a computer because the vocabulary is uncontrolled.

One of our most challenging tasks is the rationalization of vocabulary across all of bacteriology. To manage this aspect, we need a form of controlled vocabulary. A controlled vocabulary is a list which contains all allowed terms in all or, more realistically, part of a database. Thus, it acts as a check-list. The keywords often used by literature abstract services as search parameters are a common example of controlled vocabulary.

Synonymy and ambiguity are main causes of failure to communicate with a database. Synonymy is the easier of two difficult problems to manage. The synonymy problem stems from the very richness of the evolution of natural language. Among and within languages, synonyms exist because they amplify meaning and because of convergent evolution of terms. A microbial ecologist might sample a stream, brook, run, rivulet, branch, creek, bayou, etc. and still be in one place. Lists of synonyms must be built in the computer even to begin to manage the problems.

The converse problem of ambiguity, a single word having multiple meanings, is usually understood through context. A computer file is understood to be a different entity from a single file of people or a triangular file used by a metal worker. However, most computer search algorithms would not differentiate these three meanings for 'file'. This form of ambiguity is at least approachable with lists of words having multiple meanings.

Words used to describe organisms can be the same for different taxa but code for different functions or structures (e.g., the words: spore, flagellum, or sheath). This diversity of meaning is not a problem since the word is understood through context and associated specific information. A *Bacillus* spore would not often be confused with an *Aspergillus* or *Dictyostelium* spore.

Multiple words can code for the same meaning in varying taxa. On rare occasions, multiple taxon names are assigned to the same group of organisms. Such redundancy usually arises through rediscovery combined with lack of effective communication. The standards defining coding systems should provide mechanisms for coping with synonymy across taxon lines.

Just as the organic chemist combines symbols to construct a description of a compound (the structural formula), so too, must the bacteriologist be able to construct a symbolic description of a taxon. The system must include the capacity to record such information as history, utility, health and environmental risks, location, taxonomy, genotype, as well as the phenotypic features. It must be open-ended to allow for additional informational categories.

Some major advantages of a controlled vocabulary are the ability to detect errors (e.g., spelling, inappropriate entries), control of formats, table look-ups for searching, ease of sorting and reporting, and, through the use of numerical coding, better communication among users having different native languages.

Managing the Vocabulary of the Bergey's Manuals

The RKC Code (Rogosa *et al.*, 1971, 1986), a structured list of attributes of various microorganisms, uses integer numbers to represent codable concepts. More than 50 people have contributed substantively to the coding system. The RKC Code uses six-digit decimal integers to represent currently over 13,800 codable attributes of microorganisms (including their morphology, growth and cultural characteristics, metabolism, etc.). The Code covers features of bacteria, protozoa, some algae, yeasts, and some other fungi. The Code uses integers rather than abbreviations or acronyms for a variety of reasons. Arabic numerals are universally understood regardless of native language or alphabet. Standardizing on a six-digit code permits complex concepts to be represented simply (e.g., an entire metabolic pathway can be represented by a single code number). In the RKC Code, carbohydrate catabolism attributes fall within the range of 25,001 to 25,999. The range 20,001 to 20,999 represents pigment and odour features.

The present project requires, and is contributing, significant expansion of the RKC Code. The driving force behind this expansion will be the needs of the various contributors of the information. Suggested new attribute statements will be collected and reviewed by the editors under the aegis of the Bergey's Manual Trust. Bionomics International will expand the RKC Code for computer entry of standardized descriptions of microorganisms.

Standardization of information coding by taxa requires consideration of commonalities and differences among the elements coded for the various taxa. Such consideration is difficult across all bacteria since each person's knowledge of bacterial taxa, perforce, tends to be focused on the comparatively narrow range of taxa of immediate interest. A natural outgrowth of this enforced parochialism is the view that a particular group of organisms is so unique that it requires a unique coding system. This seems to be especially true of morphology. If it appears different, how can a common coding system be used? This view emphasizes differences.

Conversely, the commonality of life emphasizes similarities. Nucleic acid replication is the universal mechanism of reproduction. Essentially the same complex of monomers are the constituents of the polymers which form living structures. A current philosophical tendency in biology is to propose that we only need to record the nucleotide sequences of genes and determine homology among the nucleic acids to disclose all taxonomy. Some might add amino acid sequence information to the list. At this level of information, all life fits into the same categories to be recorded.

Both extremes are oversimplifications of the problem facing the compiler of taxonomic information. There are both parochial distinctions to be made among taxa and common threads that bind them together. The common threads extend beyond sequence coding. We are embarking on an effort that will allow all these classes of information to be entered and managed without allowing any particular philosophical bent to drive the design. The coding system, the computer programs, and the editorial process will allow eclectic entry of the information. The users of the system will be able to produce results which are driven by phenetic or genetic considerations.

An old problem of taxonomic nomenclature is still and increasingly with us. This is the linkage between the taxon member being described and taxa bearing some specific functional relationship to the member. The list of kinds of relationships is wide-ranging and important: pathogen–host, parasite–host, saprophyte–host, symbiosis, plasmid–host, inserted foreign genes, cell fusion partners, antigenic reactivity patterns, viral attachment and infectivity ranges and so on. In the same vein, linkage to other classification schema is required. These include chemical structures, geography, anatomy, literature references and sources, geometry, physical properties (e.g., colour 'standards'), material properties (especially in biodegradation and fermentation processes).

In the present effort, such important, but ancillary, information will be managed as text (character) information. The vocabulary will be controlled insofar as standards are available from the practitioners of the various disciplines involved. For example, the RKC Code already uses the Enzyme Nomenclature standards of the International Union of Biochemistry wherever possible.

The Database Management System

The software on which the electronic version of *Bergey's Manuals* is being developed is the Microbial Information System (Micro-IS) package (Portyrata and Krichevsky, 1992). Micro-IS has been used for handling microbiological data at the strain level for several years. It has been used for diverse applications in such areas as taxonomy, ecology, epidemiology, building of identification schema, regulatory microbiology, and culture collection management. The package was designed to allow for the problems outlined above and its database structure is well suited for the complexity and volume of data that will be generated.

A large proportion of strain data is yes/no/unknown. Micro-IS was designed with functions to manage such data efficiently. One can search, tabulate, generate reports, make data subsets, remove or transfer taxa among databases, based on simple or complex combinations of phenotypic attributes. The usual Boolean operators are used to construct the search criteria. Also, it is possible to search for similar taxa having a proportion of attributes in common. For example, we might wish to find strains able to utilize sedoheptulose, even though the strains in the database were not tested for sedoheptulose utilization. We speculate that strains capable of using a wide variety of other monosaccharides are more likely to utilize sedoheptulose. If ten carbohydrates were tested, we can search for strain utilization of any eight of the ten carbohydrates. This is an area of weakness in most general-purpose database systems.

Bionomics International, with active consultation by the Bergey's Manual Trust and Williams & Wilkins, is adapting the Micro-IS to perform in a variety of settings in which bacterial taxonomic information will be entered, edited, managed, and disseminated. The package runs under MS-DOS on IBM-compatible personal computers. The two main user communities of the enhanced Micro-IS will be the producers of the information for the *Bergey's Manuals* (i.e., the contributors, the editors, and the publishers) and the general community of users of the information.

All of the above is accomplished without recourse to any application programming by the user. The basic functions of the Micro-IS are described in a self-instruction manual. The tutorial manual can be worked through in a day by a reasonably computer-literate microbiologist. In a classroom setting two-day courses, including more advanced use of the system, have proved successful. It is no harder to learn and use than any of the commercial relational database programs (easier than many). However, because of the richness of the facilities available, the user must still think and learn to structure their tasks to make efficient use of the capabilities provided. In this too, there is no 'free lunch'.

The combination of the RKC Code and Micro-IS is uniquely suited to acquire, manage, analyse and communicate tabular microbial strain and species information. The *Bergey's Manuals'* structures are predominantly tabular or can be easily derived from tables. That information which is best managed as 'free text' can be integrated into the Micro-IS structure as specific alphanumeric fields which can be up to 1024 characters in length. The number of these fields can be over 100,000 in any database. This capacity far exceeds the requirements of the current project.

The system has a unique synonym capacity for feature descriptors, such that each user can have a personal synonym for any feature or share the synonyms already in the synonym file. Utilities are provided for a broad spectrum of editing functions from changing the information in a single data cell to moving and combining whole tables and databases.

During the early phases of this work, it became evident that the genus descriptions that occurred in the printed *Manuals* were effectively redundant with respect to a database, although necessary for a printed compendium. Genus statements can be created directly from the species descriptors by identifying those attributes that are constant for all species within that genus. The electronic generation of such statements will also remove ambiguities that currently arise as a result of natural language. The statement describing a genus as 'strains usually motile' could mean that all the isolates of specific species are motile whereas all isolates of other species are not motile, or it could mean that some isolates from each have been found to be non-motile. Building a genus description from species information can clarify such statements, which might resolve to 'three out of four species in this genus are motile'.

The Process of Building the Database

The unit record in each table is at the level of the species or subspecies. Although our conceptual model describes one master table, the database(s) stored in the computer is not a simple, very large, single table. The actual format of the database is quite different but transparent to the user; many tables exist, where each table represents one genus.

Conceptually, each contributor to the *Manual* fills in one or more tables of information depending on how many genera they are describing. In some cases, data from prior editions of the *Manuals* will have been entered for them and all the contributor has to do is update or add to the table. All information on higher taxa is attached to the species description to allow for competing taxonomies (whether based on phenotypic or genotypic models). Here, too, the hierarchical constructs can be generated from the species descriptions.

The table format for each contributor is provided after an iterative dialogue among the personnel of Bionomics International, the Bergey's Manual Trust editorial staff, and the contributor to map the information to be supplied by the contributor to the RKC Code. A preformed input template is generated as follows.

The manuscripts for groups in the ninth edition of *Bergey's Manual of Determinative Bacteriology*, in Wordperfect 5.1 format, were supplied by the Bergey's Manual Trust to Bionomics International. These files are 'parsed', attribute by attribute, for each species of each genus, into appropriate format for vocabulary building. Synonyms are rationalized.

Correspondence with equivalent statements in the RKC Code system is determined. Where no correspondence is found, new statements are either added to the RKC Code or deferred pending clarification. Initial feature lists for each of the taxa are established in draft form.

The Trust organizes the lists and decides on the appropriate consulting authorities for the verification process. These feature lists serve two purposes. First, they, by virtue of the process of their construction, automatically serve as the framework for converting the source documents into the database format. Second, the lists are inserted into a format to be sent to putative contributors for verification of the lists' applicability to their taxa of interest.

As agreement on the list of characteristics for each genus is finalized through the above iterative procedure, the list is entered into an appropriate electronic form for input of the species information. These lists/input forms will serve as the input mechanism for the data to be included in the database(s).

The individual tables will be maintained by the contributor, either directly or by sending the updated information to the Bergey's Manual Trust. The contributed tables are merged into a master database at appropriate intervals. The database functions in two ways: (i) the information is either modified or supplemented as new information becomes available, and (ii) reports are generated. These reports may be simple: a list of species exhibiting a specified set of features; or complex: a new version of a *Bergey's Manual*, either printed or for electronic distribution.

With intensified study of the diversity of the world's bacteria, new species will be described and the data on existing species increased. The traditional publishing process cannot cope with this information explosion. The building of a database of the new information, combined with that previously presented in *Bergey's Manuals* will resolve this problem and provide the scientists with a critical tool needed to enhance knowledge of the bacteria of the planet.

References

Bergey, D.H., Harrison, F.C., Breed, R.S., Hammer, B.W. and Huntoon, F.M. (1923) *Bergey's Manual of Determinative Bacteriology*. Williams & Wilkins, Baltimore.

Buchanan, R.E. (1916–1918) Studies in the nomenclature and classification of the bacteria. *Journal of Bacteriology* 1, 591–596; 2, 155–162; 2, 347–350; 2, 603–617; 3, 27–61; 3, 175–181; 3, 301–306; 3, 403–406; 3, 461–474; 3, 541–545.

Holt, J.G., Kreig, N.R., Sneath, P.H.A., Staley, J.T. and Williams, S.T. (1994) *Bergey's Manual of Determinative Bacteriology*, 9th edn. Williams & Wilkins, Baltimore.

Lapage, S.P., Sneath, P.H.A., Lessel, E.F., Skerman, V.B.D., Seeliger, H.P.R. and Clark, W.A. (1975) *International Code of Nomenclature of Bacteria. 1976 Revision*. American Society for Microbiology, Washington, DC.

Portyrata, D.A. and Krichevsky, M.I. (1992) MICRO-IS. A microbiological database management and analysis system. *Binary: Computing in Microbiology* 4, 31–36.

Rogosa, M., Krichevsky, M.I. and Colwell, R.R. (1971) Method for coding data on microbial strains for computer manipulation. *International Journal of Systematic Bacteriology* 21 (Suppl.), 1A–184A.

Rogosa, M., Krichevsky, M.I. and Colwell, R.R. (1986) *Coding Microbiological Data for Computers*. Springer-Verlag, New York.

Skerman, V.B.D., McGowan, V. and Sneath, P.H.A. (1980) Approved lists of bacterial names. *International Journal of Systematic Bacteriology* 30, 225–420.

Index

Abies 340, 346
Absidia corymbifera 267, 275
    glauca 164
    heterospora 278, 280, 282
    ramosa 164
Acacia 450
Acari 48
Acarina 58
Acer pseudoplatanus 236–237
Achaetomium 275
    strumarium 275
Achromobacter 144, 163–164
Acidianus 152
    brierleyi 147, 152
Acidiphilium 95–96, 152–154
Acinetobacter 163
    calcoaceticus 164
Acremonium 163, 275
    alabamense 266, 274
    furcatum 278
    strictum 266–267, 275
Acrophialophora fusispora 275
    levis 275
Actinomucor elegans 275
Actinomyces 163
actinomycetes 97–98, 100–102, 105, 293

Aderidae 60
Aerobacter 144
Aeromonadaceae 9
Aeromonas 163–164, 215
AFLP *see* Amplified Fragment Length Polymorphism
African Intellectual Property Organization (AIPO) 437
Agaricus cupreobrunneus 345
AGENDA 21 384, 427
Agrobacterium tumefaciens 217
Agrocybe paludosa 340
    pediades 340–341, 345, 350
AIPO *see* African Intellectual Property Organization
Alcaligenes 163
    denitrificans 164
    eutrophus 164
    faecalis 164
Alicyclobacillus 98
alkaliphilic microorganisms 255–263
All Taxa Biodiversity Inventory (ATBI) 74–86, 440
Allescheria 163
Alternaria 296
Alteromonas nigrifaciens 10–12
Amanita 339

ambrosia beetles 247
amoebae 39
amphibia 36
Amplified Fragment Length
    Polymorphism
    (AFLP) 374–376
Amplified Ribosomal DNA Restriction
    Analysis (ARDRA) 85
Anacystis nidulans 217
Ancalomicrobium 193
Anobiidae 60
ants 204, 246
aphids 94
Aphyllophorales 339
Apiocrea chrysosperma 338–339
Arachis hypogaea 450
Arachnida 32, 35–36, 46
Arachniotus dankaliensis 275
    ruber 279
Araucaria 127
Araucariaceae 126
ARDRA *see* Amplified Ribosomal DNA
    Restriction Analysis
Argyrodendron actinophyllum 53
arid ecosystems 199–209, 265–288
Armillaria 244, 428
    bulbosa 235
army ants 246
Arthrobacter 80, 144, 163
Ascodichaena rugosa 236
ascomycetes 58, 126, 238, 240, 268,
    274, 281–284, 338
Ascomycotina see ascomycetes
ash *see Fraxinus*
Aspergillus 271–272, 276, 277,
    281–284, 461
    alutaceus 274
    candidus 274, 277
    carneus 274, 279
    egyptiacus 274, 278–279
    fischeri 274
        var. *spinosus* 275
    flavipes 274
    flavus 274, 278–279
        var. *columnaris* 275
    fructiculosus 278, 282
    fumigatus 266–267, 274–276,
        278–280, 282, 284

glaucus 275
nidulans 274, 276, 277–280, 282
    var. *dentatus* 275
    var. *echinulatus* 275
    var. *latus* 274, 278
niger 164, 274–276, 277–280, 282,
    284
ochraceous 164
quadrilineatus 266, 274, 278–280,
    282, 284
rugulosus 266, 274
sydowii 274
terreus 274–276, 277–278, 280,
    282
    var. *africa* 275
    var. *aureus* 275
ustus 274
versicolor 274
violaceus 274
ATBI *see* All Taxa Biodiversity Inventory
Attelabidae 43
attine ants 247
Aureobasidium 163
    pullulans 296
Australia, Environmental Research
    Information Network
    (ERIN) 440, 443

Bacillus 98, 100, 102, 105, 144, 163,
    190, 256–262, 461
    alkalophilus 256–257
    firmus 257
    subtilis 217, 256, 258–259
bacteria, coryneform 198, 215
    iron 137–159
    methanotrophic 162
    systematics 371–379, 455–466
Bacteroides 100–101, 144
    fragilis 217
Basidiobolus ranarum 164
basidiomycetes 53, 56, 126, 238, 240,
    246, 268, 338, 410
    cord forming 246
Basidiomycotina see basidiomycetes
BCM *see* biologically controlled
    mineralization
BDT *see* Tropical Data Base

Index

beans *see Phaseolus vulgaris*
Beauveria 163
beech *see Fagus sylvatica*
bees 39
beetles *see Coleoptera*
Beggiatoa 219
Beijerinck, Martinus 80
Beijerinckia 164
Beneckea 163
Bergey's Manual of Systematic
    Bacteriology 457–458
    database 458–466
Bergey's Manual Trust 455–466
Berne Convention 417
Betula 242
Betulaceae 238
Bifidobacterium 374
BIM *see* biologically induced
    mineralization
BIN 21 *see* Biodiversity Information
    Network 21
Biodiversity Information Network 21
    (BIN 21) 441–445
BIOLOG 374
biological control 322–327, 332, 408
biologically controlled mineralization
    (BCM) 138–139
biologically induced mineralization
    (BIM) 138–139
BIOMASS 440
bioremediation 161–182
Biosphere Reserves 381
BIOTAS 440
Biotechnological Information Exchange
    System (BITES) 431, 435–437
birch *see Betula*
BITES *see* Biotechnological Information
    Exchange System
Bjerkandera 246
black band disease, brain
    corals 218–219, 221–222,
    224–226
black fungi 289–302
black yeasts 289–302
Blastocaulis 131
Bolbitius vitellinus 344
Boletales 339
Boletus 341

botanical gardens 399, 402, 429
Botrytis 163
Bovista nigrescens 340
Bradyrhizobium 97, 449–453
    japonicum 449, 451–453
brain corals, black band
    disease 218–219, 221–222,
    224–226
Brevibacterium 93, 163
Bruchidae 60
Budapest Treaty 18, 423, 435
Byssochlamys 275
    nivea 267, 275
    verrucosa 275

CAB INTERNATIONAL x, 383, 430
Cajanus cajan 450
Calocybe carnea 340, 345
    gambosa 341
Campylobacter 374
Candida 163
    maltosa 164
    tropicalis 164
    utilis 164
Cantharidae 43, 60
Carabidae 43, 54, 60
carbon dioxide, atmospheric 116–117,
    119–120
Caring for the Earth: A Strategy for
    Sustainable Living 382
Caulobacter crescentus 99
CCINFO (database) 393–394, 396,
    398–399
CEC *see* Commission of the European
    Community
Center for International
    Development 383
Centre for International Forestry
    Research (CIFOR) 383
Cephalosporium 163, 276
Cerambycidae 43, 60
Ceratoporella nicholsoni 212–215, 227
CGIAR *see* Consultative Group on
    International Agricultural
    Research
Chaetomium 277, 279
    atrobrunneum 275

Chaetomium contd
  aureum 278
  bostrychodes 275, 278
  coarctatum 278
  cymbiforme 278, 280
  globosum 275, 278
  indicum 279–280
  jodhpurense 278–280, 282
  olivaceum 274, 278
  perlucidum 278
  senegalensis 278–279
  subcurvisporum 275
  thermophile 266, 268, 273, 274, 281
    var. *coprophile* 274–276, 279
    var. *dissitum* 274–276
chalcocite 94
Chernobyl Atomic Power Station accident 411
chickens 401
Chlorobium vibrioforme 217
Chloroflexus 99
Choanephora conjuncta 164
Chromobacterium 163
Chrysomelidae 43
Chrysosporium 163
  indicum 280
  luteum 279–280
CI *see* Conservation International
CIFOR *see* Centre for International Forestry Research
CITES *see* Convention on the International Trade in Endangered Species
Citrus 376
Cladina 127
Cladonia 127
Cladosporium 163, 275, 296
  herbarum 275, 279
Clavaria vermicularis 345
Clavulinopsis corniculata 345
  helveola 342
  luteoalba 343
Clematis vitalba 244
Cleridae 43
Clitocybe 339
  amarescens 345
  nebularis 247

clone libraries 92–93, 95–96, 99–103, 105
Clonothrix 141
Clostridium 144
clovers 449
COBIOTECH 436
Coccinellidae 43
Coccocarpia 127, 131
Coccotrema 131
Cochliobolus 163
coconuts 277
Cokeromyces poitrasii 164
Coleoptera 32, 36, 40, 42–44, 48, 51–55, 59–60
Collema 127, 131
Collybia 339
  butyracea 344, 350
Colpophyllia natans 221
Colwellia psychroerythrus 10–12
Comamonas terrigena 10–12
Committee on Data for Science and Technology (CODATA) 392, 436
Commission of the European Community (CEC) x, 434, 436
Commission on Plant Genetic Resources 382
competitive exclusion principle 187
Conidiobolus gonimodes 164
Conocybe 339
  rickeniana 345
  siliginea 345
  tenera 345
Conservation International (CI) 383
Consultative Group on International Agricultural Research (CGIAR) 383
Convention on Biological Diversity, Rio de Janeiro 17, 23, 381–382, 385, 415, 417–425, 439
  Articles 418–422
Convention on Climate, Rio de Janeiro 19–20
Convention on the International Trade in Endangered Species (CITES) 382, 417
Coprinus 339, 341
  bisporus 345

curtus 345
echinosporus 345
ephemeroides 345
friesii 344
miser 345
patouillardii 344
picaceus 234
radiatus 345
stercoreus 345
copyrights 418
corals 19–20, 211–229
  mucus 221–226
Cordyceps capitata 239
  militaris 343
Coriolus 246
  versicolor 245
corticolous fungi *see* fungi, corticolous
Corylus avellana 244
Corynascus heterothallicus 273
  sepedonium 267, 275–277, 279
  setosus 279
  thermophilus 274
Corynebacterium 93, 144
  renale 164
cowpeas *see Vigna unguiculata*
Crenarchaeota 98
Crenothrix polyspora 141
creosote 174–179
Cronartium 236
Crustacea 32, 35, 46, 50–51
Cryptophagidae 43
culture collections 18, 358–360,
  389–403, 411–412, 418,
  422–424, 429, 431–438, 440
Cunninghamella 163
  blakesleeana 164
  echinulata 164, 275, 279
  elegans 164
  phaeospora 275
curare 422
Curculionidae 43
cyclosporin 422
Cylindrocarpon 163
Cytophaga 100–101, 163

Daldinia concentrica 242
Datronia mollis 246

Debaryomyces 163
Declaration of Belem 416
Degelia 127, 131
Deinococcus radiodurans 217
Denaturing Gradient Gel
  Electrophoresis (DGGE) 92,
  105
dendrograms 10–12, 214, 217, 307,
  313–316, 397
Dermestidae 43
desert rock varnish 270, 300
desert tortoises 204
Desertella globulifera 278
deserts 199–209, 265–288
Desulfomonile tiedje 164
Desulfotomaculum 144
  nigrificans 146
Desulfovibrio 144
  desulfuricans 146, 217
deuteromycetes 58, 338
Deuteromycotina see deuteromycetes
developing countries 415–416,
  419–420, 424, 427, 429,
  431–435
DGGE *see* Denaturing Gradient Gel
  Electrophoresis
diatoms 42
Dichomera saubinetii 236–237
Dicksonia antarctica 357
Dictyonema 131
Dictyostelium 461
Diptera 32, 36–37, 48, 59
Dipterocarpaceae 357
DIVERSITAS ix, 384, 386, 440
DNA fingerprinting 7, 82, 372–374
  hybridization 7–9, 13, 84, 90–94,
  96, 98, 101, 312, 314–315, 332,
  372–373, 375–376
  polymerase 3
Doratomyces columnaris 279
Dorylaimida 58
Douglas fir *see Pseudotsuga menziesii*
Drechslera halodes 280
  spicifera 278, 280
dried reference collections 358,
  403–414, 423
  holdings 407
Dutch elm disease 241

Dytiscidae 43

Earthwatch 382
earthworms 247
ecological dominance 187–188
Elateridae 43
elms *see Ulmus*
Emericella 275, 276
    desertorum 278, 282
    nivea 275
    similis 275
Endogonaceae 238
endophytic fungi *see* fungi, endophytic
Enterobacteriaceae 8
Entoloma 341, 346
    caesiocinctum 344, 350
    chalybaeum 343
    clypeatum 341
    conferendum 343
    hispidulum 345
    infula 345, 350
    juncinum 345
    papillatum 342
    sericellum 343, 348
    sericeum 343
    serrulatum 345
Entomophthorales 58
Epulopiscium 428
Erichsonius 45–46
ERIN *see* Australia, Environmental Research Information Network
Erioderma 127, 131
Erwinia 163
Escherichia 144
    coli 10–12, 189, 217–218, 259, 315, 329, 391
Eucnemidae 60
European Laboratory Without Walls Programme 434
Eurotium 275
Euryarchaeota 98
Exophiala jeanselmeii 296, 300
Exxon Valdez oil spill 170–174

Fagaceae 238
Fagus sylvatica 236, 243, 245

fairy rings 233, 247, 350
FAME (Fatty Acid Methyl Esters) analysis 83, 85 178, 221–222, 224–225, 374–377
FAO (United Nations Food and Agriculture Organization) 381–382, 391, 437
Fibrobacter 105
fir *see Abies*
fish hatcheries 371, 377
Flammulina velutipes 341
Flavobacterium 100–101, 163–164
fluoranthene 174–179
Fomes 339
Fomitopsis pinicola 241
food webs 37–38, 149
Foraminifera 61, 118
forest canopy ecosystems 53–56, 127–128, 130, 246, 356
forest decline 242–243
Fraxinus 242, 246
fungi, corticolous 236, 356
    endophytic 235–237, 242, 355
    litter inhabiting 246–247, 357, 359
    phylloplane 235
    rhizoplane 237
    rhizosphere 237, 356
    thermophilic 269–285
    thermotolerant 266–269, 271, 273–277, 284
    tropical plants 355–360
Fusarium 163
Fuscoderma 127, 131

Galerina 339
    atkinsoniana 343
    hypnorum 343
    mniophila 345
    pumila 344
    unicolor 343
    vittaeformis 343
Gallionella ferruginea 140, 142
Ganoderma 339
Gasteromycetes 339
GBA *see* Global Biodiversity Assessment
GEF *see* Global Environment Facility

Gemmata 97
   obscuriglobus 97
GEMMOs *see* genetically modified microorganisms
GEMS 382
gene banks 418, 423–424, 437
genetically modified microorganisms (GEMMOs) 322, 327–332, 421, 427–428
GENIE 440
Geobacter metallireducens 144, 146
Geoglossum fallax 345
   glutinosum 344
   nigritum 344
Geotrichum 163, 275
   candidum 275
Gilbertella persicaria 164
Gilmaniella macrospora 275
Gliricidia 449
   sepium 449
Global Biodiversity Assessment (GBA) 381–386
Global Biodiversity Strategy 384, 386
Global Diversity Forum 440
Global Environment Facility (GEF) 382, 384–385
Global System for the Conversion and Sustainable Use of Plant Genetic Resources 382
Glycine 451
   max see soyabeans
   soya see soyabeans
Gonytrichum 163
Gordioidea 58
Gracilicutes 102
greenhouse gases 117
Gregarinida 57–58
guanacos 204
guinea pigs 401

Haemophilus influenzae 8
Halsiomyces 270
hamsters 39
Hansenula 163
Hartig net 238
hay 268
hazel *see Corylus avellana*

heartrot 241
Hebeloma vinosophyllum 247
Hedera helix 236
Heliobacterium chlorum 217
Helminthosporium 163
Helvella 341
Hemimycena delectabilis 345
Hemiptera 36
herbaria *see* dried reference collections
Heterobasidion annosum 244
Histeridae 43, 54
Holospora 94
Holothuria 100–101
   atra 100
homeostasis 194–195, 260
Homoptera 94
Homothecium 127, 131
Human Genome Initiative 8
Humicola 163, 273, 275
   grisea 275
      var. *thermoidea* 266
   insolens 266
   lanuginosa 267
Hungary, National Collection of Industrial and Agricultural Microorganisms 435
hydrocarbons 161–182, 192–193
Hydrophilidae 43
Hygrocybe 341
   ceracea 344, 350
   conica 343
   helobia 341
   laeta 342
   miniata 345
   pratensis 343
   psittacina 342
   virginea 342
Hygrophoropsis pallida 345
Hygrophorus marzuolus 346
Hymenamphiaster cyanocrypta 215–217
Hymenochaetaceae 245
Hymenochaete corrugata 244
Hymenoptera 32, 36–37, 42, 48, 58–59
Hypholoma fasciculare 246
Hyphomicrobium 141
hyphomycetes 268, 276, 284, 357
Hypogymnia 127

IBPGR *see* International Board for Plant Genetic Resources
ICBN *see* International Code of Botanical Nomenclature
ICECC *see* Information Centre for European Culture Collections
ICRAF *see* International Centre for Research in Agroforestry
ICSU *see* International Council of Scientific Unions
ICTV *see* International Committee on Taxonomy of Viruses
INBIO 419, 424, 440
Index Herbariorum 405–406
Information Centre for European Culture Collections (ICECC) 434, 436
INFOTERRA 382
Infusoria 22
Inonotus dryadeus 241–242
insect-plant relationships 51–53
Intellectual Property Rights (IPR) 360, 415–427, 429, 437
Intergovernmental Panel on Climate Change 385
International Board for Plant Genetic Resources (IBPGR) 383, 437
International Cell Research Organization (ICRO) 392, 435
International Centre for Research in Agroforestry (ICRAF) 383
International Code of Botanical Nomenclature (ICBN) 18, 405, 409
International Code of Nomenclature of Bacteria 4, 18, 457
International Code of Zoological Nomenclature 18
International Committee on Taxonomy of Viruses (ICTV) 5, 361
International Council of Scientific Unions (ICSU) 382, 384
International Fund 382
International Geosphere-Biosphere Programme 384
International Mycological Directory 405–406, 412–413
International Organization of Biotechnology and Bioengineering 433
International Plant Genetic Resources Institute (IPGRI) 383
International Undertaking on Plant Genetic Resources 382
International Union of Biochemistry 463
International Union of Biological Sciences (IUBS) ix–x, 382, 384, 392, 437, 439, 441
International Union of Microbiological Societies (IUMS) ix–x, 382, 384, 392, 436–437, 439, 441
inventorying 74–86, 308, 355–360, 439–440
Iodophanus carneus 345
IPGRI *see* International Plant Genetic Resources Institute
IPR *see* Intellectual Property Rights
iron bacteria *see* bacteria, iron
Isosphaera 97
IUBS *see* International Union of Biological Sciences
IUBS/IUMS Committee on Microbial Diversity x
IUCN – The World Conservation Union 382–384, 392, 444
IUMS *see* International Union of Microbiological Societies

Japan, National Institute of Agrobiological Resources (NIAR) 399
  Ministry of Agriculture, Forestry and Fisheries (MAFF) 399
  Genetic Resources Centre 399
Japan Collections of Microorganisms (JCM) 399–400
Japan Federation for Culture Collections 390–391, 404
Japan Society of Laboratory Animals (JSLA) 401
JCM *see* Japan Collections of Microorganisms
JSLA *see* Japan Society of Laboratory Animals

Keeling curve 116–117
Klebsiella 163
    pneumoniae 164

laboratory animals 401
Laboulbeniales 57–60, 62, 404
Lactobacillus 137, 163, 374
    minutus 97
Lactococcus 374
Lasiobolidium orbiculoides 280
Lathridiidae 43
Lathyrus 449
Leguminosae 357, 431–432, 440, 447–454
Leioderma 127
Leiodidae 43
Lenzites betulina 246
Lepidoptera 32
Lepista nuda 345
    sordida 345
Lepolichen 127, 131
Leptogium 127, 131
Leptospira illini 217
Leptospirillum 95
    ferrooxidans 95, 147, 149–150
Leptothrix 140, 143
    cholodnii 143
    discophora 143
Leucaena 449
Leuconostoc 374
Leucothrix 163
LFRFA *see* Low Frequency Restriction Fragment Analysis
lichens, rain forest 125–135
Lieskeela 141
Linnaeus 21, 42–44, 457
Listonella aestuariana 10–12
    anguillarum 10–12
    damsela 10–12
    ordalii 10–12
    pelagia 10–12
    tubiashii 10–12
Lists of Approved Bacterial Names 457
litter inhabiting fungi *see* fungi, litter inhabiting
living reference collections *see* culture collections

Lobariaceae 130–131
Lophodermium pinastri 236
    seditiosum 236
Low Frequency Restriction Fragment Analysis (LFRFA) 374–375

MAB *see* Man and the Biosphere Programme
macrofungi 337–353, 356
    sporocarps 338–352
Macrophomina phaseoli 275
macrotermites 247
MAFF *see* Japan, Ministry of Agriculture, Forestry and Fisheries
Malbranchea cinnamomea 266, 268, 273–274
man 39
Man and the Biosphere Programme (MAB) 381
manioc 376
Marasmius 247, 339
    graminum 345
    oreades 340, 345
    wynneri 247
marine ecosystems 211–229
mealy bugs 94
Melanocarpus albomyces 266–267, 273–2745, 278, 280–282, 284
Meloidae 60
Meloidogyne incognita 326
Melyridae 43, 60
Menegazzia 127
Metallogenium 141, 143
    symbioticum 270
Metallosphaera sedula 152
Metanosaeta soehngenii 217
Methanobacterium omelianskii 195
methanotrophic bacteria *see* bacteria, methanotrophic
Methylosinus trichosporium 164
Metus 127
mice 401, 431
Microascus cirrosus 280
MICROBIAL DIVERSITY 21 ix
Microbial Germplasm Database 441, 443
Microbial Information Network Europe

(MINE) 434, 436, 441
Microbial Resources Centres
    (MIRCEN) 430–435, 448
Microbial Strain Data Network
    (MSDN) 397, 434, 436, 441
Micrococcus 144, 163, 256
Micromonospora 93
Microdochium nivale 266
microphytic soil crusts 199–209
    nitrogen fixation 210–202
    decline and recovery 204–205
Microsporida 57–58
MINE *see* Microbial Information
    Network Europe
MIRCEN *see* Microbial Resources
    Centres
Mollicutes 213
molluscs 19, 32–33, 35–36
Monilia 163
Monotropa hypopitys 239
monuments 289, 292–300
Moraceae 357
Moraxella 163–164, 220
Morchella 341
Mordellidae 60
Mortierella verrucosa 164
MSDN *see* Microbial Strain Data
    Network
Mucor 163
    circinelloides 275
    hiemalis 164
Mucorales 268, 282
mushrooms 356
mycelial networks 232–235, 246, 350
Myceliophthora 275
    fergusii 274
    thermophila 273, 274
Mycena 247, 339, 341
    avenacea 342, 345
    cinerella 340, 342
    epipterygia 345
    flavoalba 345
    galopus 343
    leptocephala 343
    pelliculosa 343, 348
    pura 345
    sanguinolenta 343
    sepia 342

stylobates 344
    tintinnabulum 341
Mycobacterium 163, 178
mycocoenology 337–353
mycoplasmas 39, 64, 213, 427
mycorrhizas 22, 203, 238–239, 241,
    248, 321, 323, 337–338, 341,
    346, 349–351, 356, 408
Mycotypha africana 275
Myriococcum thermophilum 274
Myrtaceae 238
myxobacteria 428

naphthalene 171–176
natural pruning 242
Naumanniella 141
nematodes 32–35, 37, 46, 48, 50–51,
    56–57, 64, 75, 322–327
Neotestudina rosatii 278, 280
Nephroma 127, 131
Nephromataceae 130
Neurospora crassa 164, 275
NIAR *see* Japan, National Institute of
    Agrobiological Resources
Nitidulidae 43
nitrogen fixation 321–322, 428, 432
Nitrosomonas europaea 164
Nocardia 93, 163–164
Nordic Gene Bank 437
Nothofagus 126–127
    antarctica 128
    dombeyi 128
    fusca 128
    menziesii 128
    pumilio 128
nucleic acid probes 5, 8, 82, 85, 90–94,
    96, 98, 101, 103, 105, 156, 220,
    327, 372, 374–375, 431
nucleotide sequencing 5, 7–13, 18,
    74–76, 80, 82–85, 90–96,
    98–105, 148–149, 218,
    261–262, 305–308, 310,
    312–314, 372–376, 378, 410,
    428, 431, 462

oaks *see Quercus*

oceans 48–50, 56–57, 61, 90–91, 99–100, 114, 116, 118–119, 189, 218
Ochrobium 141
Oidiodendron 163
Omphalina 339
Ophiostoma novo-ulmi 241
orchids 239
ozone 129

Paecilomyces 163
   lilacinus 323
   variotii 266, 274, 277–280, 282
paints, biodeterioration 299
Palaminus 54
Panaeolus acuminatus 344
   fimicola 343
     sphinctrinus 345
     subbalteatus 164
Pannaria 127, 131
Pannariaceae 130
Pannoparmelia 127
Paracoccus 144
parasitoids 38, 57–59, 64
Parmelia 127
Parmeliella 127, 131
Pasteurellaceae 101
patents 416–417, 420–421, 429, 431, 437
PCR (polymerase chain reaction) 3, 5, 83–85, 92–96, 98, 101–103, 218, 262, 332, 372–374, 410
peas 449
Pedomicrobium ferrugineum 141
Peltigera 131
Peltigerales 131
Penicillium 163, 271, 275, 281–284
   argillaceum 275
   crustosum 275
   dupontii 268
   egyptiacum 272
   funiculosum 279
   ochro-chloron 164
   piceum 275
   restrictum 279
Peridermium 236
Petriella setifera 280

Peziza cerea 345
Phanerochaete chrysosporium 164
   magnoliae 246
   velutina 246
PHAs *see* polyhydroxyalkanoates
Phaseolus vulgaris 376, 449–451
Phellinus 339
phenanthrene 173–177
Phialophora 163, 276
Phlyctis 127
Phoma 275, 296
Phormidium coralyticum 218–219
Photobacterium angustum 10–12
   fischeri 10–12
   leiognathi 10–12
   logei 10–12
   phosphoreum 10–12
Photorhizobium 97
phylloplane fungi *see* fungi, phylloplane
Phytophthora cinnamomi 164
Pinaceae 238
pines *see Pinus*
Pinus 236, 350
   sylvestris 338
Piptoporus betulinus 242
Pirellula 97
Pisum sativum see peas
Placopsis 131
Planctomyces 97, 105, 141
plankton 61, 84, 120, 212, 306
plant disease gardens 411
Pleurotus ostreatus 341
Podocarpaceae 126
Podospora faurelii 278
polyaromatic hydrocarbons 161–182
   structures 166
   degradation 167–176
Polychidium 131
polyhydroxyalkanoates (PHAs), degradation 371, 377
polymerase chain reaction *see* PCR
Porites 219
   astreoides 219–220
Prochlorococcus 100, 105
   marinus 99
pronghorn antelope 204
Proteobacteria 98–102, 148
Proteus 144, 163

Protousnea 127, 129
Psathyrella romagnesii 344
Pseudocyphellaria 127–128, 130–131
　homoeophylla 128
Pseudomonas 93, 144, 146, 163, 167, 328–329, 331
　aeruginosa 164
　aureofaciens 329, 331
　cepacia 164
　diminuta 99
　fluorescens 10–12, 164
　mendocina 164
　paucimobilis 174, 177
　pickettii 164
　pseudoalcaligenes 164
　putida 164
　rhodochrous 164
　testosteroni 217
　vesicularis 164
Pseudotrametes gibbosa 246
Pseudotsuga menziesii 235–236
Psilocybe inquilina 345
　montana 344
　semilanceolata 345
　strictipes 164
Psoroma 127, 131
Psoromidium 131
Ptiliidae 43
pyrene 178
Pyrodictium brockii 428
Pythium ultimum 324

Quercus 241
　robur 242–243

rabbits 401
Ramalodium 131
Ramulariopsis tenuiramosa 345
rats 401
RENEASA *see* Rhizobium Ecology Network of East and Southern Africa
reptiles 36
research and training 419, 421, 423–424, 429, 431–435, 437, 445

Restriction Fragment Length Polymorphism (RFLP) 374–375
RFLP *see* Restriction Fragment Length Polymorphism
Rhabditida 58
Rhabdocline parkeri 235–236
　pseudotsugae 236
Rhexothecium globosum 278–280, 282
Rhizobium 193, 203, 430–432, 447–454
Rhizobium Ecology Network of East and Southern Africa (RENEASA) 448–450
Rhizobium leguminosarum 449
　bv. *phaseoli* 449–451, 453
　bv. *trifolii* 449, 451, 453
　bv. *viceae* 449, 451, 453
　loti 449, 451, 453
　meliloti 451
　phaseoli 449, 452
　trifolii 449
Rhizomucor miehei 266, 273, 274, 278, 281–282
　pusillus 266–267, 273–274, 278, 281–282
Rhizophlyctis harderi 164
rhizoplane fungi *see* fungi, rhizoplane
Rhizopus 275
　chinensis 270
　microsporus var. *microsporus* 266, 276
　　var. *rhizopodiformis* 266, 275
　nigricans 266
　stolonifer 164, 278–279
　　var. *stolonifer* 275
rhizosphere fungi *see* fungi, rhizosphere
rhizosphere microorganisms 203, 321–336
Rhodococcus chlorophenolicus 164
Rhodomicrobium 144
Rhodopila globiformis 96
Rhodopseudomonas palustris 97, 164
Rhodosporidium 163
Rhodotorula 163
Ribosomal Database Project 8
rice 376
Rickenella fibula 342
　setipes 342

RKC Code 462–465
Roccellinastrum 127
rocks
    biodegradation 113–123, 202, 289–302
        scanning electron microscopy 293–298
    microflora 289–302
roe deer 338
roots, fungi 236–239, 356
Rosaceae 346
Rotifera 22
Russula 339
rust fungi 357, 411

SAB *see* Society of American Bacteriologists
Saccharomyces 163
    cerevisiae 164
Saccobolus truncatus 279
Sagenidium 127
Salicaceae 238
Salmonella 307
Saprolegnia parasitica 164
Sarcina 163
Sarcoscypha coccinea 341
scale insects 236
Scarabaeidae 43, 54
Schrödinger ratio 188
Scientific Committee on Problems of the Environment (SCOPE) ix–x, 119, 382, 384
scleractinia (hard corals) 218–226
Sclerotinia trifoliorum 345
Scolecobasidium 163, 275
Scolytidae 43, 60
SCOPE *see* Scientific Committee on Problems of the Environment
Scopulariopsis brevicaulis 275
    candida 275
Scots pine *see Pinus sylvestris*
Scytalidium thermophilum 266, 273–274
sea cucumbers *see Holothuria*
seawater 213–215, 219, 221–226, 377
Seliberia 141
Sepedonium chrysospermum 275
Septobasidiaceae 236

Shannon index 188
sheep 204
Shewanella benthica 10–12
    colwelliana 10–12
    hanedai 10–12
    putrefaciens 10–12, 144, 146
ship worms 94
Siderocapsa 141
Siderococcus 141
Silphidae 43
Siphulastrum 131
smallpox virus 8, 428
Smittium culcis 164
snails 204, 338
Society of American Bacteriologists (SAB) 456
soil, microorganisms 321–336
    Rhizobium 447–454
    thermophilic fungi 269–285
soil fungi, Middle East 271–284
soil-root ecosystems 321–336
Sordaria fimicola 278
soyabeans 449, 451
species concepts 3–15, 30–31, 34, 80, 361–368
species diversity index 188
species numbers, estimates 4–6, 29–71, 75–77, 126, 268, 338, 351, 369, 404
species-environment interactions 185–198
Sphaerobolus stellatus 345
Sphaerophorus 127
Sphaerospora saccata 275
Sphaerosporium 275
Sphaerotilus 143, 163
    natans 140, 143
Sphingomonas paucimobilis 164, 174
Spirillum 163, 192
spirochaetes 99
Spiroplasma 62
Spirulina 219
sponges 212–218
Sporotrichum carnis 275
    pruinosum 275
    pulverulentum 275
    roseolum 275
squirrels 338

Staphylinidae 43, 45–46, 54, 58–60
Staphylococcus 144
statistical methods 305–319
    binary data 306–307, 310–311
    cladistics 315
    cluster analysis 313–314
    molecular data 307
    most probable numbers
        (MPN) 307, 322, 330, 447–449
    numerical taxonomy 311–316, 372
    principal component analysis 313
    quality monitoring 308
Stereocaulon 131
Sticta 127–128, 131
Stictaceae 130
strangler figs 246
Streptococcus 190
Streptomyces 93, 96, 102, 105, 163, 297, 377
Stropharia albocyanea 345
    semiglobata 344
Suillus bovinus 350
Sulfobacillus 151
    thermosulfidooxidans 147, 151
Sulfobolus 98, 152
    acidocaldarius 152–153
    solfataricus 217
    brierleyi 152
suppressive soils 322–324
sycamore see *Acer pseudoplatanus*
Syncephalastrum racemosum 164, 275, 279
Synechococcus 99–100, 104
    lividans 99

Talaromyces 275
    byssochlamydioides 280
    emersonii 266, 274, 280, 282
    flavus 275
        var. *flavus* 279
    leycettanus 266, 275
    thermophilus 266, 268, 273–274, 280, 282
    trachyspermus 279
TAXACOM 443
technology transfer 420–421, 423–424, 429

Teloschistes 127
Tenebrionidae 43, 204
Tephrocybe tylicolor 343
Terpios granulosa 215–216
Tetrahymena 39
Thallophyta 24
Thermoascus aegyptiacus 273, 274
    aurantiacus 266, 268, 273–274, 280
    crustaceus 266, 274
Thermomicrobium roseum 217
Thermomyces lanuginosus 266–267, 273–274, 278, 280–282
    stellatus 266, 276
thermophilic microorganisms 97–99, 139, 147–148, 151–154, 265–288, 428
Thermophymatospora 273
    fibuligera 274
thermotolerant fungi see fungi, thermotolerant
Thermus aquaticus 4
Thielavia 275, 279
    arenaria 275, 278–280, 282
    microspora 275, 279, 281
    subthermophila 278–280, 282
    terrestris 266, 274–276, 280, 282
    terricola 275
Thiobacillus 95, 151
    acidophilus 152
    cuprinus 151
    ferrooxidans 95, 147–150, 152–154, 156
    prosperus 147, 149
    thiooxidans 95, 151–153
Tirmania 272
Torula herbarum 275
    terrestris 275
Torulopsis 163
Toxothrix 141
training see research and training
Trichoderma 163, 275
    harzianum 324–325
    viride 271
Trichothecium roseum 278
Trichsporon 163
Trifolium semipilosum 449
Trimmatostroma 275

Tropical Data Base (BDT) 439, 441, 443–444
tropical plants, fungi *see* fungi, tropical plants
truffles 239, 272, 356
Tubaria furfuracea 345
Tylenchida 57–58
Typhula incarnata 345

Ulmus 246
Ulocladium 296
UNCED *see* United Nations Conference on the Environment and Development
UNDP (United Nations Development Programme) 381–382, 384, 435
UNEP (United Nations Environment Programme) x, 381, 383–386, 391, 396, 435–436, 439–441, 443
UNESCO (United Nations Educational, Scientific and Cultural Organization) ix–x, 381, 384, 391, 433–437, 443
UNIDO (United Nations Development Organization) 381, 382
United Nations Conference on the Environment and Development (UNCED) 17, 381, 384–385, 415–416, 439
United Nations Conference on the Human Environment, Stockholm 381, 435
United Nations Development Organization *see* UNIDO
United Nations Development Programme *see* UNDP
United Nations Educational, Scientific and Cultural Organization *see* UNESCO
United Nations Environment Programme *see* UNEP
United Nations Food and Agriculture Organization *see* FAO
UNU 391
USA, National Biological Survey 440
Usnea 129

Verticillium chlamydosporium 324–326
viable but non-culturable (VBNC) microorganisms 5–6, 61, 64, 79–80, 82, 89–110, 218, 226, 311, 371
Vibrio 9, 100–101, 103, 144–146, 163, 215, 371, 377
  adaptatus 10–12
  alginolyticus 10–12
  campbellii 10–12
  carchariae 10–12
  cholerae 9–12, 306–307, 428
  cincinnatiensis 10–12
  costicola 10–12
  cyclosites 10–12
  diazotrophicus 10–12
  fluvialis 10–12
  furnissii 10–12
  gazogenes 10–12
  harveyi 10–12
  hollisae 10–12
  marinus 10–12
  mediterranei 10–12
  metschnikovii 10–12
  mimicus 10–12
  natriegens 10–12
  nereis 10–12
  nigripulchritudo 10–12
  orientalis 10–12
  parahaemolyticus 10–12
  proteolyticus 10–12
  splendidus 10–12
  vulnificus 10–12
Vibrionaceae 9
Vicia 449
Vienna Convention on the Ozone Layer 382
Vigna 449–450
  unguiculata 449
viruses 361–369
  classification 362–367
  quasispecies 369

wallpaper, biodeterioration 299
Wardomyces anomalus 279
WCMC *see* World Conservation Monitoring Centre

WDCM *see* WFCC World Data Center on Microorganisms
weevils 54
WFCC *see* World Federation for Culture Collections
WFCC World Data Center on Microorganisms (WDC) 389, 391–401, 430, 433, 440–441, 443
whales 39
WHO *see* World Health Organization
whole cell probes 103–104
WIPO *see* World Intellectual Property Organization
wombats 204
wood decay fungi 239–247, 356
  compartmentalization 244
wood wasps 247
woodland ecosystems 231–251
  fungi 231–251
woodlice 204
World Bank 383–384
World Climate Research Programme 384
World Congress on National Parks and Protected Areas, Caracas 384
World Conservation Monitoring Centre (WCMC) 382
World Conservation Strategy 382
World Data Center on Microorganisms *see* WFCC World Data Center on Microorganisms
World Federation for Culture Collections (WFCC) 389, 391, 393, 396–397, 432, 434, 436, 439, 441
World Health Organization (WHO) 19, 391
World Intellectual Property Organization (WIPO) 434
World Resources Institute (WRI) 382–384
World Wide Fund for Nature *see* WWF
WRI *see* World Resources Institute
WWF (World Wide Fund for Nature) 382–383

Xanthobacter 164
Xanthomonas 163, 371, 374, 376
  campestris 372, 376

Zambia, Regional Genetics Resources Centre and Gene Bank 437
zoos 402, 429
Zygomycotina 58, 238, 283, 357